THE PHYSICS OF TIRE TRACTION
THEORY AND EXPERIMENT

PUBLISHED SYMPOSIA

Held at the
General Motors Research Laboratories
Warren, Michigan

Friction and Wear, 1959
Robert Davies, *Editor*

Internal Stresses and Fatigue in Metals, 1959
Gerald M. Rassweiler and William L. Grube, *Editors*

Theory of Traffic Flow, 1961
Robert Herman, *Editor*

Rolling Contact Phenomena, 1962
Joseph B. Bidwell, *Editor*

Adhesion and Cohesion, 1962
Philip Weiss, *Editor*

Cavitation in Real Liquids, 1964
Robert Davies, *Editor*

Liquids: Structure, Properties, Solid Interactions, 1965
Thomas J. Hughel, *Editor*

Approximation of Functions, 1965
Henry L. Garabedian, *Editor*

Fluid Mechanics of Internal Flow, 1967
Gino Sovran, *Editor*

Ferroelectricity, 1967
Edward F. Weller, *Editor*

Interface Conversion for Polymer Coatings, 1968
Philip Weiss and G. Dale Cheever, *Editors*

Associative Information Techniques, 1971
Edwin L. Jacks, *Editor*

Chemical Reactions in the Urban Atmosphere, 1971
Charles S. Tuesday, *Editor*

The Physics of Opto-Electronic Materials, 1971
Walter A. Albers, Jr., *Editor*

Emissions From Continuous Combustion Systems, 1972
Walter Cornelius and William G. Agnew, *Editors*

Human Impact Response, Measurement and Simulation, 1973
William F. King and Harold J. Mertz, *Editors*

The Physics of Tire Traction, Theory and Experiment, 1974
Donald F. Hays and Alan L. Browne, *Editors*

THE PHYSICS OF TIRE TRACTION
THEORY AND EXPERIMENT

Edited by
DONALD F. HAYS and ALAN L. BROWNE
General Motors Research Laboratories

PLENUM PRESS • NEW YORK–LONDON • 1974

Library of Congress Cataloging in Publication Data

Symposium on the Physics of Tire Traction, Warren, Mich. 1973.
The physics of tire traction, theory and experiment.

"Presented by the [General Motors Corporation] Research Laboratories."
Includes bibliographical references.
I. Tires, Rubber—Traction—Congresses. I. Hays, Donald F., 1929- ed.
II. Browne, Alan L., 1944- ed. III. General Motors Corporation. Research Laboratories. IV. Title.
TS1912.S9 1973 678′.32 74-10536
ISBN 0-306-30806-1

Proceedings of the Symposium on The Physics of Tire Traction held at the General Motors Research Laboratories Warren, Michigan, October 8-9, 1973

A Division of Plenum Publishing Corporation
227 West 17th Street, New York, N.Y. 10011

United Kingdom edition published by Plenum Press, London
A Division of Plenum Publishing Company, Ltd.
4a Lower John Street, London, W1R 3PD, England

Printed in the United States of America

PREFACE

This book contains the papers, panel discussions, and general discussions that comprised the formal sessions of the Symposium on The Physics of Tire Traction: Theory and Experiment, which was held at the General Motors Research Laboratories on October 8 and 9, 1973. This symposium was the seventeenth in an annual series presented by the Research Laboratories. Each symposium has covered a different technical discipline that was timely and of vital interest to General Motors as well as to the technical community at large.

The subject of the 1973 symposium was considered to be an appropriate one in view of the growing awareness and concern in the private sector that the available tire traction can drastically affect safety in vehicle operation and in view of the (then) impending government regulations concerning acceptable levels of pavement skid resistance and tire traction performance. In recent years, a great number of tests have been run in an effort to measure tire traction. Although large quantities of data exist, there has been a lack of theoretical work that would give meaning to the data that has been recorded. Today there is a need for a predictive capability in the design of both the tire and the road for good all-weather tractive capability.

In this situation it was particularly appropriate to provide a forum for the key individuals from the tire industry, the automobile industry, university research groups, independent research institutions, and government agencies where the newest theories and experimental results could be presented and discussed.

The small group of engineers and scientists invited to the symposium were chosen because of their expertise and current activity in the field of tire traction physics. The number was limited both by the desire to keep the group intimate and by limitations of the physical facilities. Attendees came from 13 foreign countries as well as the United States. For each of the four sessions, papers were presented by currently active authorities in the particular technical areas. To assure competent direction during the technical sessions, men of recognized international reputation in the field of tire traction research acted as session chairmen. Time was allotted during the meeting for the presentation of 15 technical papers, three formal panel discussions, and for informal discussions on the papers, panel topics, and allied investigations of the attendees.

A summary of the symposium proceedings was presented by one of the symposium chairmen. The proceedings were recorded on tape that was subsequently transcribed and edited lightly. These edited accounts were then submitted to the discussers for corrective action and approval. The submitted manuscripts of the papers, the formal panel sessions, the informal discussions, and the symposium summary have been reproduced in this volume. In addition, a critical review of the literature on tire traction through the year 1970, which was not presented at the meeting, has been included so that the present book can serve as a more complete reference volume on the subject of tire traction.

Although this book records the original papers of the authors and discussions that they elicited, a major contribution of the symposium could not be recorded here. The two-day meeting offered many opportunities for personal interaction between attendees, many of whom were meeting together for the first time. The new channels of communication that were opened may allow an increase in the awareness of the concerns and objectives of others and contribute greatly to the direction and success of future tire traction research.

This symposium could not have been held or these proceedings published without the valuable assistance of many people: Mr. R. L. Scott, who efficiently handled the myriad details involving the physical arrangement of the symposium; Mr. D. N. Havelock, for assistance in editing the symposium volume; Miss M. E. Johnson, Mrs. K. A. Kirksey, Mrs. B. M. Lavender, and Mrs. D. A. Stafford, for not only typing but acting as symposium hostesses; and Mrs. M. A. Alexander, Mrs. M. J. Arnold, and Mrs. J. M. Rascano who so competently handled the microphones during the symposium discussion periods.

The chairmen would also like to acknowledge the considerable contributions of the symposium organizing committee members who gave unselfishly of their time and talents. These included Messers. V. E. Caste, R. C. Collister, R. A. Daskivich, A. O. DeHart, C. W. Griffin, K. H. Huebner, R. G. Lenz, E. R. Maki, R. C. Rosenberg, J. J. Smith, S. R. Tiderington and E. G. Trachman.

Finally, thanks are extended to the symposium advisory committee consisting of Messers. F. Smithson, K. C. Ludema, and S. K. Clark, and to the meeting participants for creating a stimulating and worthwhile symposium.

A. L. Browne
D. F. Hays
January 1974

CONTENTS

SESSION I

TIRE TRACTION
A MANY FACETED PROBLEM

Session Chairman
H.C.A. van ELDIK THIEME

Delft University of Technology
Delft, The Netherlands

INTRODUCTORY REMARKS (Van Eldik Thieme)

To review today where we stand technologically in the subject of tire traction and to assure an adequate treatment seems impossible in the available time. I, therefore, will only touch upon some factors affecting tire traction. In general, the role of the tire can be divided into two modes of performance. The tire supports the vertical load and transmits longitudinal braking and driving forces. It also supplies the lateral cornering and camber forces which are necessary for the directional control of the vehicle. In practice, routine investigation of the behavior of the combination car-tire is usually restricted to the purely experimental part. The effect of tire construction is tested in carefully chosen automobile maneuvers. The tire engineer wants to know what is the best compromise in tire construction regarding carcass stiffness, tread pattern, tread compound, etc., to meet the conditions set by the vehicle suspension engineer.

With these kinds of tests, it is the overall behavior of all four tires which counts because a number of important quantities have to be considered, such as moments of inertia, lateral and longitudinal load transfer, steering system elasticity, roll steer and roll camber, self aligning torque, etc. Although a variety of test methods have been devised, all methods are subject to many factors which may adversely affect the repeatability and accuracy of data. Even when control tires are used for establishing the friction coefficient rating of the test surface, this practice does not always provide the basis for comparison of test data among different tests.

It may be evident that a deeper insight into the function of the tire can hardly be achieved by means of the above mentioned practical methods, unless the vehicle is highly instrumented so that the relevant motions and forces can be measured. Due to the necessitously complex instrumentation of automobiles, tire testers were developed for measuring tire characteristics on road surfaces at different speeds and loads. Numerous tests have been carried out to find the influence of braking, traction and cornering behavior both on dry and wet roads.

DRY TRACTION

Although the problem has been studied in detail, the physics of tire traction on so-called "dry" surfaces is far from clear. Some results on dry traction physics will be reported in Session II of this symposium.

To describe the tire traction of different tire constructions on several roads with different surface textures, we want to know more about the detailed mechanisms that control the frictional coupling at the tire-road interface. We often do not know enough about the distribution of the actual slip velocities at the different points in the contact area or about the normal force distribution. And what about the shape of the contact area or the deformation of the tread elements at different speeds as a function of tire geometry, tire construction and tire wear? Because all these data relate to the longitudinal and lateral shear force/slip relationships, publications and discussions of possible new research findings are very important.

WET TRACTION

As a result of numerous tests, it was found that on wet roads the most important factors are speed, road surface texture, water depth and tread depth. Because of the interaction among these and other variables, a multifactor program was conducted on behalf of the Institute for Road Safety Research (S.W.O.V.) by our Vehicle Research Laboratory. The object was to find the dominant parameters of the road surface texture influencing skid resistance.

To formulate the measuring program, statistical design techniques were used to select the sequence and type of necessary tests. In this program the type of tire, speed, road surface texture, water depth, and tread depth (among other variables) were varied, and the braking and cornering coefficients were measured on six road textures on a special test track. The tire tread drainage capacities and certain tread compound values were investigated as well as the road drainage capacities and road wet skid resistance values (S.R.T.). In an earlier similar program carried out on main roads, it was found that, in general, the influence on the friction coefficient of the inflation pressure and tire load was rather small. Therefore, in the second program these variables were kept at constant levels.

Another multifactor program was not so much directed to road surface texture evaluation as to the determination of tire parameters dominant in wet skid resistance.

Because variation of tire parameters at the desired level is a rather difficult problem, these tests were carried out with a large number of normal production tires. To ensure sufficient variation in these parameters, tires of different brands were used. With statistical correlation techniques, the influence of several tire parameters was evaluated. We found, for instance, (by using a correlation matrix) that there was a significant correlation between cornering stiffness and cornering coefficients on wet roads. Obviously, there will also be a strong correlation between footprint void area and tire drainage capacities.

HYDROPLANING

The situation in which there is almost complete loss of tire traction on a wet pavement is usually described as hydroplaning. The speed at which the loss of braking and steering capabilities first occurs is called the hydroplaning speed. Dynamic hydroplaning can be said to be an inertial phenomenon. The lifting effect due to the hydrodynamic pressure in a fluid film of density ρ accelerated from rest to the velocity v is dependent upon the square of the velocity v. A complete separation of the tire surface and pavement by a critical fluid film depth is observed in the footprint area. Because these fluids cannot develop reasonably high shear forces, tire traction drops to very low values. Records of rainfall intensities (mm/h) over long periods at a number of sites give the calculated water depths for a certain drainage length (10 m) and slopes of 1:40 and 1:100. On tire proving grounds, the installation of a glass plate and high speed photography is used with water containing a fluorescent material to give spectacular results, where through progressively increasing speeds, the water wedge can be seen extending under the tire until it has no further contact with the road. The details of the pressure distribution for a fluid film in the footprint region have been described by A. Browne, and it has been shown that hydroplaning is very much dependent upon lateral inward buckling of the tire in the footprint region.

A different type of loss of traction on wet surfaces is called viscous hydroplaning which requires a thin fluid film to be present on the pavement. The scale and sharpness of pavement texture in relation to its potential for penetrating the squeeze film under the rolling tire illustrates its paramount importance in preventing viscous hydroplaning. The ability of the tread pattern to transport fluid from the contact patch also results in a decrease of traction loss. Tread rubber reversion in the skidding footprint due to overheating of the rubber sometimes accompanies hydroplaning and has been observed by Allbert on car and truck tires.

SNOW TRACTION

Having touched upon hydroplaning we come to traction on snow, and it is well known that the heavily profiled treads of winter tires are designed to provide an increased tractive capability of tires on snow.

The interaction of the tire and snow calls for descriptions of the properties of different types of snow, in order to discuss the functions of the tread pattern.

Because it is often very difficult to explain the reason for the large spread of different tread patterns in driving force and braking force coefficients at large slips, further research on the important parameters influencing the tire-snow system should be carried out.

The effectiveness of studded tires on ice and the influence of temperature on their effectiveness should also be mentioned. Due to the damage of the road surface, studs are not permitted in several European countries.

CONCLUSION

I have tried in this very short introduction to show that there are many aspects of the tire traction problem, and that a wide range of topics must be covered by the four speakers.

TIRE WET TRACTION: OPERATIONAL SEVERITY AND ITS INFLUENCE ON PERFORMANCE

A. G. VEITH and M. G. POTTINGER

The B. F. Goodrich Company, Brecksville, Ohio

ABSTRACT

Tire wet traction, both cornering and braking, depends on several factors. The primary ones are speed, pavement texture, and water depth. The dependence of tire traction on these is complex due to the interaction among these variables. This paper examines some aspects of this complex behavior and offers suggestions on how to overcome part (hopefully a large part) of this complexity.

A concept of Operational Severity is defined. Operational Severity is directly proportional to tire (vehicle) speed and water depth and inversely proportional to tire and pavement water drainage capacities. The logic for this concept is presented.

The difficulties involved in assessing tire wet traction performance at a particular speed are outlined. To overcome this problem, an additional factor, the Ultimate Performance Rating or UPR, is defined. This is based on the full information content of the linear data plot of wet cornering coefficient, μ_c, vs. speed. Tires rated on the basis of UPR under moderate Operational Severity yield performance values that closely correlate with actual available traction (μ_c values) under severe Operational Severity.

Data showing the UPR correlation with high operational severity traction are used to illustrate how the actual wet cornering traction coefficient, μ_c, and relative ratings based upon μ_c, change as water depth varies. This is done for two different pavement textures. Both actual μ_c and *relative ratings* are dependent upon water depth at 60 mph. This change in relative rating with varying water depth brings into question the "control tire" concept normally used in this type of testing. This problem is discussed and suggestions made on ways to alleviate some of the difficulties.

References pp. 19–20

INTRODUCTION

It has been well established that tire traction, both braking and cornering, is quite speed dependent on pavements typical of most real road surfaces (1, 2, 3, 4, 5, 6). For the small fraction of real road surfaces which possess a very high macro-texture, only a minimal speed dependence is obtained.

A surface with large macro-texture is best described as being rough. It has aggregate particles or asperities as large as several millimeters in diameter. On such rough surfaces, the probability of loss of traction is quite low.

Surfaces or pavements that have moderate to low macro-texture (1–2 mm or less) contribute to loss of traction when they are wet from moderate to heavy rainfall. On such surfaces as these, the tire alone must efficiently discharge water from the contact area and thereby promote good pavement "grip."

In this report we examine the question of the speed dependence of tire traction on low to moderate macro-texture pavements. Specifically examined is the question of how to realistically rate tire wet traction performance. The data presented were obtained with a special instrumented trailer designed to measure free-rolling wet cornering traction. The trailer has been described in a previous publication (6). The test procedure is briefly outlined below to provide some background information on how the data was obtained. Although the remarks and discussion specifically concern wet cornering traction, the conclusions reached should also apply to wet braking traction.

TEST PROCEDURE AND DATA ANALYSIS

The procedure used for each "test cycle" is shown in Fig. 1. A test cycle is comprised of a rapid slip angle input from zero to the selected maximum angle, in this

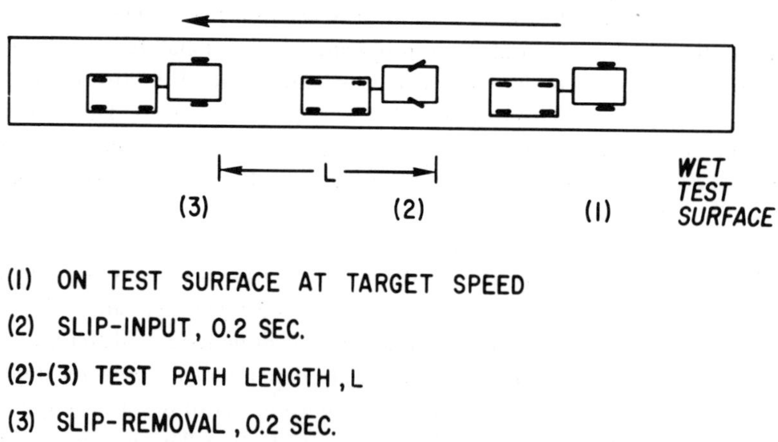

Fig. 1. Cornering wet traction test sequence.

case 12°, a hold at that angle for about 1 second (or longer at lower speeds), and a reduction of slip angle to zero.

Tests are conducted with both tires in the "toe-in" slip angle configuration. In the configuration, the left tire cornering force is balanced by the right tire force, and an appropriately situated load cell measures the average "single tire" cornering force for the two tires.

The overall procedure is as follows. Two or three 75 ft. long sections are selected on the wet test surface and marked by appropriate markers. A series of target speeds is selected for testing, usually 30, 40, 50 and 60 mph. Test tires are mounted and warmed-up for 3 miles at 60 mph; and the air pressure is adjusted to 28 ± 0.5 psi or other specified values. A test run is made going in each direction. A run consists of 2 or 3 test cycles, one cycle on each selected section. The speed (to ± 0.5 mph) of each test cycle is recorded. The average cornering force for each test cycle is obtained, and from this the cornering wet traction coefficient is calculated.

The cornering wet traction coefficient μ_c, is defined as follows:

$$\mu_c = \frac{F_c}{L} \qquad (1)$$

where F_c is the average cornering force (lbs.), and L is the static tire load (lbs.).

Average values for μ_c and speed are obtained for each direction. Plots of μ_c vs. speed are linear, and the basic μ_c - speed data are subjected to linear regression analysis to calculate the slope and intercept. These represent averages for both directions where a significant directional effect exists.

$$\mu_c = b_o + b_1 V \qquad (2)$$

where b_o is the intercept or value of μ_c when the speed, V, (in mph) is zero and b_1 is the slope which equals $d\mu_c/dV$.

Correlation coefficients and standard deviations of μ_c about the regression line are also obtained from the analysis which is performed on a digital computer.

RATING TIRE PERFORMANCE FOR WET TRACTION

The basic problem of how to realistically rate tires for their wet traction performance is conveniently set forth in Fig. 2. In this figure, cornering wet traction coefficient data are plotted vs. speed for tires with two tread pattern "types" and for two pavement textures. In Part (A) are plots for a rough or high macro-texture surface. The closed and open tread patterns (representing tires with identical internal construction and tread compounds) are, respectively, a typical seven rib regular highway tire with 20% void or groove area in the contact patch and a block tread pattern with 40% void area.

References pp. 19–20

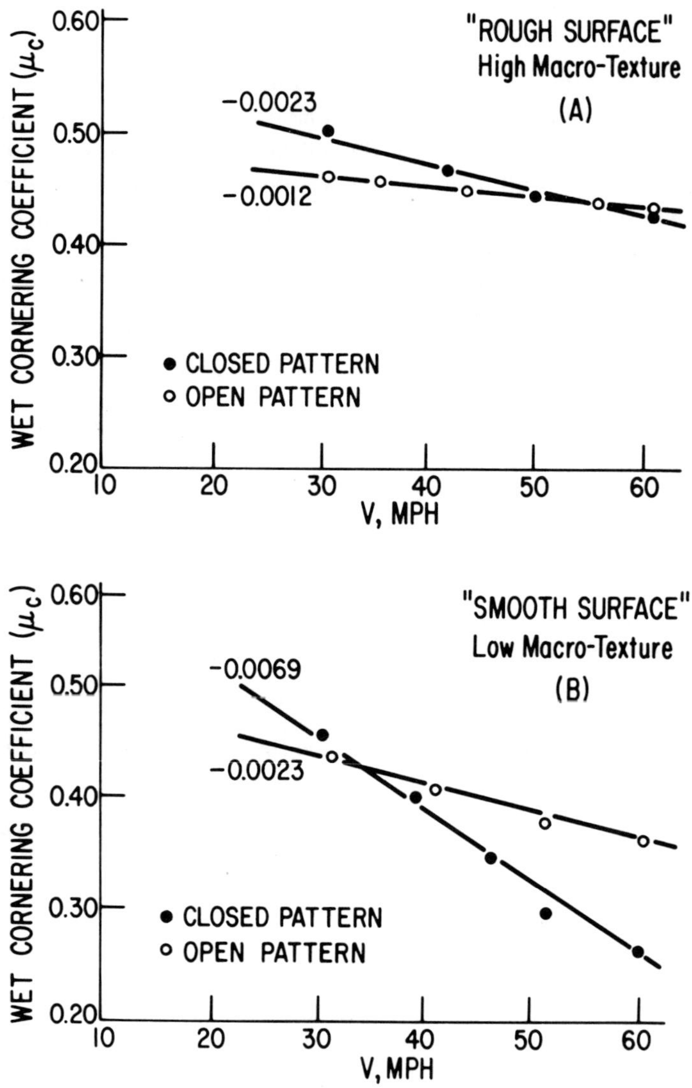

Fig. 2. Cornering wet traction performance of an open pattern tire and a closed pattern tire on two surfaces.

In Part (A) the closed pattern gives higher μ_c values compared to the open pattern in the 25–50 mph speed range. At 60 mph it is slightly inferior. Appended to each curve is the velocity gradient of the cornering coefficient in mph^{-1}.

In Part (B) the situation is quite different. The closed pattern is now superior (higher μ_c values) to the open pattern only below 35 mph and at 60 mph a considerable difference exists in favor of the open pattern. If one asks the question – What is the relative rating of these open and closed pattern tires? – an unequivocal answer cannot be given. One must specify pavement texture and speed (an additional factor is water depth which will be discussed later). If pavement texture is specified, a large variation in relative rating or performance still exists including outright reversals. Clearly, some considerations must be made as to priority or order of importance. Otherwise, one may choose any rating desired simply by selecting the appropriate speed and/or pavement texture.

Both the stopping distance (on wet pavements) and the lateral force required to negotiate a curve of fixed radius increase with the square of the vehicle velocity (initial velocity in the case of braking). Available traction decreases with increasing vehicle velocity. Therefore, high speed wet traction performance is often given high priority, as should be done, in assessing overall tire performance.

One could arbitrarily adopt 60 mph as a high speed and assess tractive performance at this speed. Certainly, this approach to tire rating would be superior to a rating assessment in the 20 to 40 mph range which is often seen in the literature. However, Fig. 2 shows that the arbitrary selection of 60 mph would still leave something to be desired. Ratings at 60 mph would be different from those at 70 mph and also different from those at 50 mph, in spite of the fact that these test speeds are relatively close together. In the next section, we will attempt to resolve this problem by adopting a procedure which, while not perfect in every aspect, has much to offer in providing a realistic "worst possible case" rating assessment. The term "worst possible case" implies an emergency motoring situation at high speed on low texture or slippery pavement in heavy rain.

OPERATIONAL SEVERITY CONCEPT

A better understanding of wet traction performance can be gained if one views the situation from the standpoint of "operational severity." Operational severity is defined as follows:

$$\text{Operational Severity} = \frac{k}{WD_P} \times \frac{V}{WD_T} \tag{3}$$

where V is the tire (vehicle) speed or velocity, WD_P is the capacity for dynamic water drainage or venting through the contact patch due to pavement texture

References pp. 19–20

drainage paths, and WD_T is the same capacity for water drainage due to drainage paths provided by the tire tread pattern.

For any given tread pattern/tire design, this latter drainage capacity is defined as being equivalent to the following ratio.

$$WD_T = h_G/h_W \tag{4}$$

where h_G is the depth of the tread grooves and h_W is the water depth on the pavement.

The constant k of Eqn. 3 encompasses various geometric factors of tread pattern, other tire design factors such as aspect ratio, and tread material or compounding variables.

Thus, tires will operate at high "operational severity" when speeds are high, when pavement texture is low, (smooth, polished pavement), when water depths are high, and when groove depth is low. This is the situation for any given tire design.

Ideally, one should design tires to give maximum performance at high operational severities consistent with other tire performance requirements. Does routine testing under some arbitrary test site conditions (pavement, water depth, speed) give realistic information about high severity performance? The answer to this question will vary from one situation to the next. Rewriting Eqn. 3 in these terms, we have

$$\begin{aligned} \text{Operational Severity} &= \frac{k \quad V}{WD_P \; h_G/h_W} \\ &= \frac{k \; V \; h_W}{WD_P \; h_G} \end{aligned} \tag{5}$$

It will be readily appreciated that the variables in Eqn. 5 are the ones that change both with time and location as a motorist operates a vehicle in normal usage.

A plot of μ_c vs. speed (velocity) contains two important pieces of information, and it is necessary to use both of these to fully utilize the data. These are the slope or velocity gradient of μ_c and the general location (in a vertical sense) of the curve or line. This latter quantity is specified by the value of μ_c at some selected speed. We postulate that the high operational severity rating of a particular tire design is given by how much traction capability or reserve, a tire has above a critical speed. For the linear plots of μ_c as a function of velocity, this is defined as the area bounded by the extrapolated curve, the vertical line at 60 mph and the abscissa. This is shown in Fig. 3 as Area A.

The Area A, multiplied by 10 to obtain convenient numbers, is called the "Ultimate Performance Rating" abbreviated by UPR. This name is selected to provide a unique term. We realize that the rating of tires based on this calculated parameter is

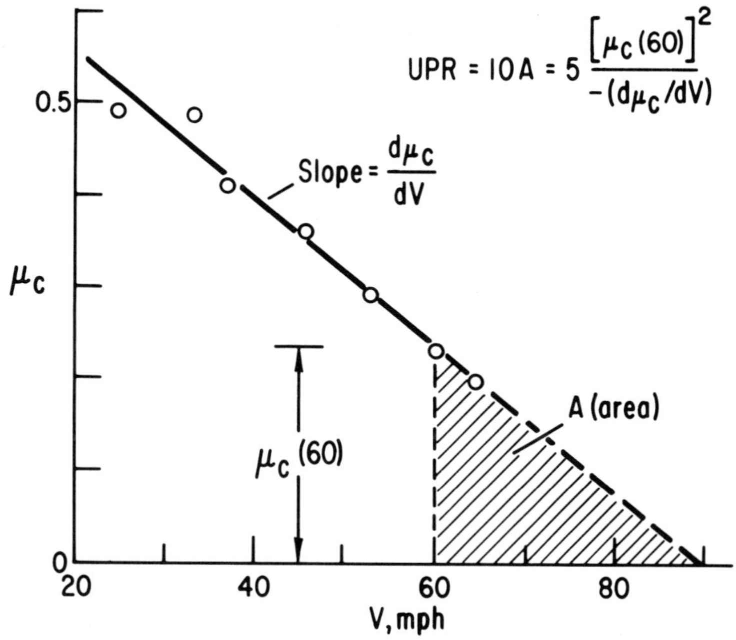

Fig. 3. The ultimate performance rating, UPR, concept.

only a first order approximation to their ultimate performance. It is, however, a much better indicator of high operational severity performance than any other simple parameter, especially values of μ_c selected at some arbitrary speed.

It may be shown from geometry that the Area A or UPR is given by

$$\text{UPR} = 10\text{A} = 5\mu_c{}^2 \Big/ \frac{-\,d\mu_c}{dV} \tag{6}$$

where μ_c is the value of μ_c at 60 mph.

In calculating UPR, we assume all such plots will smoothly go to $\mu_c = 0$ at some speed normally above the upper limit testing speed of 65 mph. With only a handful of exceptions, *all* plots of wet μ_c vs. speed are linear. Well over 500 such plots have been made with a variety of tires, surfaces, and water depths. Values of μ_c as low as 0.04 have been obtained and all evidence to date indicates that the plots are linear down to this μ_c limit.

Evidence will now be presented to show that UPR measured under low to moderate operational severity predicts actual available (cornering wet traction coefficient) *traction forces* under high severity conditions. Two factors that govern operational severity will be selected, h_G, groove depth and h_W, water depth.

References pp. 19–20

In Table 1 three tire designs are considered and are designated A, B, C: these represent different patterns and aspect ratios with identical tread compounds.

TABLE 1

Correlation of UPR and High Operational Severity Available Traction, μ_c

Operational Severity	h_G	Measured Parameter	Rating		
			A	B	C
1) Moderate	Full Non-Skid[a] Groove depth	μ_c (60 mph)	100	83	75
2) Moderate	Full Non-Skid[a] Groove depth	UPR	100	37	28
3) Severe	30% Non-Skid[b] remaining	μ_c (60 mph)	100	30	21

a = 0.37 inch
b = 0.12 inch

When tires B and C with full tread depth are rated compared to A by using the measured μ_c, at 60 mph, we get ratings of 83 and 75. Taking the same data and using $d\mu/dV$ in addition to μ_c at 60 mph and calculating UPR we get the rating values of 37 and 28 for B and C. This is based on UPR alone under full non-skid depth conditions.

When these tires are 70% worn (30% remaining), the *measured* μ_c at 60 mph gives the ratings 30 and 21 for B and C. The UPR full non-skid calculated ratings of 37 and 28 are much *closer* to the measured μ_c (60 mph) ratings (under severe conditions) of 30 and 21 than are the ratings 83 and 75. Thus, UPR ratings from tires with full tread depth are good predictors of relative tire performance (μ_c values) after 70% of the tire tread depth is worn away. Table 2 lists the actual μ_c and $d\mu/dV$ values used to compute the ratings shown in Table 1.

TABLE 2

Actual Values – Tire Series A, B, C

Surface 1 (see Table 3)			
Tire	h_G	μ_c at 60	$\frac{-d\mu}{dV} \times 10^3$
A	0.37	0.38	1.9
	0.12	0.28	6.5
B	0.37	0.31	3.5
	0.12	0.074	9.2
C	0.37	0.28	3.8
	0.12	0.054	11.4

We now turn to another factor in the operational severity equation, the pavement water depth, h_W. Additional evidence to support the UPR concept is presented in Table 3. These data are for two tire sets (different from those of Table 1) that are identical except for non-skid or groove depth; the groove depth of tire B is 50% that of tire A.

Table 3

Correlation of UPR and High Severity Traction

Surface*	Operational Severity	h_W	Measured Parameter	Rating A	Rating B
(I) Smooth[a]	Moderate	0.02 in.	μ_c at 60 mph	100	75
	Moderate	0.02 in.	UPR	100	42
	Severe	0.16 in.	μ_c at 60 mph	100	40
(II) Textured[b]	Moderate	0.01 in.	μ_c at 60 mph	100	79
	Moderate	0.01 in.	UPR	100	28
	Severe	0.08 in.	μ_c at 60 mph	100	38

[a]Smooth concrete (low macro-texture)
[b]Asphalt aggregate (moderate macro-texture)
*In the remainder of this paper, Surface (I) and Surface (II) are the same as in this table.

The two different test surfaces represent a well-worn concrete road and a typical asphalt aggregate road worn to some degree by traffic. The asphalt road has much more texture than the smooth concrete.

Under moderate operational severity conditions, B has ratings of 75 and 79, respectively, on the two surfaces on the basis of the 60 mph μ_c values. From the same basic data, using both the slope and 60 mph μ_c, the calculated UPR gives ratings (B compared to an A value of 100) of 42 and 28, respectively, on the two surfaces. If conditions are now altered to produce a very severe operational situation by testing in deep water, the ratings based on actual measurements of μ_c are now 40 and 38, respectively. Thus, the UPR rating obtained under moderate severities gives a much closer prediction of the rating of tire B under severe conditions than does the measured μ_c under moderate conditions. The actual available traction is, of course, a direct function of μ_c.

The water depths for this test series were measured *above* the tops of the asperities by a special water depth meter that was previously described in (6). Due to this, water depths on the textured surface (II) cannot be directly compared to those on the smooth surface. A depth of 0.01 inch on the textured surface represents more actual water than an equal depth on the smooth surface. From the standpoint of good tire

References pp. 19–20

traction or grip, the total water content should be the important factor in dynamic water drainage from the tire contact area. This effect partly offsets the 2 to 1 ratio in "above asperity" water depths reported in Table 3 and is, we believe, primarily responsible for the lower B ratings on the textured surface.

TIRE PERFORMANCE AND WATER DEPTH – FURTHER DISCUSSION

Part of the data used to illustrate the ability of moderate condition UPR to predict severe condition traction or μ_c was obtained in a test series where four water depths were used in conjunction with two tire sets. The sets A and B were bias ply tires with a full tread depth of 0.35 inch and a one-half non-skid depth of 0.17 inch, respectively. The four water depths are outlined below.

Test	Water Depth, * inch Surface I	Surface II
1	0.005	0.005
2	0.036	0.017
3	0.050	0.036
4	0.25	0.06

*-above asperity depth

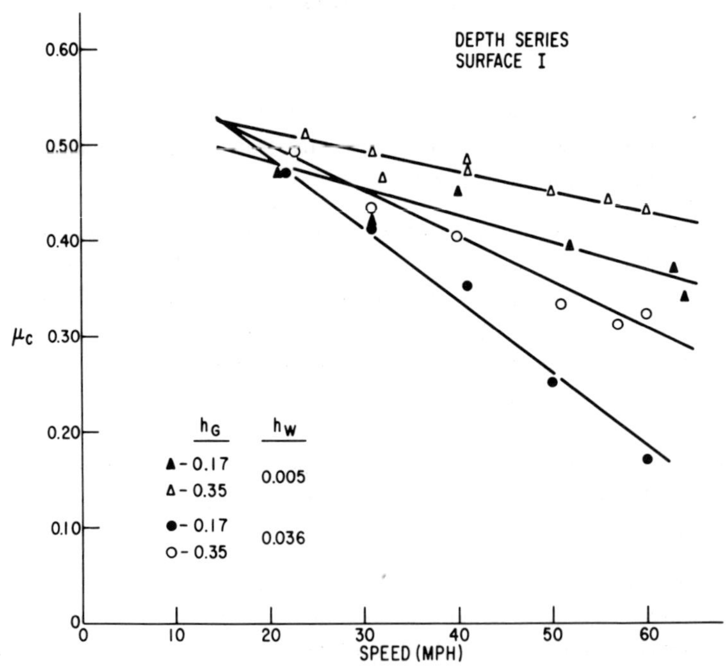

Fig. 4. Cornering wet traction performance of tires of the same tread pattern but two different groove depths in two water depths.

The depth of 0.005 was obtained by applying water to the surface and then brooming off the "excess" water with a power broom. Such brooming gives a "just wet" condition. The depth of 0.005 inch is an estimated value. Other water depths were obtained by sprinkler pressure adjustment.

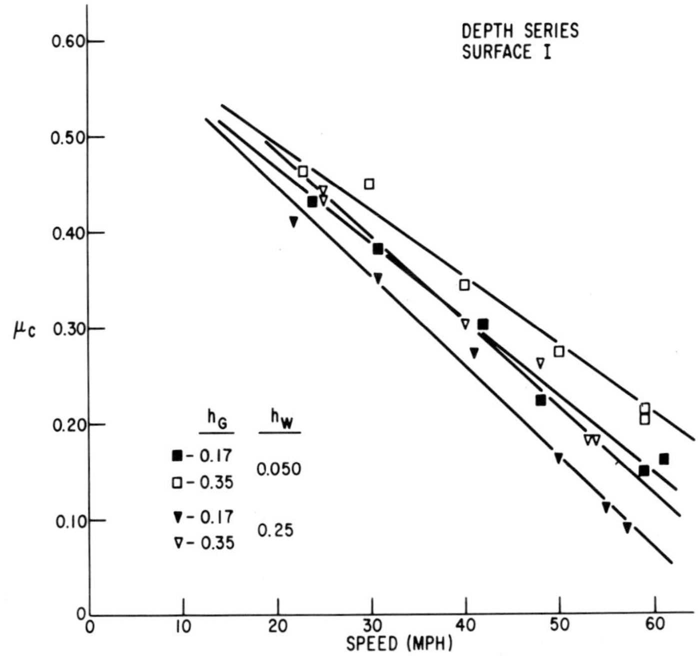

Fig. 5. Cornering wet traction performance of tires of the same tread pattern but two different groove depths in two water depths.

Figs. 4 and 5 show plots of μ_c as a function of speed on Surface I at the various tread groove and water depths. The data on this surface encompass the widest range of water depths, and they are typical of a "low coefficient" surface. Similar results were obtained on the more textured surface (II).

The relationship between μ_c (at any speed) and the water depth is depicted in Fig. 6. Plots of μ_c at 20, 30 and 60 mph as functions of h_W are shown. The first items of interest in this figure are the linear dependence of μ_c on log (h_W) and the essentially numerically constant differential between tires with full and half-tread depth at any selected speed. It is fortunate, in a sense, that the logarithmic h_W relationship exists. Traction requirements in ordinary motoring would not be met if a sharper depth dependence existed. The accumulated experience of motorists confirms this somewhat gradual dependence of μ_c on water depth.

The linear-logarithmic dependence of μ_c and h_W should not be interpreted as having a minor or inconsequential influence on the values of μ_c measured in tire

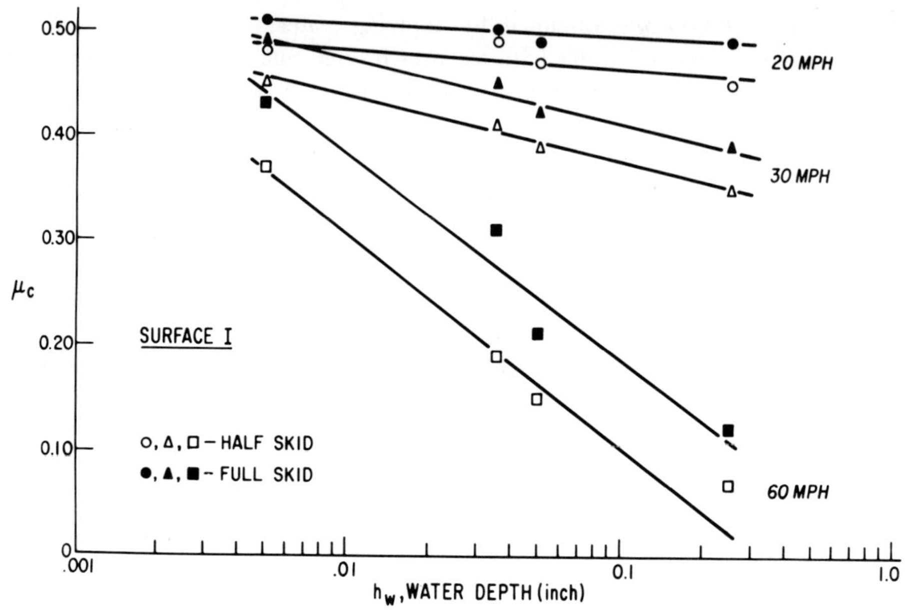

Fig. 6. The influence of water depth on tire cornering wet traction coefficient for tires of the same tread pattern in two different groove depths.

testing. On the usual test surfaces with lateral slope in the 1-2% range, ordinary (sometimes hourly, sometimes daily) wind velocity – direction variations can change water depth from 0.01 to 0.03 inch or more. This is a factor of 3 or greater. Fig. 6 shows how such changes materially affect the relative rating. A major source of water depth variation is the effect of wind velocity and direction. For lateral sloped test surfaces, the lateral wind velocity vector across the test strip can have a marked influence. Measurements at the Automotive Proving Grounds Inc., Pecos, Texas, where the tests described herein were conducted, show smooth surface film depths varying by a factor of three for wind velocities of 10-15 mph blowing east vs. west. The track is oriented north-south and such east-west variations either thicken water film depths or thin down these films. This is a very important factor for high speed testing.

The second item of interest is the virtual lack of dependence of μ_c on water depth at 20 mph and the very slight dependence at 30 mph. This change in water depth dependence of μ_c is illustrated in Table 4. The difference in this dependence when comparing values of μ_c taken at 20 and 60 mph is a 20 to 1 ratio (20 mph = 1). The ratio of the two speeds is 3 to 1.

TABLE 4

Effect of Speed: μ_c Dependence on h_W

Speed	Slope*
20	0.01
30	0.05
60	0.20

(* = $\Delta\mu_c$ per decade)

The near independence of μ_c and h_W in the 20–30 mph range derives from the mode of frictional lubrication that exists at these speeds. It is, for all practical purposes, boundary lubrication as defined by Dowson (7). Such lubrication is produced by a water layer of approximately molecular dimensions between contacting surfaces. Excess lubricant plays no part in this mode of lubrication. At 60 mph a substantial part of the lubrication mode is hydrodynamic. Under such conditions the mass of water that must be moved (squeezed out) by the tire is a major determinant of the frictional force developed. For a fuller discussion of this, see (6, 7) and the references contained herein.

Fig. 6 illustrates one of the reasons why there has been scant mention of water depth in the literature on tire traction testing. By and large, the majority of wet traction has been done at low to moderate speeds (20-40 mph) and under these conditions, water depth really is not that important. The practical experience of people doing such testing confirms this. When the important high speed range is approached, water depth is quite important. This was the impetus for the development of the water depthmeter previously described (6).

Although the slopes for tire sets A and B in Fig. 6 are essentially the same, the relative rating or ratio of B to A is not invariant with water depth because of the decreasing base or reference μ_c value (which is the value for A). Since such testing is comparative, this variation is a rather serious matter. The change in relative rating is illustrated in Fig. 7. The relative rating or $[\mu_c\,(B)\,/\mu_c\,(A)]$ x 100 is plotted as a function of water depth for two speeds, 30 and 60 mph.

The invariance of the relative rating at 30 mph is to be expected from the results of Fig. 6 which were discussed previously. However, the marked dependence of the *relative rating* at 60 mph on water depth is a serious matter from the standpoint of tire wet traction testing technique and philosophy. It clearly shows the fallacy of the "control tire" concept, expecially at high speeds, that is almost unanimously employed by all engaged in this sort of testing. In using the control tire concept, one assumes that external factors that are difficult or impossible to control, affect the test or experimental tire in the same way and to the same degree that they affect the control tire. This "equal effect assumption" than renders logical the use of the

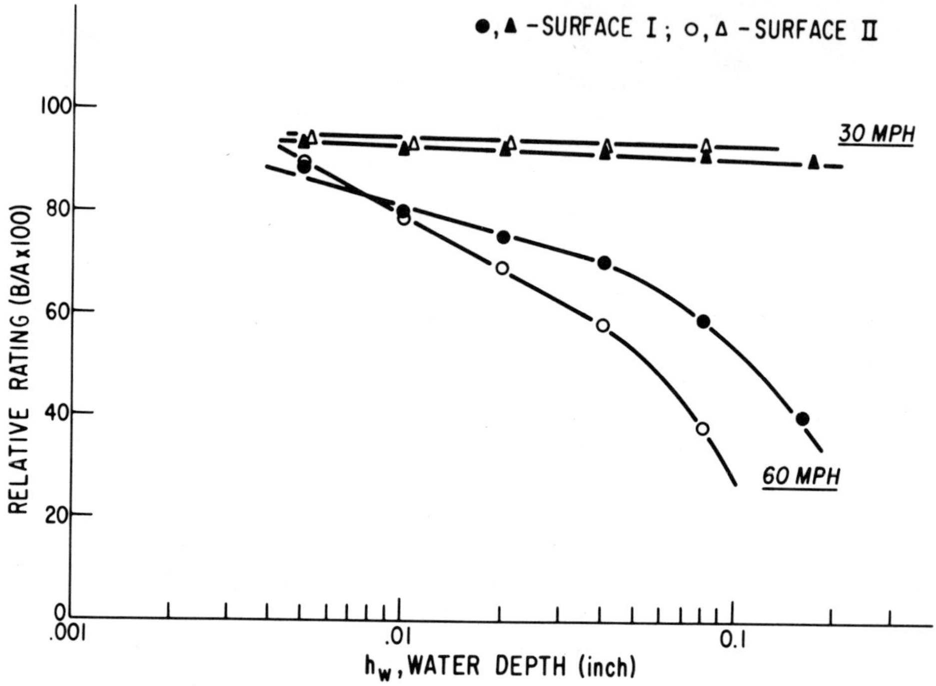

Fig. 7. Relative wet traction rating of tires with the same tread pattern but two different groove depths as a function of speed and water depth.

relative rating approach. If one can show, as has just been done, that this "equal effect assumption" cannot be made for such a simple tire performance or operational severity factor as tread depth, then there is serious doubt as to whether it can be applied in the very broad manner normally used.

The data plotted in Fig. 7 certainly show it cannot be applied in this particular case. And there is serious doubt, on our part, that it can be applied in an omnibus manner without considering the *particulars* of the test situation in question. Its application is permissable only in those cases where the difference between the experimental and the control is minimal.

What is minimal and what are the differences in question? No firm answer can be given. Answers depend upon particulars and the particulars change from case to case. The core problem is due to the *interaction* of "response characteristics" of the control and experimental tire. "Response characteristics" dictate how a tire's performance will change as certain design features are changed i.e., tread pattern, tread compound, and construction type. When external factors are varied, the influence of the external factors is not the same for each of these design features. Thus, the response effects are not simply additive. Such non-additivity or failure of superposition underlies practically all plysical and chemical phenomena, and one should not expect any different behavior for tire performance testing.

If two tires (types of tires) are quite close in their overall design, their nonlinear response can be closely approximated by a linear relation; and thus, the control tire concept is applicable. The use of this "approximate linear" approach for ordinary testing requires numerous "different" controls; each not far removed in performance – design features from its accompanying experimental or candidate test tires. The idea of numerous controls or a "variable constant" is closely akin to a contradiction in terms and involves difficult judgements on when a tire is a control and when it is not. The use of such an approach is also inefficient as a substantial testing volume would be engaged in non-productive control testing. This is a fundamental basic problem in all wet traction testing, and those doing such testing must give serious thought to this problem.

It seems self-evident that two criteria must be met for any real solution. First, a rigorous control of external testing factors must be maintained. The two most important are water depth and surface texture. These must be maintained as nearly constant as possible; hour to hour, day to day and week to week. The micro-texture (asperities in 0.1 mm range) of a test surface is most subject to seasonal and weather induced changes. Macro-texture is ordinarily reasonably constant.

The micro-texture of a test surface changes due to the polishing action of tires during actual testing. A plateau or equilibrium "state of polish" must be maintained for proper testing. The use of power broom-type polishers should be investigated. The normal nylon bristles of such a broom would be replaced by a series of rubber paddles or blades to give a fairly high pressure wet-polishing action. One half hour of such polishing could do much more than a similar period of actual car or trailer "run-in" testing.

Second, a consensus must be reached as to what type of pavement or pavements should be employed to evaluate the wet traction performance of tires. Without either of these criteria, the test data produced by the wet traction test community will be contradictory and confusing. The origin of this state of affairs is the same as the cause of the variation of B/A relative rating with water depth that was shown in Fig. 7, – the interaction of tire design factors with external factors. Once such agreements and decisions are reached, real progress can be made toward eliminating conflicts and producing realistic tire test data for wet traction performance.

REFERENCES

1. *G. Maycock, Rubber Chemistry and Technology, Vol. 41, p. 780, 1968.*
2. *B. J. Allbert, J. C. Walker, Rubber Chemistry and Technology, Vol. 41, p. 753, 1968.*
3. *A. Schallamach, Journal of Institute of Rubber Ind., Vol. 4, p. 251, 1970.*
4. *W. B. Horne, Astronautics and Aeronautics, August, 1967.*

5. *"Skidding Accidents" brochure published by Institute for Road Safety. Research – SWOV Voorburg, 2110, The Netherlands.*
6. *A. G. Veith, Rubber Chemistry and Technology, Vol. 44, p. 962, 1971.*
7. *D. Dowson, chaper in "Boundary Lubrication – An Appraisal of World Literature" Ed. by E. Kling, E. Klaus, R. Fein, ASME Special Publication, N. Y., 1969.*

DISCUSSION

Prof. Thieme

We now call for questions on this very interesting paper.

G. G. Balmer

I would like more detail on the measurement of the water depth.

Pottinger

The water depths noted in the paper refer to "static" water depths measured at 18-20 locations under equilibrium lateral water flow conditions on the test surface. They are, thus, average "static" depths above the tops of the asperities.

The water depth-meter consists of 10 brass pins each insulated from the case and arranged in a circular pattern with a 100 mm diameter. The first pin is located 0.01 inch (or 0.4 mm) above the base plane of the depth-meter which is the plane of the surface asperities. The base plane is also defined by the bottom base legs as shown in the diagram. The additional pins are arranged in a stair-step pattern each pin incremented 0.01 inch farther away from the base plane than its preceding adjacent pin.

Water contact with any pin is evidenced by a signal light that indicates current flow through a transistor amplified circuit. The depth is defined by the highest numbered pin that gives a water contact reading. Measured values are, thus, read to the nearest $\pm$ 0.01 inch.

Water depth measurements are made after equilibrium water application has been attained. This implies that there is no change of wind velocity/direction factors during testing.

Balmer

Is this a water depth measured from the top of the asperities?

Pottinger

It is from the top.

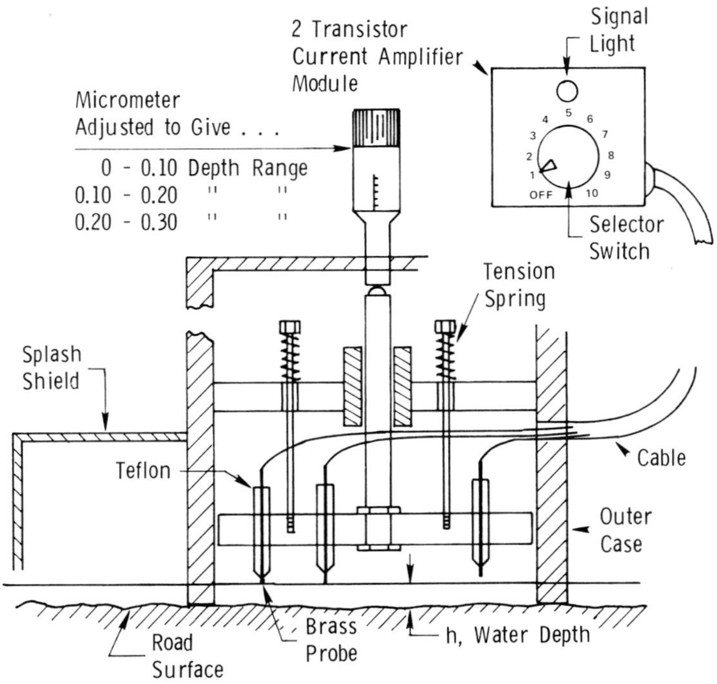

Fig. 1. Schematic of water depth meter.

Balmer

Thank you.

K. A. Grosch

I wonder whether it would not be more appropriate to control the rate of water application to the surface rather than to insist on having the same water depth. You may have a very close texture surface which will easily take a thick water film and you may have an open textured one which drains the water well. On such a surface it would be very difficult to maintain a water film of constant thickness. A rainfall does not take any notice of well draining or close surface; one should control the rate of application rather than the film thickness.

Pottinger

We do control the rate of application, but to study the effect of static water depth, this static depth must be measured. In southwest Texas the influence of wind velocity and direction is of great importance especially for high speed testing on low macro-texture surfaces. Lateral wind barriers greatly reduce the effect of wind velocity on water depth variations, but improved barriers are required.

J. D. Walter

Could you elaborate on the rationale for picking the 12° slip angle for the coefficient; and if you worked with a lower slip angle, would this change relative rankings?

Pottinger

The rationale for selecting the 12° slip angle is fully outlined in the referenced paper (6). Briefly, in the 10-14° range a maximum or plateau wet cornering or lateral force is developed on all but the most abrasive, highest macro-texture surfaces. The desire, therefore, is to measure the maximum wet cornering or lateral force that the tire is capable of producing.

Walter

Would you get crossovers if you were to work with some other angle?

Thieme

For instance 〉8°.

Pottinger

Testing at a 6° slip angle would not change the relative results of this paper, but below this, say 2-3°, relative results may change somewhat.

Thieme

We use values 〉8°. It almost seems that you get to the top and then you don't run into that kind of problem.

F. R. Davis

I would like to ask about the operational severity relations that you have developed. What factors go into the quantity K? Are there any exponents on those drainage terms, e.g. 1/3 or 1/2?

Pottinger

I am afraid I am unable to answer you for several reasons. Some are ignorance, some are proprietary.

L. R. Hamkins

I have two questions relating to the water depth. To what extent does the tow vehicle become involved in changing the water depth at high vehicle speeds? And secondly, what effect or how can you functionalize the amount of water brought into the contact area by the tire itself, that is, residual water coming back around on the tire?

Pottinger

The tow vehicle does influence the actual dynamic water depth experienced by the test tires, but the test tires do not "track" in the same path as the tow vehicle tires, they are offset by 16 inches and, thus, intercept a water film undisturbed by the partial "wiping dry" effect of the tow vehicle tires. The aforementioned influence of the tow vehicle on "dynamic" water depth is due to a spray induced by the vehicle. We operate on the premise that an increase in "static" water depth yields a corresponding increase in the dynamic water depth, which of course cannot be easily measured. This situation is analogous to actual real world wet driving conditions.

Hamkins

Does the mass transfer of water into the contact area from the tire itself begin to have an effect on the water depth as the speed goes up? And how then do you functionalize that effect, if in fact you can, or if it ever has been?

Pottinger

We are aware that this happens. We have done some photographic studies of the way the water is transferred around the tire, sort of entrained by the tire, if you will, but at the present moment we have no definite statement on this subject.

A. Chiesa

Have you considered only the slide position, i.e. locked wheel or have you considered also the peak value position, transient condition? The transient condition seems to correlate better with actual behaviour of the car's performance.

Pottinger

The act of measuring the cornering force of a tire on a wet surface at 12° slip angle is analogous to measuring the wet peak braking force which of course is a transient condition. The actual sliding velocities of rubber vs. pavement in both of these measuring modes, i.e. cornering vs. braking, is very similar. The tire is rotating in both modes and maximum forces are being generated. This is, in fact, the reason for selecting a 12° slip angle. The sliding is mainly lateral in cornering, of course, and longitudinal in braking testing; but the contact patch sliding phenomena are similar. We have data that shows that μ_c and μ_p (peak braking coefficient) are highly correlated. See reference (6).

Thieme

Thank you gentlemen for contributing to the discussion, and as chairman I would like to thank both authors for their paper describing wet tire traction. We have heard that due to the difficulty involved in assessing wet tire traction performance at particular speeds, a "UPR" factor was defined, the Ultimate Performance Rating. Operational severity, road and tread drainage, and water depths were discussed and how much the traction capability is influenced say at 60 km/hr. It was particularly interesting to see that for different pavement textures, how the actual wet cornering coefficients μ_c and relative ratings were influenced as water depth changed. You also brought into question the control tire, which is a legitmate question I personally believe.

I think, gentlemen, we all feel very grateful to Dr. Veith and Dr. Pottinger for having produced their paper, and to Dr. Pottinger personally, for having so ably put forward before the audience their paper. I am sure Symposium members would like me to congratulate these gentlemen. On their behalf I propose a vote of thanks.

TIRE HYDROPLANING: TESTING, ANALYSIS, AND DESIGN

ROBERT W. YEAGER

The Goodyear Tire and Rubber Company, Akron, Ohio

ABSTRACT

Water film action between the surface and the tire was photographed for passenger and truck tires by driving and/or pulling a locked tire over a glass plate.

These results were correlated with vehicle stopping performance on specially developed water depth controlled surfaces.

Normal and tangential unit shearing forces were measured in the footprint under braking and cornering modes on surfaces that were dry or had simulated wet and flooded conditions.

A rain laboratory was developed to measure water depths on various road surfaces as a function of cross slope, drainage length, and rainfall intensity. These results were correlated with results reported by the Texas Transportation Institute and The Ministry of Research.

Expected rainfall intensity, frequency, and distribution was gathered for all parts of the United States.

Road design data were obtained to determine proper drainage lengths and cross slopes at various points in highway configurations.

A subjective analysis was made of horizontal visual range during several brief thundershowers as a function of measured rainfall intensity.

The data obtained from these various studies were compiled to gain an insight into the tire hydroplaning phenomenon and its effect on wet traction in general.

References p. 57

INTRODUCTION

For proper vehicle control when driving at high speeds, maneuvering, or decelerating, maximum frictional levels must be available upon demand. Tire hydroplaning is one condition adversely affecting friction levels. This condition occurs when the tire footprint is separated from a wet road surface, to a varying degree, by a film of water. To nearly the same degree that the footprint is separated, the stopping, cornering, or tractive ability of the tire is degraded.

This report describes the varying conditions of tire hydroplaning, the method and probability of hydroplaning occurring, and the influencing factors.

PHOTOGRAPHING A DYNAMIC TIRE FOOTPRINT

As a tire is driven, rolled, or pulled across one of Goodyear's three glass plate facilities, a photograph can be taken of the tire footprint from either a movie camera or a single-photo still camera. A fluorescein dye solution is added to the water to indicate flow and define tread-to-surface contact. The black portion of the tread footprint is that area in contact with the glass. As the proportion of the green dye solution is increased, thin layer definition increases.

VISCOUS TIRE HYDROPLANING

When removal of the water film from beneath the tire footprint is resisted by internal friction within the fluid layer, the resulting action is referred to as viscous tire hydroplaning. The viscous properties are relatively low for water and this action generally is related to thin films. The formation of this fluid layer can occur on smooth or macro-textured surfaces at any speed above 20 mph. Removal of this layer from between the tire footprint and the road surface requires discontinuities in the tread or road surface and a relative localized sliding between the interfaces. This type of action is best shown for a free rolling smooth truck tire at 40 mph (see Fig. 1). A thin film of water has penetrated the complete tire footprint area and has separated the tire from the surface.

DYNAMIC TIRE HYDROPLANING

In the region where the water depth is relatively deep and rapid changes in flow direction occur, the inertial effects of the fluid usually are dominant. Resulting changes in momentum cause a reaction force normal to the tire tread. As vehicle speed increases, the reaction force increases and a hydrodynamic lift begins at the forward edge of the footprint and tends to separate the footprint from the road. This separation penetrates gradually through the length of the footprint, excluding the shoulder rib region, with increasing vehicle speed.

See color plate, page 63.

Fig. 1. Viscous hydroplaning of smooth truck tire at 40 mph in 0.080 inches of water (free rolling).

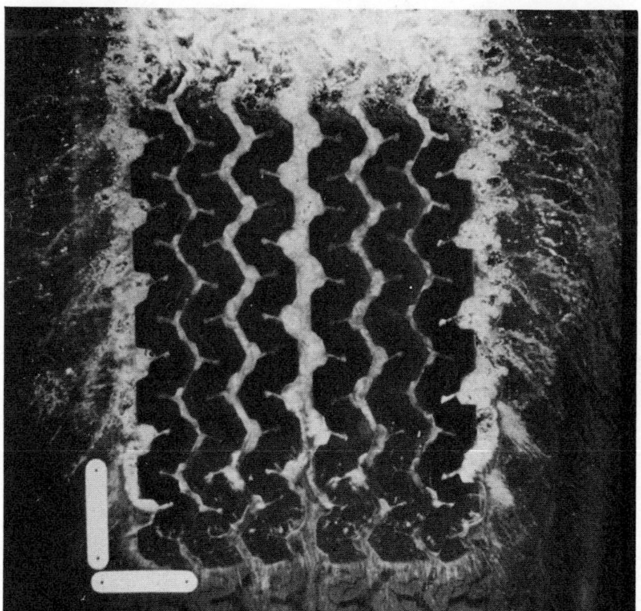

Fig. 2. Dynamic partial tire hydroplaning of passenger tire at 60 mph in 0.080 inches of water (free rolling).

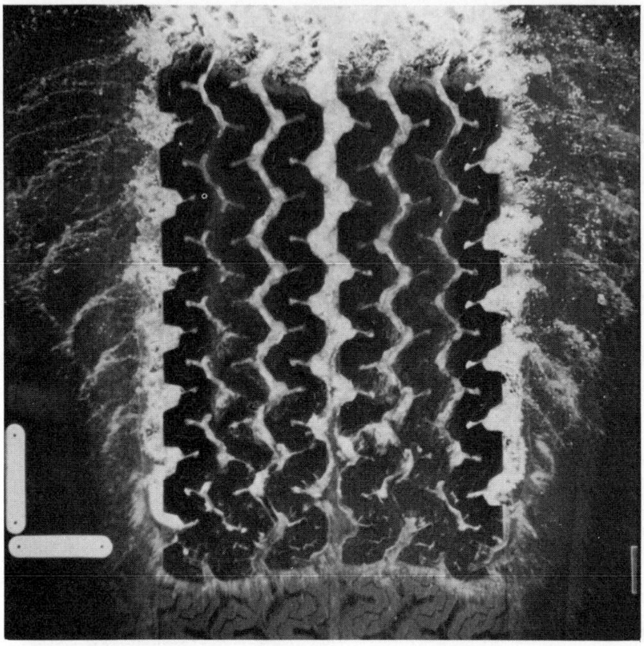

Fig. 3. Dynamic tire hydroplaning of passenger tire at 70 mph in 0.080 inches of water approaching a condition of full tire hydroplaning.

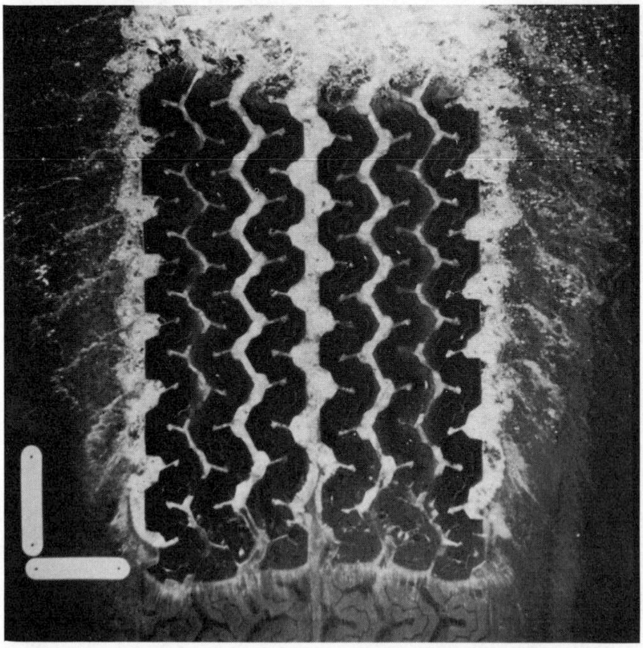

Fig. 4. Dynamic tire hydroplaning of passenger tire at 70 mph in 0.050 inches of water.

The progression of road-to-footprint separation occurs over a varying speed range depending on many factors, eventually separating the major portion of the footprint area. This type of action is defined as "dynamic tire hydroplaning"; in the early stages of separation, it is referred to as "partial tire hydroplaning."

In Fig. 2 a passenger tire is in the early stages of partial tire hydroplaning at 60 mph. The second and fifth rib are separated from the road surface in the forward edge of the footprint. As the vehicle speed is increased to 70 mph as shown in Fig. 3, the degree of partial tire hydroplaning increases and the tire is approaching a condition of "full tire hydroplaning."

Dynamic tire hydroplaning is affected by the water depth standing on the surface and the surface texture. Fig. 4 illustrates the decrease in the extent of dynamic tire hydroplaning as the water depth is decreased to 0.050 inches.

Dynamic tire hydroplaning can occur to any pneumatic tire traveling on a wet surface in a free rolling, locked, and/or cornering mode.

FREE ROLLING AND LOCKED-WHEEL DYNAMIC TIRE HYDROPLANING

There are many influencing factors that determine the vehicle speed at which dynamic tire hydroplaning occurs. For a free rolling tire, the factors are:

- The ability of the pavement to rapidly drain water horizontally and vertically under pressure from the tire footprint.
- The ability of the tire tread to absorb and move water rapidly out of its path.

In addition to the factors listed above, a sliding tire depends on:

- The size of the wedge that is built up in front of a sliding tire – This wedge is caused by a plowing action of the tread that contacts a given water volume and is greatly affected by the surface texture and of course by the standing water depth on the surface. For a rough texture, the wedge will not occur on normally drained surfaces except for the possible condition of a completely smooth tread. As the surface becomes smooth, the wedge size will increase, especially as the tread nonskid depth is reduced.
- The forward spray increasing the standing water depth – The amount of forward spray occurs as a maximum between 30 and 40 mph. The amount will increase generally until partial tire hydroplaning begins and then gradually decrease. It is affected by the tread design and the amount of tire wear. In Fig. 5, this forward spray can be observed for a sliding tire crossing the glass plate during a locked wheel hydroplaning test.
- The ability of the fluid to flow longitudinally through the tread grooves.

References p. 57

Fig. 5. Forward spray occurring for sliding tire.

RESEARCH FACILITIES AND PROGRAMS

Requirements – To evaluate the speed at which flow through the footprint creates hydrodynamic lift and to study the flow characteristics of the fluid, three glass plate facilities were installed with the necessary testing area and photographic equipment. To correlate the results of the glass plate with those from actual road surfaces, seven specially developed water depth controlled surfaces were installed at the Akron and San Angelo test track.

The absorption of the contacted water through the footprint, as well as the resulting lift and removal of a portion of the contacted water out of its path, depends on the normal tread and shearing forces acting in the footprint. To measure these forces, a tri-axial force transducer was installed in the test track and in a laboratory mill with elaborate and rigidly controlled test conditions.

An estimation of the water depth that may be encountered during the rain-showers is important because the depth of water for a given surface and tire determines the speed at which dynamic hydroplaning will occur. A rain facility was constructed to determine water depths on actual road surfaces as a function of cross slope, drainage length, and rainfall intensity.

The water depth study included the probability of expected rainfall intensities and driver visibility during a rainfall based on subjective tests.

Glass Plate Facility – A complete description of the glass plate facility is given in Reference 1. At present there have been nearly 5,000 tire hydroplaning tests completed on the glass plate facility.

These tests have been conducted primarily to evaluate tread design, but have included effects of inflation, load, tread width, water depth, and tire construction at test speeds from 2 to 140 mph.

Fig. 6 illustrates a vehicle crossing the glass plate at a speed of nearly 140 mph.

Fig. 6. Vehicle crossing glass plate at nearly 140 mph.

Added features to the facility are the enclosing of this portion of the track with a monolithic sunshade and the installation of a high-intensity-strobe, single-photo camera system for increased resolution.

Fig. 7. Lighting system for high-speed movie camera and single-photo camera.

References p. 57

Fig. 8. Glass plate facility view of cameras and lighting.

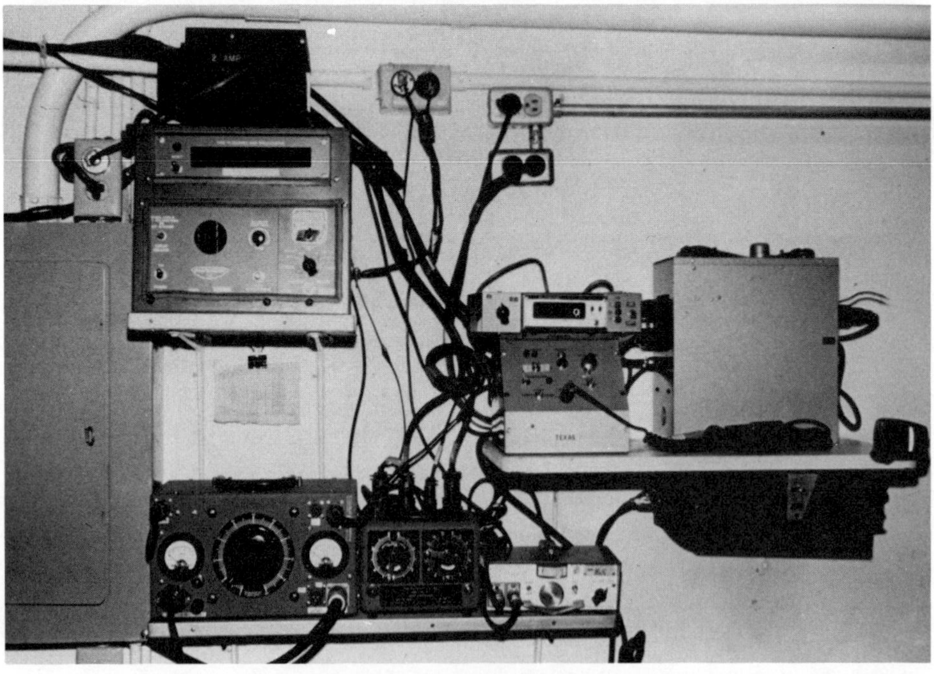

Fig. 9. Instrumentation for glass plate facility.

The versatility of the glass plate complex is shown in Figs. 7 and 8. In Fig. 7, the two larger 2,000-w floodlights located at the underside of the glass plate are used with a high-speed movie camera that stops the action of the tread. For greatly increased resolution, the center lights are high-intensity strobes with a flash duration of 1/30,000 sec. Fig. 8 shows the installation and setup of the movie camera and the high-resolution single-action camera.

Instrumentation for the camera, lighting, and speed timing are shown and located in the laboratory below the glass plate (see Fig. 9). The instrumentation includes a two-way radio, a power supply for the strobe lights and the high-speed motion camera; interval timers for shutter delay and vehicle speed; and control units for both cameras. The single photo method of photographing a tire allows excellent definition of tread geometry and the determination of specific flow characteristics which are helpful in the design of efficient groove shapes and paths.

An enlargement of a 16-mm frame taken by the high speed movie camera is shown in Fig. 10. The tire was traveling at 70 mph in 0.080 inches of water.

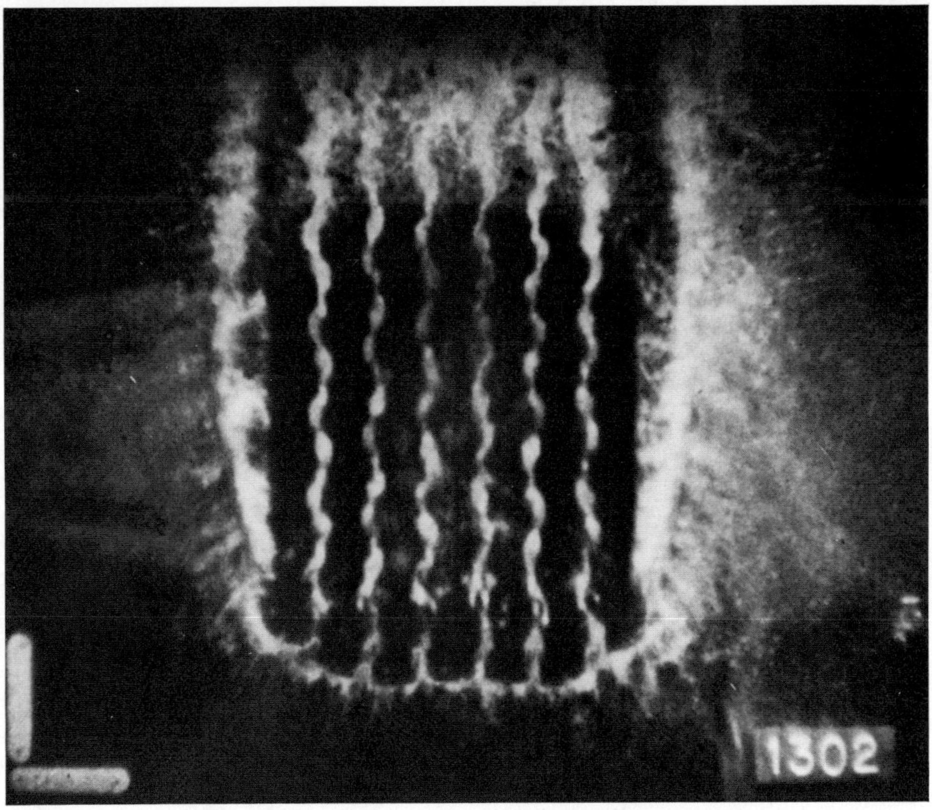

Fig. 10. Enlargement of 16-mm frame of passenger tire (free rolling) at 70 mph in 0.080 inches of water; camera speed is 7,200 frames per second.

References p. 57

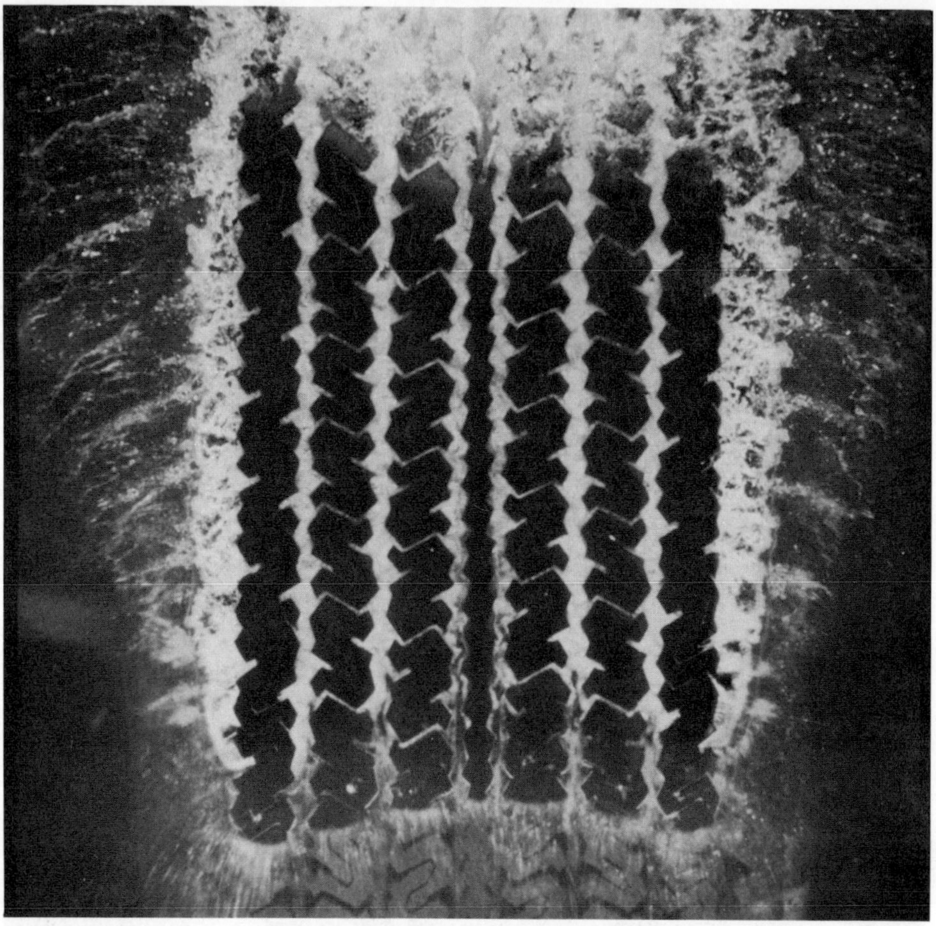

Fig. 11. Enlargement of a 4 x 5 format 120 size frame from a single-photo camera with strobe firing at 1/30,000 sec. (free rolling); speed is 70 mph in 0.080 inches of water.

In contrast, the photo shown in Fig. 11 was taken by the single-action camera at 70 mph. With this system, water depths between the tire footprint and the glass plate are being obtained and studied with highly developed Moiré fringe techniques.

Water Depth Controlled Surfaces – The need for test surfaces with precisely controlled water depth was found to be necessary in past glass plate hydroplaning tests. Five such surfaces were constructed in Akron to determine the feasibility of control techniques. Each surface was 3 ft. wide by 250 ft. long (see Fig. 12). Based on preliminary results from these surfaces, two water depth controlled surfaces were installed on the edge of the glass plate road at the San Angelo, Texas, facility.

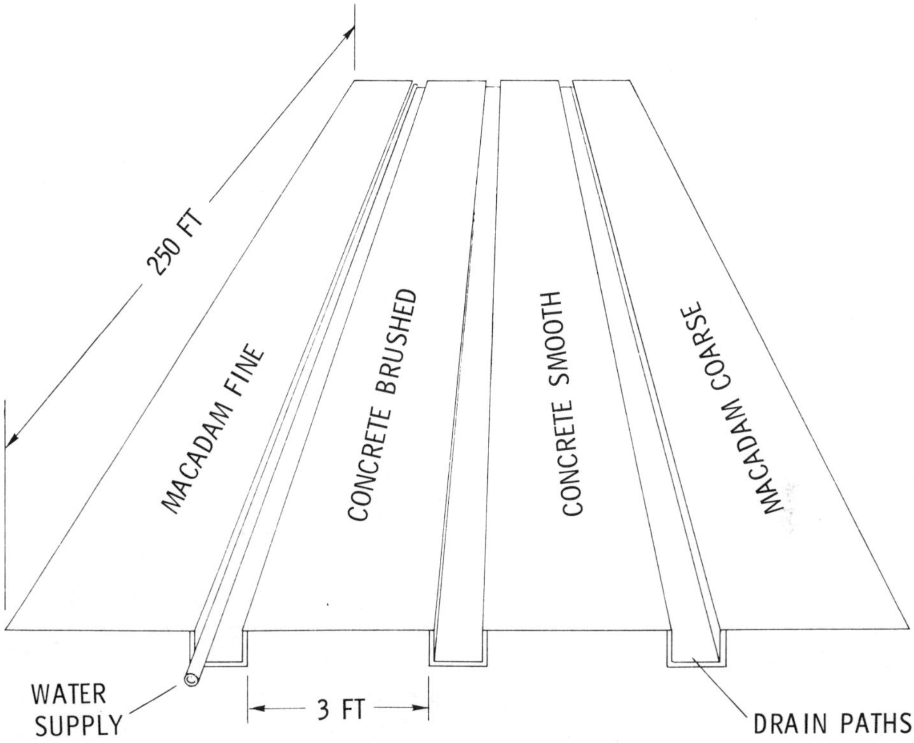

Fig. 12. Water depth controlled surfaces.

After completion of these surfaces, over 5,600 test runs were made to compare load, inflation, speed, and water depth using a bias/belted Polyglas test tire and an SAE straight ribbed control tire. Additional testing utilized various production and control tires to speeds of 70 mph.

A noticeable advantage of these surfaces over identical wide surfaces in the same testing area was the increased repeatability of wet skid trailer test results due to a more uniform water depth. Engineers also found that when surfaces contain a high degree of macro-micro texture and are free of puddling and lightly wetted, there is not any great change in either skid trailer or slide wet skid braking coefficients with increased test speed.

Fig. 13 illustrates some important points that were observed during many of the skid trailer braking tests. This figure shows peak and slide coefficients measured by a skid trailer for two surfaces with a water depth of 0.020 inches which varied, however, to 0.060 inches. The first surface is brushed concrete with a wet skid number (SN) of 35 and the second surface is lightly textured macadam with a wet SN of 50.

References p. 57

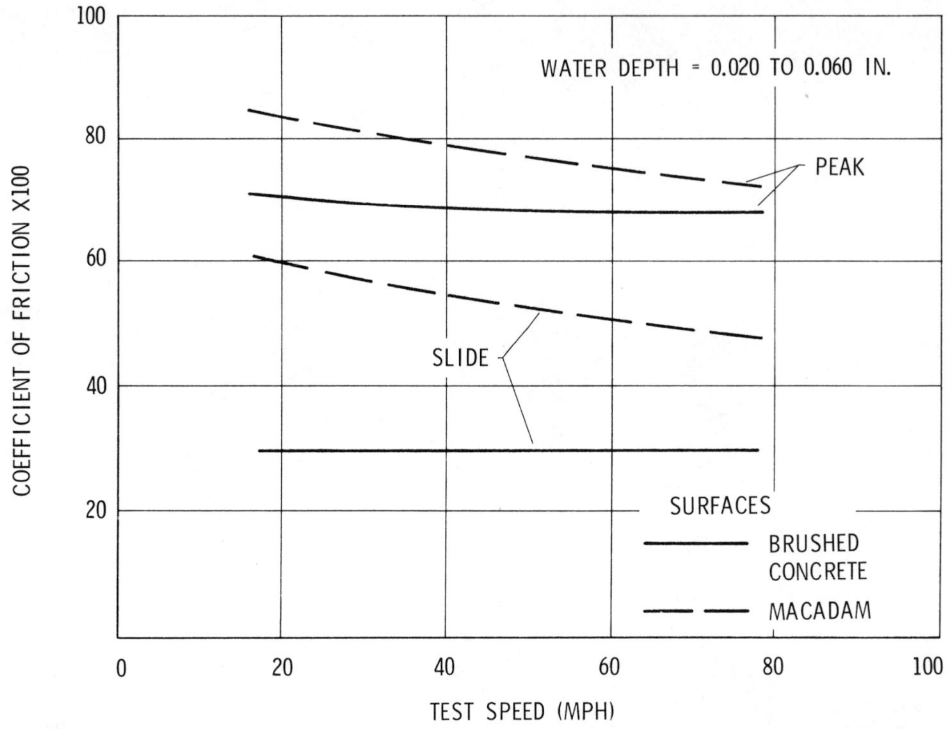

Fig. 13. Peak and slide values for L60-15 bias tire vs. test speed.

The fact that the peak and slide braking coefficients do not decrease with speed for a wide bias tire with extra wide grooves suggests that any decrease in these coefficients with increased speed is largely a result of tire viscous and dynamic hydroplaning. Also, the fact that the same tire when tested on a higher SN surface did degrade with speed suggests that the SN of a surface does not totally reflect the ability of that surface to drain water.

Finally on relatively smooth surfaces, excluding ultrasmooth surfaces such as Jennite or trowled concrete, there is a rank correlation between peak and slide braking coefficients obtained from a skid trailer and tire hydroplaning tests obtained from the glass plate.

Normal and Tangential Unit Shearing Forces – A tire rolling on a road surface produces two primary forces acting in the tire footprint at the tread-road interface (see Fig. 14). The force normal to the road surface is the tread pressure, F_z and the force in the plane of the road is the shearing force, F_{xy}. These forces are measured by a tri-axial force transducer flush with the road or flat test surface having a known surface area as shown in Fig. 15. The resultant shear force F_{xy} is measured as two components, F_x and F_y. F_x is the longitudinal unit force exerted in the direction of

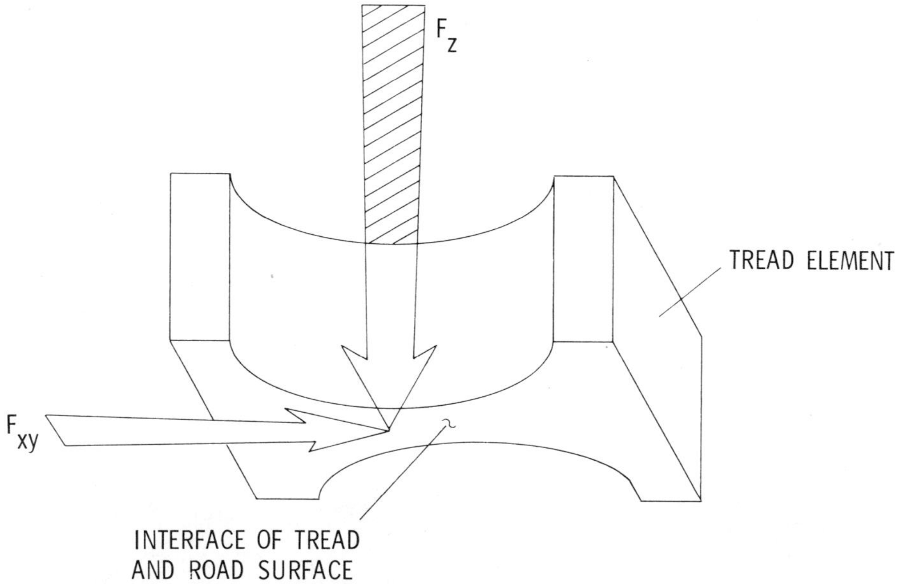

Fig. 14. Normal and tangential forces acting in footprint.

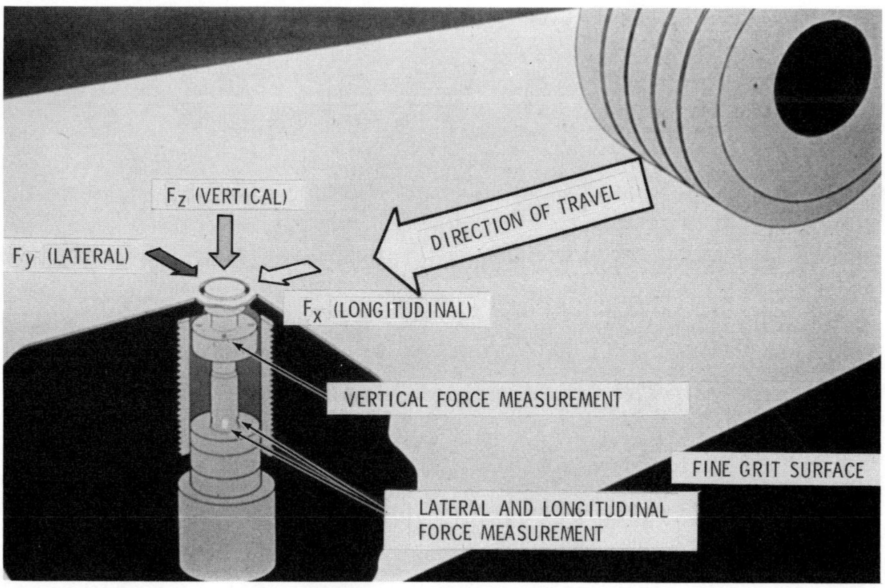

Fig. 15. Tri-axial load pin.

travel and F_y is the lateral unit shearing force, perpendicular to and in the plane of F_x. These forces have been measured for many types of passenger, truck, and aircraft tires.

References p. 57

Fig. 16. Tri-axial force pin mill.

Fig. 17. Passenger tire about to roll over a tri-axial force transducer at 1/10 mph.

Fig. 16 is a photograph of a specially adapted tri-axial force pin mill. Numerous tests were conducted at slow speed with precise control of camber, slip angle (1/8-deg increments to $\pm$ 2 deg) and torque (to 400 ft.-lb.).

Fig. 17 shows a test in which a passenger tire is about to roll over the imbedded tri-axial force transducer that is flush with the test surface. The forces are supplied

Fig. 18. Amplifying and recording instrumentation for tri-axial force measurement.

LOCATION IN U.S.	MAX AVG INCHES		MAX RATE (IN./HR)
	6 HR	1 HR	5-MIN DURATION
SOUTHEAST	4.7	2.9	10.1
NORTHEAST	2.1	1.3	5.5
AKRON	2.3	1.4	5.1
WEST	1.7	1.0	4.2

Fig. 19. Rainfall in United States.

directly in analog form to a tape recorder (see Fig. 18) and simultaneously read out visually by a Visicorder. Some testing, involving only passenger and light truck tires, was completed at moderate and high speeds and under torque loads to 1600 ft.-lb. on the test track.

Water Depth Study on Pavements – The goal of the water depth study on pavements was to determine the probability of expected water depths that could be encountered during wet driving conditions.

References p. 57

The conterminous United States was divided into sectors with a representative rainfall intensity selected over various return periods. The assigning of probabilities to peak intensity rates was based on intensity rates within a 10-sq.-mi. area of a shower for various return periods (References 2 and 3). Generally, peak intensities occur during the first five minutes of a thundershower and can be up to three times the intensity measured for the first hour.

The data shown in Fig. 19 have been selected to indicate relative expected intensities and are for a return period of five years.

A subjective analysis of horizontal visual range to rainfall intensity was made from the driver's position in three separate rainfalls while simultaneously measuring incremental rainfall intensities. The results are shown in Table 1.

TABLE 1

Results of Water Depth Study

Rainfall Condition	Return Period	Maximum Rainfall Intensity: First 5 Min. If App. (In./Hr.)	Horizontal Visual Range From Drivers Position (Ft.)	Mean Water Depth* (In.)	Range of Mean Among Conditions
Drizzle	Frequent	Less than 1/2	1000 +	.014	.010–.020
Light Rain	Frequent	1/2	1000 +	.033	.024–.043
Extended Shower	More than Once/Yr.	1	500 to 800	.045	.032–.060
Brief Thunder-Shower	More than Once/Yr.	2	200 to 400	.060	.040–.082
Heavy	Once per Yr.	4	Less than 100	.083	.055–.112
Severe	Once per 10 Yrs.	10	Near Zero	.128	.085–.172

**Mean depth calculated from 16 conditions of surface configurations from 12 ft. wide, 1.56 percent slope to 36 Ft., 1.8 percent slope. 3 percent grade, 1740 ft. radius turn. Water depth measurements within any single condition varied by ±10 percent (±20) for 0.080 inches water depth.*

Highway Design Parameters – Roadway conditions considered for estimating water depth were:

- Level and straight three 11-ft. lanes with a 1.56-percent standard crown.
- Level and curved three 11-ft. lanes with a 2.1 percent cross slope on a 1910-ft. radius curve at a 70-mph design vehicle speed; also, a 1.8-percent slope on a 1450-ft. radius curve with a design vehicle speed of 60 mph.
- Inclined and straight three 11-ft. lanes and a 1.56 percent cross slope; a varying hill slope of 2 to 4 percent; and a single condition involving one lane at 10 percent.

Development and Installation of Rain Facility – A rain facility was constructed on a concrete slab 12 ft. long and 6 in. wide (see Fig. 20). The slab could be tilted from 0.5 to 10 percent to determine the effect of cross slope. A reservoir was installed 4 ft. above the surface with a heavy wire mesh screen lined with perforated vinyl. A sheet of vinyl was draped under the reservoir to allow the water to run off adjacent to the concrete slab prior to the test. Various heights of the water depth in the reservoir could be maintained to control rainfall intensities from 0.5 inches per hour to over 10 inches per hour. At the start of the test, the drape was removed and the water from the reservoir was allowed to rain onto the concrete slab (lightly brushed concrete with a texture depth of 0.010 inches).

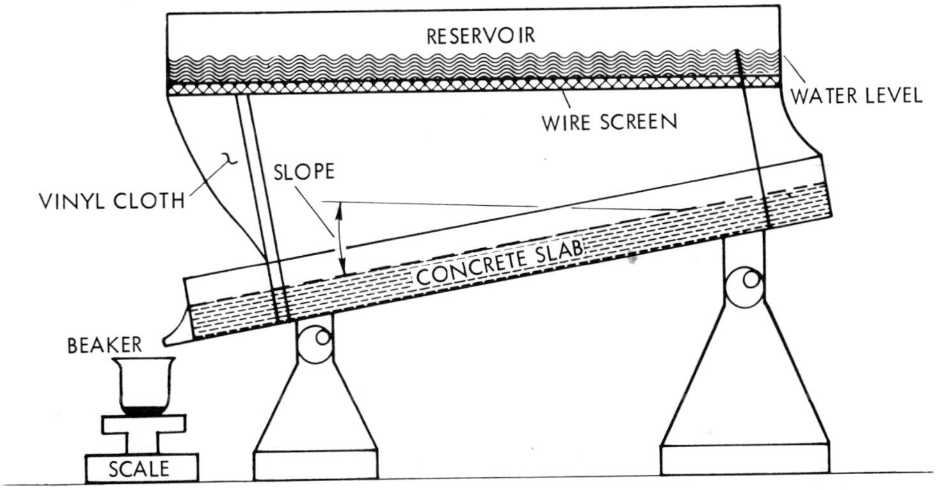

Fig. 20. Rain facility.

The runoff at one end of the slab was collected and weighed each 5 sec. until the runoff weight became constant and a steady state was assumed (see Fig. 21). With a known runoff rate and rain intensity, the water depth then was measured at each 6 inches of concrete test slab length by an electronic water depth indicator. The area of the curve shown in cross hatching was equated to the depth measurements. The area between a projected straight line tangent to the steady state condition and the rainfall intensity line is that water lost to absorption, splash, and evaporation and amounted to 15 percent.

Different curves were established for each angle of cross slope and rainfall intensity.

Results of Water Depth Study – Table 1 represents the results of the water depth study. For example, in the second line, a rainfall condition of light rain may occur frequently and produce maximum intensity rates for the first five minutes of up to 0.5 inches per hour. During this condition, visibility is not a problem to the driver and

References p. 57

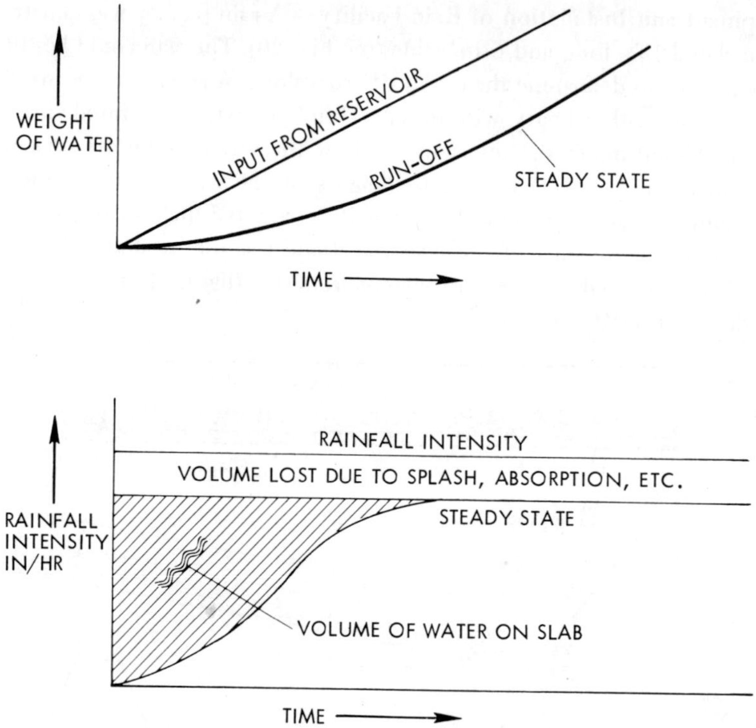

Fig. 21. Run-off vs. rainfall.

the average water depth to be expected is 0.033 inches. For the range of highway conditions, the water depth can vary from 0.024 to 0.043 inches.

In a brief thundershower, which can occur more than once per year with peak intensities of 2 inches per hour, driver visibility is lowered, but probably will not affect the driving speed. During this condition, the average water depth that may be expected is 0.060 inches with a range of 0.040 to 0.82 inches. Rainfall rates in excess of this amount generally impair the driver's visibility.

Similar studies of water depth on road surfaces have been conducted by the Texas Transportation Institute and the Ministry of Research (References 4 and 5).

ESTIMATED HYDROPLANING SPEED

The estimated minimum full dynamic hydroplaning speed for a free rolling passenger tire can be obtained by knowing the unit groove capacity (UGC). The UGC is obtained empirically and defined as equal to:

$$\frac{K_1}{W_T} \sum_{i=1}^{n} w_i d_i$$

where W_T = effective width of the tread,

wi and di = individual widths and depths of each groove measured perpendicular to the longitudinal axis in a loaded tire footprint (they must be obtained dynamically on a wet glass surface),

n = rib number, and

K_1 = weighted factor to consider lateral grooving and blading.

This equation has been discussed in a recent SAE paper (Reference 1).

The estimated free rolling hydroplaning speed is obtained from Fig. 22 as a function of water depth. The conditions assumed are a relatively rounded footprint and normally encountered loads and inflations as well as relatively smooth surfaces such as brushed concrete and fine textured asphalt.

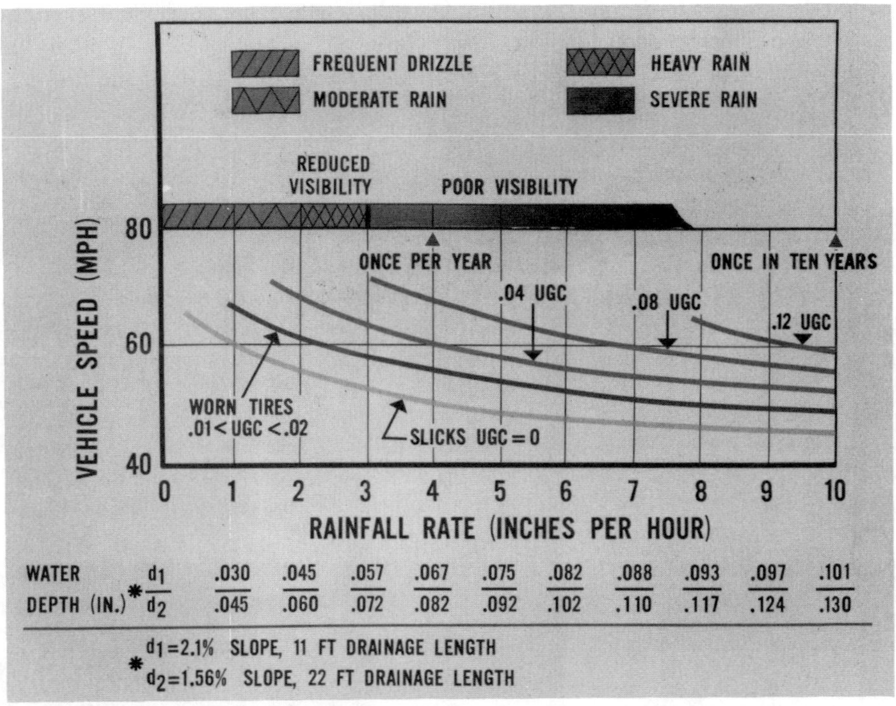

WATER DEPTH (IN.)*										
d_1	.030	.045	.057	.067	.075	.082	.088	.093	.097	.101
d_2	.045	.060	.072	.082	.092	.102	.110	.117	.124	.130

*d_1=2.1% SLOPE, 11 FT DRAINAGE LENGTH
d_2=1.56% SLOPE, 22 FT DRAINAGE LENGTH

Fig. 22. Estimated free rolling minimum full dynamic hydroplaning speed for passenger tires (conditions: relatively smooth surface, rounded footprint, and rated inflations and loads).

Most commercial tires have a UGC of 0.04 to 0.06 with some radial tires as high as 0.08.

The effect of a high UGC can be seen in that during a moderate rainfall of 3 inches per hour, a radial tire with a UGC of 0.08 will not be hydroplaning severely until a speed of 70 mph. On the other hand a smooth tire will be hydroplaning at a speed of approximately 50 mph.

References p. 57

FACTORS AFFECTING TIRE HYDROPLANING

Groove Closure – Groove closure is a condition related to the structural properties of the tire footprint. This condition is dynamically affected by the rotational speed of the tire and hydrodynamically affected by the inertial forces of the fluid in the footprint.

As a tire rolls at slow speed (see Fig. 23), a high lateral force, F_y, occurs in the footprint of the shoulder ribs of many tires acting toward the longitudinal centerline of the footprint. The force is the greatest for bias tires and decreases rapidly when a belt is applied such as with a radial tire. This force increases with tire load and causes the shoulder grooves to decrease in width. On a dry surface, the coefficient of friction of the surface is high enough to resist closure; however, on wet surfaces or at any time the tire is sliding, groove closure occurs and greatly reduces the UGC of the tire.

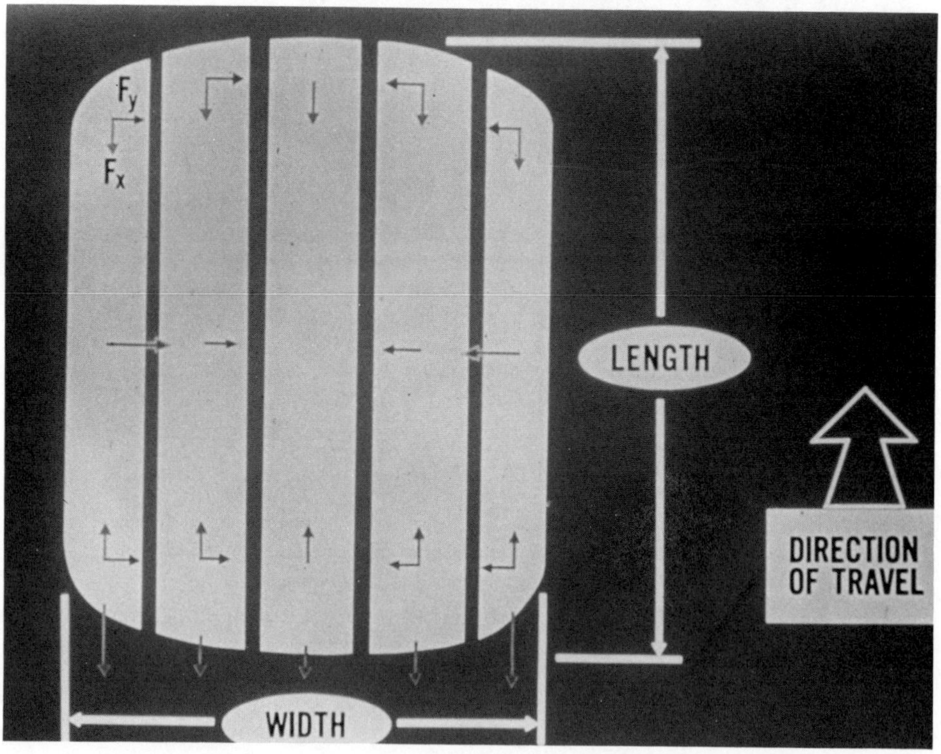

Fig. 23. Forces in the footprint.

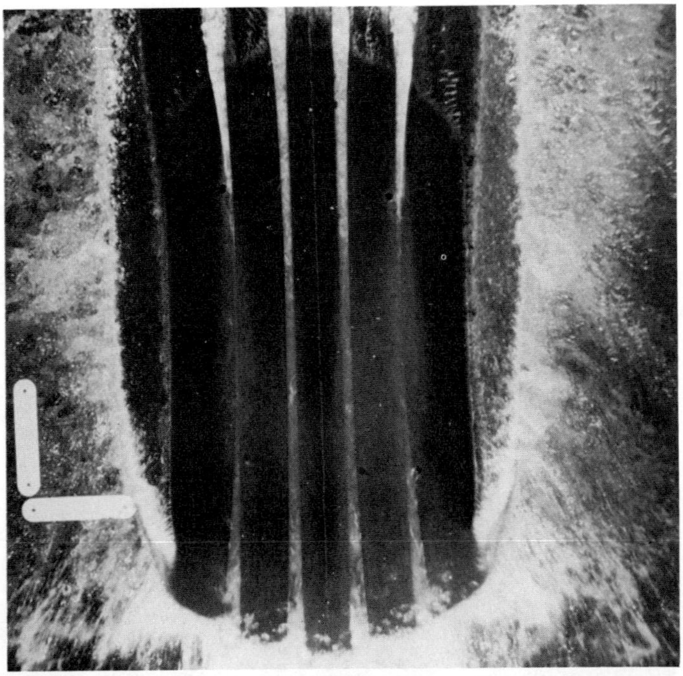

Fig. 24. Groove closure for bias tire at 60 mph free rolling in 0.050 inches of water.

Fig. 25. Bias tire at 60 mph in 0.080 inches of water (free rolling).

References p. 57

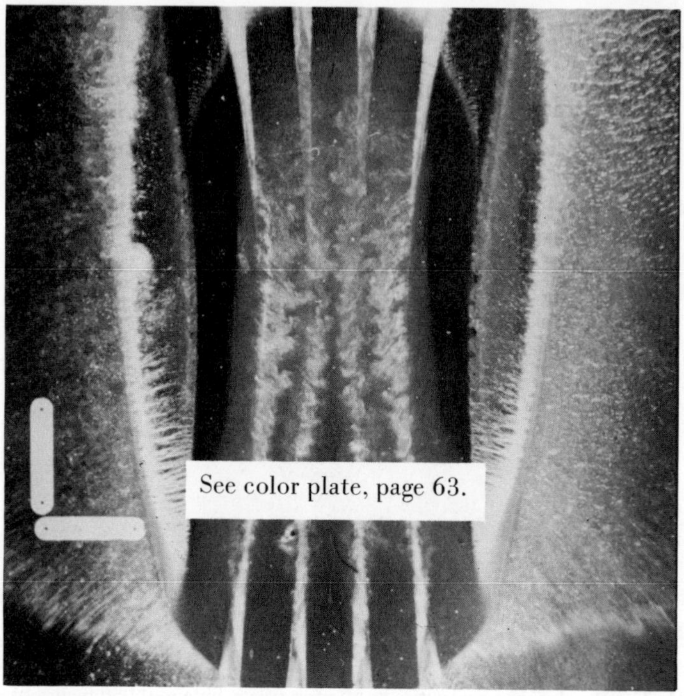

Fig. 26. Bias tire at 100 mph in 0.080 inches of water (free rolling).

Fig. 24 indicates the severe groove closure for a bias tire, which can reduce the UGC by over 50 percent. The tire is an SAE control tire free rolling in 0.050 inches of water at 60 mph.

In Fig. 25 the water depth is increased to 0.080 inches. The groove closure is slightly aggravated by the film of water lifting the center portion of the tire and increasing the lateral force, F_y in the shoulder ribs.

At 100 mph (see Fig. 26) centrifugal effects have elongated the footprint; the amount of film penetration has increased under the footprint which greatly increased the amount of groove closure.

In contrast (see Fig. 27), a steel belted radial tire free rolling at 60 mph in 0.080 inches of water indicates a groove closure of less than 20 percent. This is primarily because the lateral force, F_y, acting in the shoulder ribs is one-third the magnitude of the lateral force in bias tires.

Even at 100 mph (see Fig. 28) in 0.080 inches of water, groove closure is virtually unaffected because the UGC of the tire is great enough to keep the film of water from penetrating into the footprint.

Fig. 27. Radial tire at 60 mph in 0.080 inches of water (free rolling).

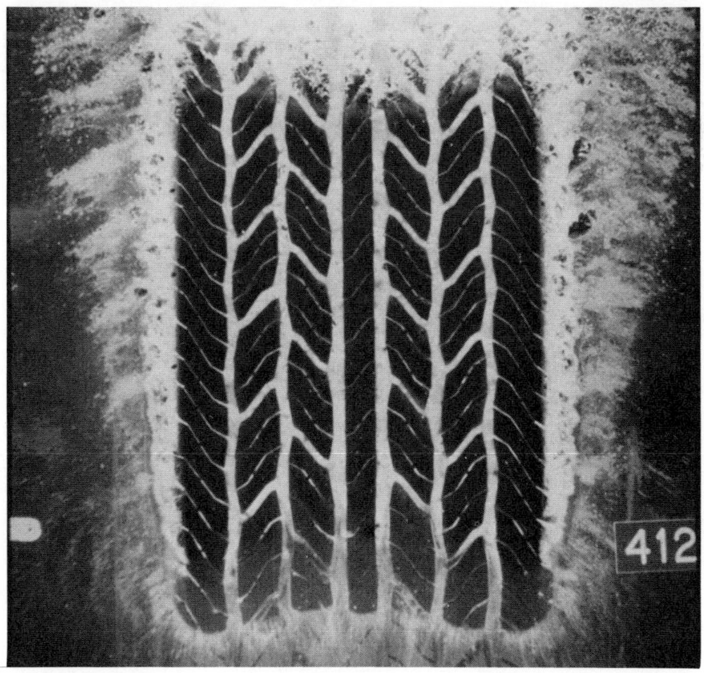

Fig. 28. Radial tire at 100 mph in 0.080 inches of water (free rolling).

References p. 57

Load and Inflation – Load and inflation play a complex role in tire hydroplaning. Increased inflation will:

- Increase the tread pressure of the interior ribs.
- Increase the unit groove capacity of the tire by lowering groove closure, but decrease the tire footprint length.
- Have the greatest effect on smooth surfaces when the water depth is greater than the UGC of the tire.

Under ideal conditions the minimum hydroplaning speed will increase 1 mph per 3 psi of increased inflation.

An increase in load will:

- Increase the tread pressure in the shoulder ribs, which is a negligible effect since the shoulder ribs do not hydroplane except in deep water films.
- Decrease the unit groove capacity by increasing groove closure.
- Increase the footprint length.

The net result is a negligible improvement.

Groove Shape – The shape and design of the longitudinal grooves is nontrivial and complex. In Fig. 29 the footprint is viewed upside down with a longitudinal groove exposed. The components of fluid velocity (U_1, U_2, and U_3) represent the axial, vertical, and transverse directions, respectively, relative to the tire tread. V_1 is the axial velocity of the tread, and V_2 is the tire forward axial velocity.

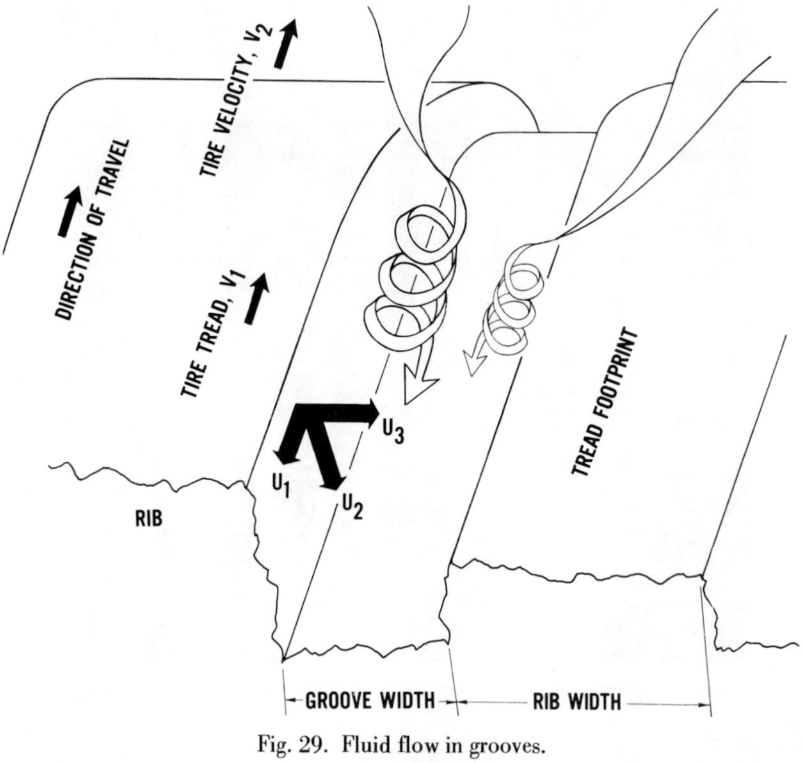

Fig. 29. Fluid flow in grooves.

When the tire is free rolling, V_1 is essentially zero. Fluid flow is confined to the forward portion of the footprint and to the grooves. The vortex is created by the fluid flowing into the grooves at the forward edge of the footprint. The flow of the fluid in the grooves is low and laminar except in the vortex region. For the sliding tire, V_1 and V_2 are equal. U_1 is stationary at the wall of the groove and increases rapidly in the free stream region. The boundary layer increases in width along the length of the groove and eventually becomes turbulent, especially at higher velocities.

Fig. 30 represents a photograph of a tire with wide shallow grooves. The longitudinal grooves were carved straight and wide to observe the flow. At the bottom of the figure two vortices are formed at the leading edge of the footprint of a free rolling tire traveling at 40 mph in 0.080 inches of water.

See color plate, page 63.

Fig. 30. Experimental tire at 40 mph in 0.080 inches of water (free rolling).

The size of diameter of the vortices remains constant through the footprint with respect to time. This fact suggests that the fluid is inviscid or at least that viscosity is of minor importance. The axial velocity of the vorticity is zero, which means that the axial pressure gradient is zero and the flow is two dimensional.

In the equation of motion in two dimensional flow, the radial variation of pressure simply supplies the force necessary to keep the fluid elements moving in a circular path. Since the axial pressure gradient is zero, there is no hydraulic head at the

References p. 57

leading edge of the groove. This fact is compatible with pressure measurements taken in the footprint of a free rolling tire and the lack of the wedge observed.

As the speed is increased, however, a slightly different condition of flow exists. In Fig. 31 the tire is free rolling at 60 mph in 0.080 inches of water. While flow in the grooves has not changed, horizontal vortices are visible under the ribs. These vortices are not steady and suggest that they have a viscous effect. Inertial effects are present also and create a hydrodynamic pressure on the ribs.

For a free rolling tire, the shape of the longitudinal grooves is not important. The primary role is the UGC, which must be great enough to absorb the contacted water. Diagonal grooves and blading is important to reduce viscous effects on smooth surfaces.

Fig. 32 illustrates the flow for a sliding tire in 0.080 inches of water at 40 mph. The twin vortices in each groove expand in size through the length of the groove. The flow in the groove is three dimensional with a boundary layer at the sides of each groove. The boundary layer is laminar at low speeds and is independent of the vortex region.

As the sliding speed of the tire is increased to 50 mph (see Fig. 33), the boundary layer becomes turbulent and mixes with the increasing size of the vortex to choke off the flow of the fluid. Thus, a head exists at the leading edge of the grooves and an axial pressure gradient also exists in the groove channel.

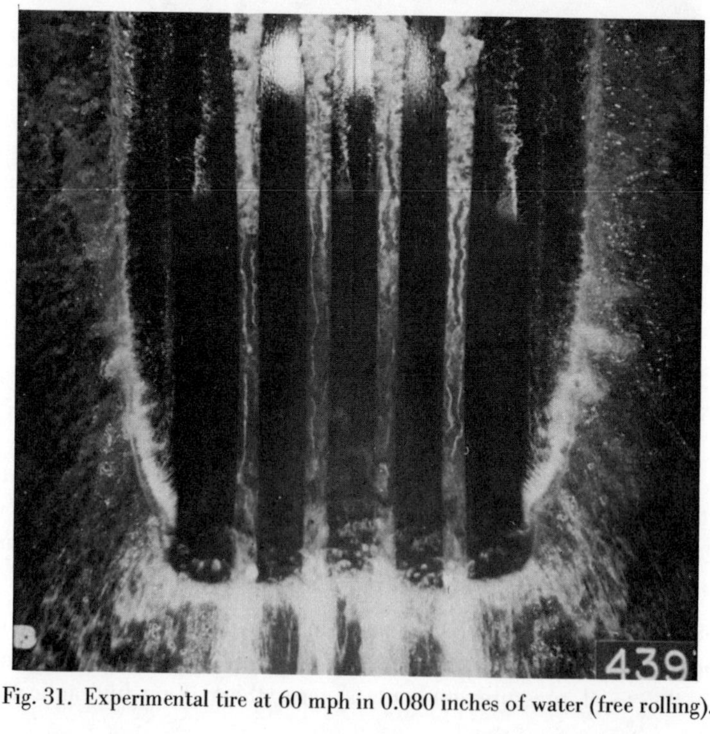

Fig. 31. Experimental tire at 60 mph in 0.080 inches of water (free rolling).

Fig. 32. Experimental tire at 40 mph in 0.080 inches of water (sliding).

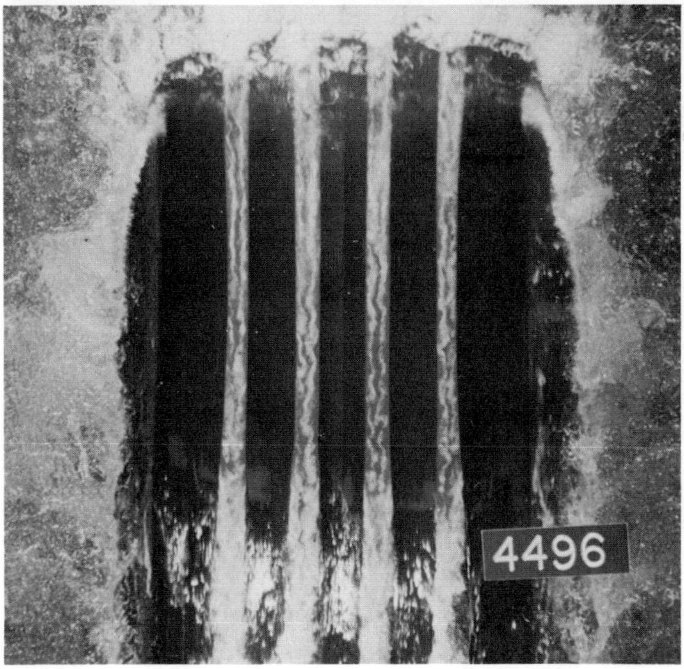

Fig. 33. Experimental tire at 50 mph in 0.080 inches of water (sliding).

References p. 57

As the speed is further increased to 60 mph (see Fig. 34), for the sliding condition, the flow is choked and becomes fully turbulent.

Fig. 35 shows a tire with narrow zigzag grooves at 40 mph in 0.080 inches of water. The boundary layer is laminar in the forward length of the grooves and is undergoing separation and adding to the turbulence.

Boundary layer separation can be observed in Fig. 36 for this tire with an increased amount of zigzag even at a speed of 40 mph in 0.080 inches of water. The result is an early choking of the fluid flow and an increased wedge at the leading edge of the footprint.

The flow in a sliding commercial-type tire is highly complex (see Fig. 37). Boundary layer separation is apparent and the blading and diagonal grooves indicate considerable improvement in removing viscous and inertial fluid effects. Choking of the flow is reduced by the diagonal grooves for this tire in a sliding mode at 60 mph. The flow study for a sliding tire clearly indicates that groove shape is important. Wide grooves are better than narrow grooves because of better flow characteristics and reduced effect of groove closure. Straight grooves also are more efficient; however, tests have shown that slight amounts of zigzag with diagonal grooves are equally efficient.

Fig. 34. Experimental tire at 60 mph in 0.080 inches of water (sliding).

Fig. 35. Experimental tire at 40 mph in 0.080 inches of water (sliding).

Fig. 36. Experimental tire at 40 mph in 0.080 inches of water (sliding).

References p. 57

Fig. 37. Commercial tire at 60 mph in 0.080 inches of water (sliding).

Fig. 38 illustrates the effect of superficial wear on the hydroplaning characteristics of a tire. When a bias tire has just 20 percent of its nonskid depth removed, the sharp edges of the individual tread elements are rounded. This rounding is due to the high rate of scrubbing and flexing of the tread elements. The result is the inability of the tread to adequately remove the film layer from these regions.

Fig. 39 shows a free rolling truck tire with an inflation of 75 psi at 70 mph in 0.080 inches of water. The high tread pressures, low tire deflections, and large groove capacities are advantageous in resisting tire hydroplaning.

Tire-to-Surface Friction during Hydroplaning – The influence of the water film in the footprint has been considered both from its effect on the road surface and on the tire tread.

The road surface is a relatively lasting, stable, rigid platform and is conducive to resisting tire hydroplaning if properly designed. These qualities are absent in the tire. The improvement that a tire tread potentially offers is governed by the level of drainage and friction capability of the surface.

Fig. 38. Effect of hydroplaning on bias tire with slight wear.

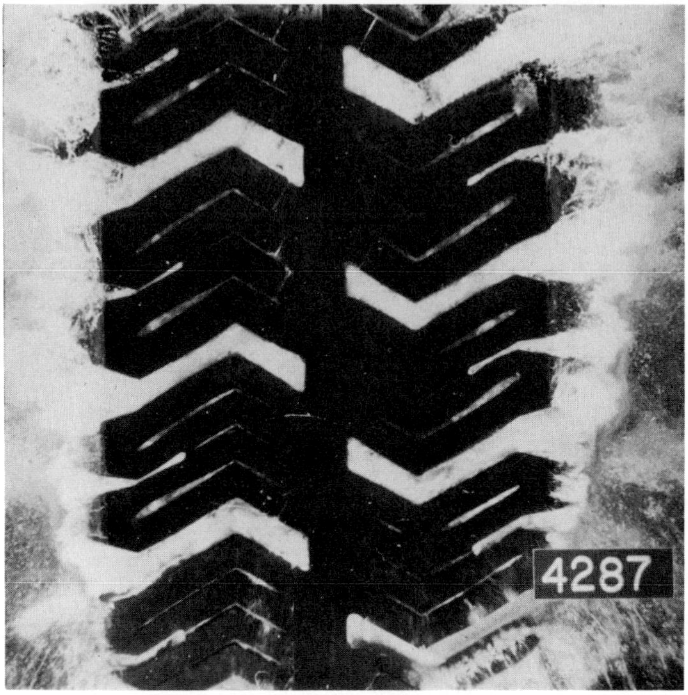

Fig. 39. Truck tire at 70 mph with 75 psi inflation in 0.080 inches of water (free rolling).

CORRELATION BETWEEN METHODS

There is a good rank correlation between peak data obtained by a skid trailer on wet surfaces with precise water depth control and results obtained by glass plate free rolling hydroplaning tests. This correlation study was obtained from glass plate free rolling hydroplaning tests of many passenger tires and subsequent testing on the brushed concrete and textured macadam water depth controlled surfaces.

On the brushed concrete surfaces, correlation was good with equal amounts of water on both the concrete surface and the glass plate. As the surface becomes more textured, correlation requires a greater water depth on the test surface. Exceptions to this are with surfaces that have high macro structured textures, surfaces with a wet SN exceeding 50, and significant changes in compound. Ultrasmooth surfaces, such as Jennite or trowled concrete, show some degree of rank correlation between slide coefficients obtained from a skid trailer and glass plate locked-wheel hydroplaning tests, but with the same exceptions.

CONCLUSIONS

After conducting more than 5,000 glass plate tire hydroplaning tests and nearly 10,000 skid tests on water depth controlled surfaces, Goodyear tire engineers have arrived at the following conclusions:

- Wet weather tire performance cannot be made independent of tire hydroplaning.
- There is good rank correlation between glass plate tire hydroplaning tests and wet skid trailer tire testing.
- Tire hydroplaning is significantly dependent on many primary nonlinearly related variables.
- Simplified solutions quantify many of these variables and seriously limit the results.
- To adequately define wet weather tire performance, surfaces must be defined by a single number denoting sliding resistance and drainage capability.

REFERENCES

1. *R. W. Yeager and J. L. Tuttle, "Testing and Analysis of Tire Hydroplaning." Akron, Ohio, The Goodyear Tire and Rubber Co., 1972.*
2. *D. M. Hershfield, "Rainfall Frequency Atlas of the United States for Duration from 30 Minutes to 24 Hours and Return Periods from 1 to 100 Years." Washington, D.C., U.S. Weather Bureau, 1961.*
3. *C. V. Davis and K. E. Sorensen, "Handbook of Applied Hydraulics." 3rd ed. New York, McGraw-Hill, 1969.*
4. *R. M. Gallaway, R. E. Schiller, Jr., and J. G. Rose, "The Effects of Rainfall Intensity, Pavement Cross Slope, Surface Texture, and Drainage Length on Pavement Water Depths." Research Report No. 138-5. College Station, Texas, Texas Transportation Institute, Texas A&M University, 1972.*
5. *N. F. Ross and K. Russam, "The Depth of Rain Water on Road Surfaces." RRL Report LR 236. Crowthorne, Berkshire, England, Road Research Laboratory, Climate and Environment Section, RG11 6AU, 1968.*

DISCUSSION

Thieme

Thank you very much Dr. Yeager. We have some time for discussion.

T. J. Brown

I was interested in your triaxis transducer in the laboratory. At what speed can the tire be run on the laboratory equipment and can cornering forces be introduced?

Yeager

In the laboratory the test tire is rolled over the tri-axial force transducer at a speed of 0.2 mph with variably controlled camber and/or slip angles up to 2 degrees in 1/8 degree increments. This is in addition to numerous testing outdoors with a similar device embedded in a test track.

Thieme

Do the dimensions of the pin have any influence? Did you try that out?

Yeager

The dimensions of the pin are very influential. The present surface of the tri-axial force transducer is circular and is 1/2 inch in diameter.

Thieme

Do you have any reason for selecting that particular size?

Yeager

Quarter inch and three quarter inch diameter cylindrical pins have been used. The smallest diameter pin requires more runs and is susceptible to a larger amount of irregularities and transient profiles. Larger diameter pins are too severe in minimizing these irregularities and overlap the ribs.

B. D. A. Phillips

I would like to ask what the practical limitation is in the unit-groove capacity of a tire and whether in order to obtain a large unit-groove capacity it is preferable to have a few wide grooves or a large number of narrow grooves and what are the problems encountered when you design tires like this?

Yeager

The unit-groove capacity as defined in SAE paper 720471 has a practial limit. Large unit-groove capacities can adversely affect dry tire performance as related to vehicle handling, noise and tire wear, depending on the type of tire construction. An increase in the number and/or size of the grooves of a new passenger tire will require a very close observance of the change in all tire properties. The optimum unit-groove capacity of a passenger tire can vary from 0.04 to 0.10 depending on the type of tire and the controlled water depth used to determine wet traction performance.

The optimum number of grooves can vary from 4 to 6 for a 78 aspect ratio tire to 8 to 10 for a 50 aspect ratio tire.

Generally, the shoulder rib can be the widest rib for bias tires as it has the highest tread pressure. The width of the shoulder rib can usually be determined by the slope of the lateral tread pressure gradient. The higher the slope, the narrower the rib requirement. Lateral grooves in bias tires or in the shoulder ribs of radial tires can adversely affect tire wear.

R. M. Gallaway

I was interested in knowing what faith you had in the correlation between the performance of the tread under glass compared to what it would do on the real-world surfaces, since the friction of the two is entirely different.

Yeager

Correlation studies obtained between the glass plate hydroplaning photos and braking tests on the water depth controlled surfaces indicate that the coefficient of friction that is measured on a relatively smooth surface does not affect the dynamic hydroplaning characteristics of a tire.

Thus, the amount of groove closure observed on the glass of a sliding tire is identical on a smooth surface regardless of its skid number rating. However, as the road surface texture increases, the free rolling tire condition as observed on the glass surface becomes conservative.

Gallaway

My question was directed toward those surfaces with large coarse surface roughness where the roughness itself might tend to direct the opening and closing of the tread under stress.

Yeager

As the surface texture on a wet road increases, a free rolling tire may not experience the amount of groove closure as observed on the glass plate. Thus, the actual unit-groove capacity would be larger. Surface texture roughness does affect the tread pressure, causing very high local pressures. Also, surface texture affects the drainage capability of the water in the tire footprint.

The correlation between glass plate free rolling tests and a skid trailer measuring peak values on a road surface and the correlation between glass plate sliding tests and a skid trailer measuring slide values on a road surface decreases with increasing road surface texture. This is reflected by the improvement and the narrowing of the differences of all tread designs as the surface texture increases. However, if the water depth is proportionally increased, within limits, as the texture increases, the differences in tread designs and correlation with the glass plate increases.

Thieme

What do you mean by very high pressures?

Yeager

Normally a passenger tire will experience tread pressures of 80 to 100 psi under the shoulder ribs and 20 to 50 psi under the remaining ribs. Macro surfaces could increase these pressures by at least twice depending on the rib and size of asperity.

Harry A. Smith

Have you studied the effect of water depth on cornering; lateral forces as opposed to longitudinal forces?

Yeager

A tire operating at a slip angle on the wet glass plate only denotes trends. This is because the smooth wet surface allows gross sliding of the ribs in the tire footprint, which would not normally occur on road surfaces. These trends are important, however, in that they show the change in footprint shape and relative groove closure. It is more useful as a development tool than as an absolute measurement.

Thieme

Did you try to see some difference in the shape of the footprint area at different speeds, because it also might be interesting?

Yeager

The footprint area is significantly affected by vehicle speed. The tire is a very elastic structure and deforms readily under the influence of inertia, braking and fluid dynamic forces. For example, at 60 mph the footprint contact area of a 60 series tire, with ample longitudinal grooves to resist dynamic hydroplaning, can vary over 50% by just changing the water depth 0.030 inch.

Significant changes can also occur due to the effects of slight variations in velocity at high vehicle speeds, braking and changes in slip angle.

A. D. Roberts

Have you made any attempt to measure the water film thickness within the contact area? Was this by the Moiré fringe technique you briefly mentioned?

Yeager

We have completed a study using a wide smooth tire. By using a Moiré fringe technique a water film thickness of 0.080 inch was measured in the center of the footprint as the tire was traveling at 40 mph in 0.120 inch water depth. These results were obtained at low tire inflations and by increasing the reflective quality of the tire.

Thieme

Gentlemen, we all know the large number of excellent research reports of Goodyear Laboratories and now a new paper has been added to the long list of publications. Mr. Yeager we thank you very much for your most interesting and excellent paper on hydroplaning. I think most particularly interesting were the forces measured in the footprint on the braking and cornering modes of dry, wet, and flooded areas. These and other data such as triaxial force facilities and water depth studies, offer an insight into tire hydroplaning and its effect on wet traction in general and we feel that Mr. Yeager suceeded in presenting today's paper in an excellent manner, and this is a great service to us all. I feel sure that countless members would like me to congratulate Mr. Yeager on their behalf, and I therefore propose a vote of thanks.

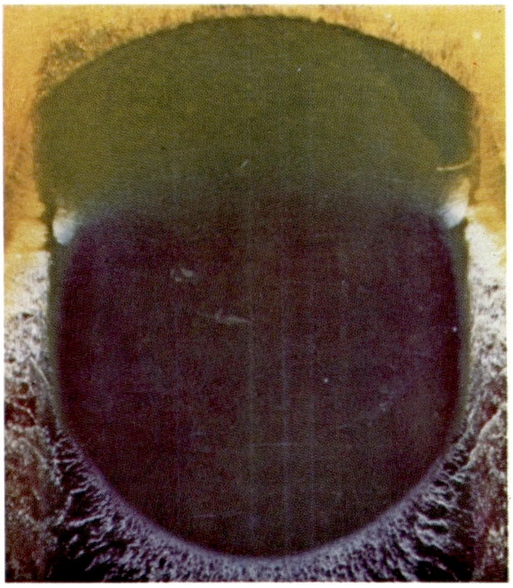

Fig. 1. Viscous hydroplaning of smooth truck tire at 10 mph in 0.080 inches of water (free rolling).

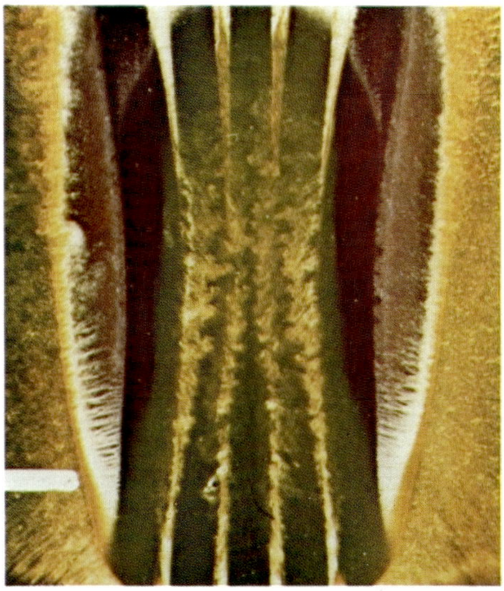

Fig. 26. Bias tire at 100 mph in 0.080 inches of water (free rolling).

Fig. 30. Experimental tire at 40 mph in 0.080 inches of water (free rolling).

Fig. 38. Effect of hydroplaning on bias tire with slight wear.

TIRE TRACTION ON DRY, UNCONTAMINATED SURFACES

L. SEGEL

University of Michigan, Ann Arbor, Michigan

ABSTRACT

Tire traction on dry surfaces is discussed with emphasis given to the viewpoint of the vehicle-tire system engineer. The paper indicates that, in this context, the mechanics of shear force generation assumes greater importance than the detailed mechanisms that control the frictional coupling at the tire-road interface. After postulating a working definition to distinguish between the concept of "tire traction" and the mechanics of shear force generation, the latter process is described in phenomenological terms. Initial, early, and current efforts to define and measure the shear force mechanics of tires in their normative operating condition are briefly reviewed prior to presenting test results obtained recently in several research programs. These findings relate to (1) the manner in which the longitudinal and lateral components of shear force depend on the longitudinal and lateral components of slip, (2) the shear force/slip relationship as a function of tire geometry and construction, and (3) the sensitivity of peak lateral force to small amounts of tread shoulder wear.

INTRODUCTION

One of the major reasons that the rubber treaded, pneumatic tire has become the success story that it has is its ability to generate large shear forces in relationship to the supported load. This statement is particularly true when the tire is being operated on a dry uncontaminated surface; it is less true when a tire-vehicle system is being operated on a contaminated or lubricated surface. Given that tires exhibit this behavior, it seems appropriate that dry traction (both the process and the results) be

References p. 94

discussed within a frame of reference that is different from that applicable to a discussion of the shear forces generated by tires on contaminated or lubricated surfaces.

Accordingly, this discussion of the dry traction phenomenon adopts as its frame of reference the demands for shear force that derive from the maneuvers executed by a tire-vehicle-operator system. In this systems context, it is necessary that the mechanism by which shear forces are generated be viewed as a slip-dependent process in addition to being viewed as a mechanism that is dependent on the frictional coupling existing between tread and road at the contact surface. Preliminary to presenting some recent experimental findings on shear force dependency on slip as measured on dry uncontaminated surfaces, a working definition of tire traction is offered in the hope that this definition will be helpful in sorting out the myriad of variables that are involved in the shear force generation process. An overview of the mechanics of shear force generation follows, prior to citing the requirements of a dynamometer for measuring the shear/slip properties of full-scale tires. The primary objective is to present some dry traction data that defines the ability of the tire to generate, simultaneously, longitudinal and lateral components of shear force on typical road surfaces. Findings will be presented with respect to the following measured relationships, viz.:

1. shear force dependency on road surface and speed,
2. shear force dependency on tire geometry and construction,
3. shear force dependency on slip, and
4. sensitivity of peak lateral force to small amounts of tread shoulder wear.

In the real world, there is no such thing as an uncontaminated road surface. Consequently, for purposes of this discussion, reference to a dry, uncontaminated surface shall mean the normative operation of tire-vehicle systems on road surfaces that appear both dry and clean to the eye of an observer. This definition of what is meant by "dry" and "uncontaminated" is admittedly nonscientific and would never suffice in the laboratory if, for example, one were to study the frictional forces generated by deformable polymers in contact with nondeformable surfaces. It is, at best, a pragmatic definition that serves technologists in a reasonably satisfactory fashion as long as the presence of contaminants, either liquid or solid, is sufficiently miniscule that the lubrication produced by the contaminant can be disregarded.

As indicated earlier, tires are designed and fabricated such that the maximum value of the shear-force vector that can be generated on an "uncontaminated" road surface is of the order of the vertical load carried by the tire. Under these conditions, tire and vehicle technicians are inclined to view the available "traction" as being quite adequate for purposes of maneuvering a tire-vehicle system. This situation contrasts with the operating mode in which liquid or solid contaminants lubricate the tire-road interface such that the maximum shear force vector that can be produced is significantly less than the load carried by the tire. In the latter instance, one commonly observes that the frictional coupling between the tire and the road has been degraded; or one states, in lay terms, that the tire has lost its ability to "grip" or "hold" the road.

The above remarks suggest that tire "traction," when viewed as a measure of the peak interfacial shear forces that can be developed, may be insufficient for purposes of maneuvering a vehicle only when lubricating elements are present. By and large, this conclusion is warranted. Thus, the question of whether tires have adequate traction on dry, uncontaminated surfaces appears to be an academic issue as long as lubricated road conditions can be encountered which result in reduced traction levels. It follows that the primary motivation for studying the dry traction characteristics of tires does not stem from a concern with the tire's ability to generate sufficient forces for maneuvering purposes but rather from a need to evaluate and predict the maneuvering characteristics of a tire-vehicle system under the control of a human operator.

Consequently, the measurement of the dry-traction characteristics of tires centers around the role that the tire plays as the mechanism for generating the forces required to maneuver a motor vehicle. In this role, the compliant character of the carcass and tread has a substantial influence on the manner in which the vehicle responds to road irregularities and to steering, braking, and acceleration commands. To the extent that these commands are of a low magnitude, the structural compliance of the tire is the only tire property of significance. In the event that large accelerations are demanded, tire compliance and the frictional coupling at the tire-road interface jointly determine how the vehicle responds to driver commands. Accordingly, in this paper, the traction performance of a tire is examined in terms of

1. the departure from a state of free rolling, as designated by the kinematic variable called "slip," and
2. a number of the variables that limit the shear forces produced by tire "slip."

A WORKING DEFINITION OF TIRE TRACTION

It will be helpful to consider, in functional terms, the relationship between the forces and moments that derive from the shear and normal pressure distributions existing at tire-road contact and the orientation and so-called slip of the tire. (We shall use the standardized terminology proposed by SAE. See Fig. 1.)

References p. 94

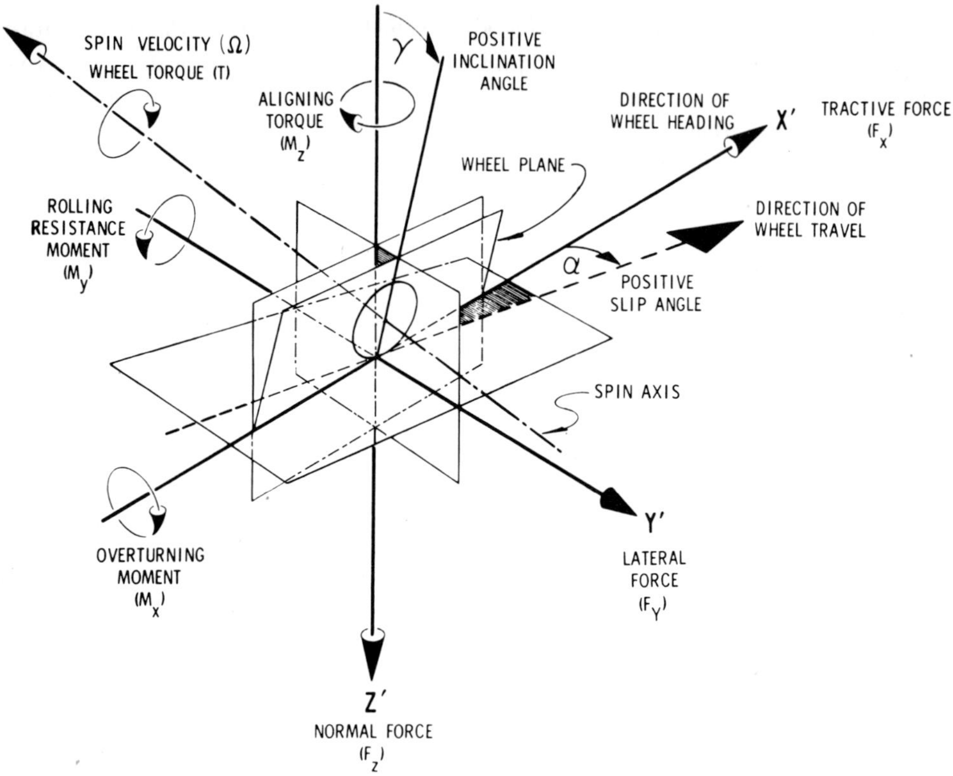

Fig. 1. SAE tire axes and terminology.

For a given tire operating on a specific road surface, the geometric and kinematic variables that completely define the steady motion and orientation of the tire relative to the road are given on the right-hand side of the following equation, viz.,

$$\begin{Bmatrix} F_x \\ F_y \\ F_z \\ M_x \\ M_y \\ M_z \end{Bmatrix} = f(R_\ell, \Omega, \gamma, \frac{1}{\rho}, \overline{V}) \tag{1}$$

where R_ℓ = distance between the spin axis and the tire contact center (also referred to as the rolling radius of a loaded tire), Ω = angular velocity of the tire about its spin axis, γ = inclination of tire center plane with respect to the vertical z axis, ρ = radius of the curved path of the tire ($1/\rho$ = curvature of path), and $\overline{V}$ = the velocity vector of the center of tire contact in the xy plane.

Note that Eqn. 1 implies that steady forces and moments must be applied to the wheel on which the tire is mounted in order to provide static equilibrium with the forces and moments acting at the tire-road interface, as expressed by the column of variables on the left-hand side of Eqn. 1.

Experimental measurements of the forces and moments represented by the left-hand side of Eqn. 1 show that there is a small disturbance operating regime, as defined by the variables on the right-hand side of Eqn. 1, in which the indicated force/motion relationship is primarily dependent on the structural compliance of the tire. These data also show that when the vector sum of the lateral force, F_y, and the longitudinal force, F_x (called "tractive force" in Fig. 1), reaches and exceeds approximately thirty percent of the normal force, F_z, the frictional coupling between the tread and the road begins to influence the indicated relationship to a marked degree. Test experience shows that the shear forces generated by tires are characterized by an upper limit, which limit is commonly interpreted as a measure of the frictional coupling prevailing between the tire and the road.

Since the mechanical characteristics of tires, as expressed by Eqn. 1, are a major determinant of the maneuvering behavior of the tire-vehicle system, the dry traction of tires will be defined and examined herein in the context of Eqn. 1. To facilitate this task, it is desirable to define some auxiliary kinematic variables. To this end, we define the rolling velocity, $\overline{V}_r$, of a tire as the velocity vector directed along the positive x axis with its magnitude given by

$$|\overline{V}_r| = R_e \Omega \tag{2}$$

The radius, R_e, is an "effective" rolling radius, conventionally defined as the rolling radius given by the expression

$$R_e = \frac{|\overline{V}_{fr}|}{\Omega_{fr}} \tag{3}$$

where $|\overline{V}_{fr}|$ = magnitude of the velocity vector of the tire contact center when this vector lies along the x axis and there is no external torque applied around the spin axis of the tire. Ω_{fr} = spin velocity of the tire when $\overline{V}$ lies along the positive x axis and the external torque applied around the spin axis of the tire is zero.

The constraints imposed on Eqn. 3 are such that the expression has meaning only when the tire is in the "free, straight-rolling" state, namely, that state in which the only forces applied to the tire from the wheel hub to achieve a state of equilibrium are a vertical load applied at the axis of rotation in the positive z direction and a longitudinal force in the positive x direction (see Fig. 2). Thus, we can write that

$$|\overline{V}_r| = \frac{\Omega}{\Omega_{fr}} |\overline{V}_{fr}| \, , \tag{4}$$

and we see that the rolling velocity, $\overline{V}_r$, as defined, is equal to the velocity of the "free, straight-rolling" tire when the spin velocity, Ω, is equal to the spin velocity of the "free, straight-rolling" tire.

References p. 94

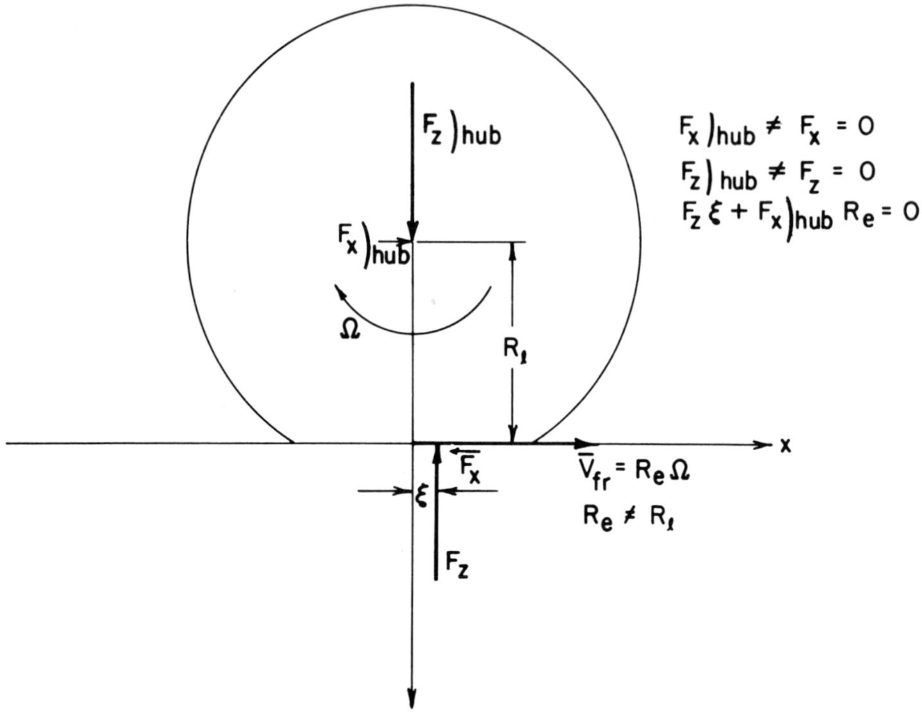

Fig. 2. Free body diagram of the "free, straight-rolling tire".

The operating condition in which

$$\overline{V} = \overline{V}_r$$

has been arbitrarily defined by tire mechanicians as constituting a state of zero translational "slip." Notwithstanding the small longitudinal force, F_x, that must act in the negative x direction to produce equilibrium in the "free, straight-rolling" state (see Fig. 2), the tire is defined to be in a zero-slip state whenever the inclination, γ, and the radius of path, ρ, are zero and the velocity of the tire contact center, $\overline{V}$, is equal to and coincident with the rolling velocity, $\overline{V}_r$. Correspondingly, whenever the velocity vector, $\overline{V}$, is not equal to the rolling velocity, $\overline{V}_r$, the velocity vector representing the difference between the vectors, $\overline{V}$ and $\overline{V}_r$, can be defined as the translational slip velocity, $\overline{V}_s$, of the tire. Figs. 3(a), 3(b), and 3(c) show the slip velocity, $\overline{V}_s$, that will exist when $\overline{V}$ has a component along the positive y axis and $\overline{V}_r$ is respectively less, equal, and greater than $u\bar{i}$. The slip conditions diagrammed in Fig. 3 will exist whenever a force is applied to the axle hub in the positive y direction and negative, zero, and positive wheel torques, T, are applied as per the SAE definition given in Fig. 1.

This slip-velocity vector, expressed as

$$\overline{V}_s = V_{sx}\,\bar{i} + V_{sy}\,\bar{j},$$

can also be written as

$$\overline{V}_s = u(S_x \bar{i} + S_y \bar{j}) \tag{5}$$

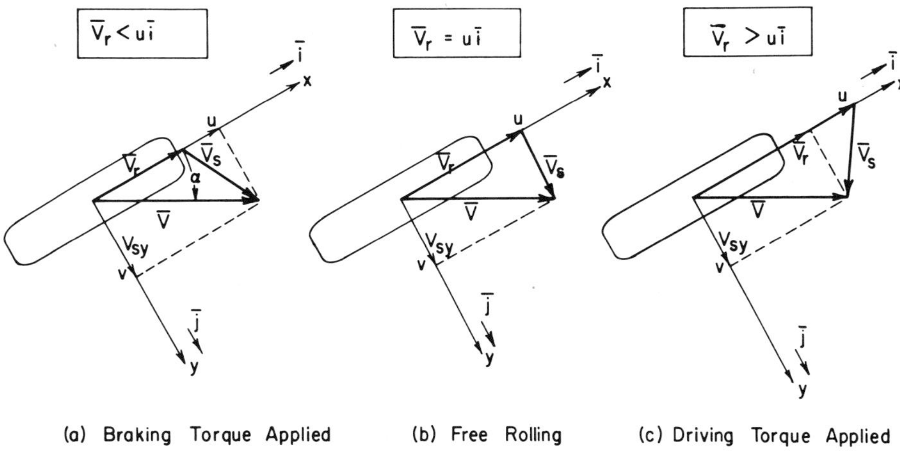

Fig. 3. Definition of the translational slip vector, $\overline{V}_s$.

where $S_x = \dfrac{V_{sx}}{u}$, a nondimensional quantity defining the longitudinal slip of a tire,

and $S_y = \dfrac{V_{sy}}{u} = \tan\alpha$, a nondimensional quantity defining the lateral slip of a tire.

Note that for small values of the angle, α, $S_y = \alpha$, meaning that the lateral slip of a tire is equal to the angle that has traditionally been called the "slip angle" of the tire.

With the above definitions to describe the translational slip of a tire, we find that Eqn. 1 can be transformed to a form that is more easily interpreted in terms of the physics that are involved. Specifically, we note that

$$\overline{V}_r = R_e \Omega \bar{i}$$

and

$$\overline{V} = \overline{V}_s + \overline{V}_r$$

such that Eqn. 1 can be written as

$$\begin{Bmatrix} F_x \\ F_y \\ F_z \\ M_x \\ M_y \\ M_z \end{Bmatrix} = f(R_\ell, \gamma, \frac{1}{\rho}, \overline{V}_s, \Omega) \tag{6}$$

References p. 94

Since the slip velocity, $\overline{V}_s$, can be expressed in terms of u, S_x and S_y, and since

$$|\overline{V}_r| = u - S_x u = R_e \Omega,$$

Eqn. 6 can be written as

$$\left\{\begin{matrix} F_x \\ F_y \\ F_z \\ M_x \\ M_y \\ M_z \end{matrix}\right\} = f(R_\ell, \gamma, \frac{1}{\rho}, S_x, S_y, u). \tag{7}$$

It must be emphasized that the functional relationship given by Eqn. 7 identifies only the independent motion variables that are necessary and sufficient to describe the attitude of the wheel and the motion of the center of tire contact relative to the road. In theory, each of the interfacial forces and moments are a combined function of the six identified geometric and kinematic variables. In practice, however, tires are generally tested to determine their mechanical properties as a function of a less complete set of motion variables than are contained in Eqn. 7. Nevertheless, Eqn. 7 should be viewed as constituting a complete definition of the mechanical characteristics of the pneumatic tire. As indicated earlier, these characteristics are determined by (1) the mechanical properties of the carcass and tread and (2) the frictional coupling that is developed at the tire-road contact.

In this discussion, we shall be concerned exclusively with tire *traction*, namely, the shear forces, F_x and F_y, and the shear moment, M_z. Accordingly, Eqn. 7 should be written in the form in which traction measurements are usually made. Consider first the case in which the inclination and path curvature of the tire are zero (viz., $\gamma = 1/\rho = 0$). Note further, that, to first order, the loaded radius, R_ℓ, is a function of the load supported by the tire which is equal and opposite to the normal force, F_z. Thus, we may write that

$$\left\{\begin{matrix} F_x \\ F_y \\ M_z \end{matrix}\right\} = g(F_z, S_x, S_y, u). \tag{8}$$

Eqn. 8 constitutes a definition of the "traction field" that is generated by a loaded, translating tire whose motion relative to the road is completely defined by the variables S_x, S_y, and u. For a specified loading, F_z, and a specified velocity vector component along the x axis, u, we see that the traction field of a tire can be defined as

$$\left\{\begin{matrix} F_x \\ F_y \\ M_z \end{matrix}\right\} = g(S_x, S_y). \tag{9}$$

Note that the traction field defined by Eqn. 9 involves both longitudinal slip, S_x, and lateral slip, S_y, in combination. However, we frequently measure the shear force/slip relationship of a tire (at a given load, F_z, and speed, u) to define

$$F_x = k(S_x) \tag{10}$$

$$\left\{ \begin{array}{c} F_y \\ M_z \end{array} \right\} = j(S_y) \tag{11}$$

where Eqn. 10 expresses the longitudinal force or traction caused by longitudinal slip ($S_y = 0$), and Eqn. 11 expresses the lateral force or traction caused by lateral slip ($S_x = 0$).

THE MECHANICS OF SHEAR FORCE GENERATION IN PHENOMENOLOGICAL TERMS

Prior to examining the traction fields generated by different tires on different surfaces, it will be helpful to examine the mechanism by which a "frictional coupling" between the tire and road presumably influences the traction field defined by Eqns. (9), (10), or (11). A complete understanding of the shear-force generation mechanism is not currently available although investigators are pursuing the topic, both from an analytical and experimental point of view (1, 2, 3, 4, 5). Accordingly, it will behoove us to limit this discussion to a description of shear force generation in terms of the phenomena that are either known to occur or can be presumed to occur when a tire is loaded against a road surface and forced to roll in a manner such that $\gamma \neq 0$, $1/\rho \neq 0$, and $\overline{V} \neq \overline{V}_r$.

Experiments (6) show that for small values of γ, $1/\rho$, and $\overline{V} - \overline{V}_r$:

1. the shear (or traction) forces and moments are linearly related to γ, $1/\rho$, S_x, and S_y;
2. the tire tread and carcass are distorted such that the actual center of the tire contact print (or patch) is not located at the center of tire contact; and
3. both the tread and the carcass deform, within and without the area of contact, such that differential elements of tread remain in essentially non-sliding contact with the road from the time that a differential element of tread rolls into contact with the road and is later lifted from the road surface.

Experiments have also shown (6) that as $\overline{V} - \overline{V}_r$ becomes large, it is no longer possible for the tread and carcass to distort sufficiently to maintain the tread in non-sliding contact with the road. It may be presumed that there is only a finite level of static shear force possible between a differential element of tread and the road. Sliding will occur if a force value greater than the maximum static level is demanded.

References p. 94

The above observations form the basis for conceiving the phrase "frictional coupling" between tire and road. We can give it sharper meaning by noting that within the area of tire-road contact there is a distribution of force per unit area (or pressure) normal to the road surface as well as a distribution of shear forces per unit area tangential to the road surface. If we designate the normal pressure distribution as $\sigma(x,y)$ and the shear force distribution as $\tau(x,y)$, as shown in Fig. 4, the traction forces and moments, F_x, F_y, and M_z, are given by the following integrations:

$$F_x = \int_A \tau_x (x,y) dA$$

$$F_y = \int_A \tau_y (x,y) dA$$

$$M_z = \int_A [x\tau_y (x,y) - y\, \tau_x (x,y)] dA$$

where τ_x, τ_y = components of the shear force distribution in the x and y directions, respectively, A = total area of tire-road contact.

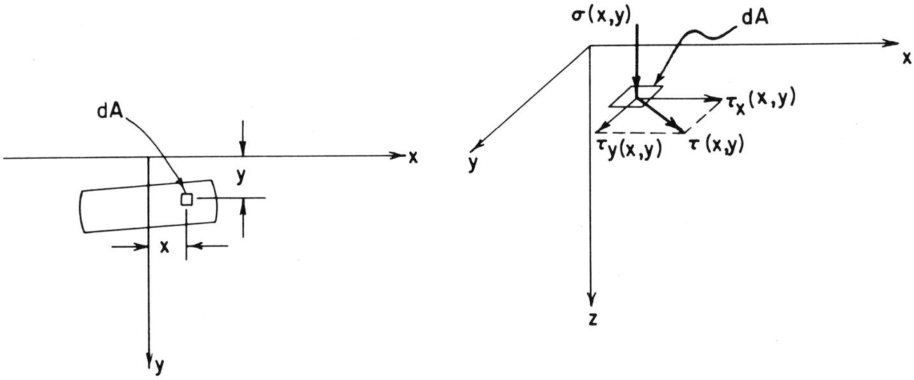

Fig. 4. Normal and shear forces acting on a differential element of tread.

The experimental observations suggest that elements of tread begin to move relative to the road surface when

$$\tau(x,y) \geq \tau_{limit} (x,y)$$

These observations also show that the traction field exhibited by a given tire-road combination is sensitive to the limiting value of differential shear force, $\tau_{limit}(x,y)$, that can be sustained before tread elements begin to slide relative to the road. The key questions are: "What are the determinants of $\tau_{limit}(x,y)$?" and "What are the determinants of $\tau(x,y)$ after tread rubber begins to slide relative to the road?"

If it is assumed that the classic concepts of friction can be invoked to explain these empirical observations, we can write that

$$\tau_{limit}(x,y) = \mu\, \sigma(x,y)$$

where μ = classic coefficient of static friction.

Although this assumption is frequently invoked in the various efforts (1-5) that have been made to model the shear-force generation process, there is empirical evidence indicating that it is incorrect to characterize the frictional coupling between tire and road by a space-independent value of μ. For example, there is evidence (6) that, in the case of contact between rubber and hard surfaces (such as road pavements), the friction coefficient is a function of (among other things) the normal pressure. Further, a directional character of the road profile at the macroscopic or microscopic level may cause the frictional coupling to be anisotropic. Thus, in the general case, we would have to write that

$$\tau_{x_{limit}}(x,y) = \mu_x(x,y)\ \sigma(x,y)$$

$$\tau_{y_{limit}}(x,y) = \mu_y(x,y)\ \sigma(x,y)$$

With these assumptions as to the character of the frictional coupling involved, the integrals yielding the components of total shear force assume the following form:

$$F_x = \int_{A_{ns}} \tau_x(x,y)dA_{ns} + \int_{A_s} \mu_x(x,y)\ \sigma(x,y)dA_s \tag{12}$$

$$F_y = \int_{A_{ns}} \tau_y(x,y)dA_{ns} + \int_{A_s} \mu_y(x,y)\ \sigma(x,y)dA_s \tag{13}$$

where A_{ns} = area of non-sliding contact and A_s = area of sliding contact.

In Eqns. 12 and 13, the first integral yields the component of the traction force that derives from tire distortion as it depends on tire compliance and the components of slip, S_x and S_y. The second integral yields the component of traction force that derives from the area of contact in which the shear force distribution is limited by a process assumed to be frictional in character. To the degree that $\mu_x(x,y)$ and $\mu_y(x,y)$ are relative-sliding-velocity dependent as well as being dependent on the normal pressure distribution, the component of traction force created within the area of sliding contact is also dependent on tire compliance and tire slip, S_x and S_y, as is the non-sliding component of traction force.

We should observe that, to first order, the following functional dependencies exist among the variables identified in Eqns. 12 and 13, viz.:

$$A_{ns}, A_s = f \left\{ \begin{array}{c} \sigma(x,y) \\ S_x, S_y \\ \mu_x(x,y), \mu_y(x,y) \\ \text{Tire compliance in the x and y directions} \end{array} \right\}$$

$$\tau_x(x,y), \tau_y(x,y) \mid A_{ns} = g \left\{ \begin{array}{c} S_x, S_y \\ \text{Tire compliance in the x and y directions} \end{array} \right\}$$

References p. 94

$$\sigma(x,y) = h \left\{ \begin{array}{l} F_z \\ \text{Tire compliance in the z or radial} \\ \text{direction} \\ \text{Tire carcass and tread cross section} \\ \text{Geometry of tread pattern} \\ \text{Net area/gross area of contact} \end{array} \right\}$$

$$\mu_x(x,y), \mu_y(x,y) = j \left\{ \begin{array}{l} \text{Physical properties of tread material} \\ \text{Temperature of tread material and road} \\ \text{Profile of road surface} \\ \sigma(x,y) \\ S_x, S_y, u \\ \text{Tire compliance in the x and y directions} \end{array} \right\}$$

In addition to the above-stated relationships, which have been postulated to be first-order relationships, there are many second-order effects. For example, the normal pressure distribution, $\sigma(x,y)$, produced by a "free, straight-rolling" tire is altered by the lateral and longitudinal distortion of the tire tread and carcass accompanying the generation of shear forces (7). We can conclude that, in addition to the dependency of the tire's traction field on normal load and the kinematic variables of motion, as expressed by Eqn. 8, there are a large number of tire design and environmental variables that govern the shape and form of this dependency. It appears that the number of variables is so large that a generalized understanding of the influence of these variables cannot be derived from empirical studies alone. Rather, it will be necessary to make additional progress in analyzing the physics that is involved. In this paper, we shall be content to examine some empirical findings in the hope that subsequent research, both analytical and experimental, will serve to explain these findings.

REQUIREMENTS FOR THE COLLECTION OF DRY TRACTION DATA

Measurements of the mechanical properties of tires, as defined by Eqn. 8, have been performed both in the laboratory and the field (8, 9, 10, 11, 12). If the primary objective is to determine the limit shear-force capability of the tire, in contrast to the forces and moments produced at small slip levels, these measurements should be conducted in the field. By "in the field", we mean testing a tire on a specific pavement such that the measured traction forces and moments reflect the influence of $\tau_{limit}(x,y)$ that derives from a specific combination of tire and pavement.

The difficulty of making field measurements as opposed to laboratory testing has been discussed elsewhere (12, 13). The major difficulties in obtaining valid and consistent data derive from three sources. First, there is the difficulty of establishing a truly steady condition of normal force in that all test surfaces have some finite irregularities in vertical profile that will excite a dynamic response on the part of the tire support system. Second, there is the difficulty of obtaining a pavement of fixed character over the entire test space with negligible change in pavement conditions over time. Third, the test agent is unable to control the temperatures of the pavement and the tire as influenced by the ambient air temperature and the amount of sunlight that is present. (The test agent may, however, be able to control the temperature increase that derives from the rolling process and the heat produced by frictional processes.) Note that these sources of error are inherent to field testing and always exist, irrespective of the tire dynamometer device that the test engineer may have at his disposal.

Since the tire dynamometers capable of measuring the complete traction field of a tire, namely, capable of defining

$$\left\{ \begin{matrix} F_x \\ F_y \\ M_z \end{matrix} \right\} = g(F_z, S_x, S_y, u)\,, \tag{8}$$

are few in number, it is logical to conclude that the design and production of satisfactory tire-vehicle systems can be accomplished without the existence of such dynamometers. On the other hand, if requirements exist for predicting the limit maneuvering performance that can be achieved by a tire-vehicle system, field-test dynamometers are required that are comparable to those described in References (12) and (14). Note that the distinguishing features of these two dynamometers are their ability to control the longitudinal slip, S_x, of the test tire.

Although the number of field devices available to measure (in a controlled manner) the complete traction field of tires is small, a substantial number of devices either exist or are under development to measure either the longitudinal traction of a straight-rolling tire or the lateral traction of a free-rolling tire. (Devices do exist, however, to measure the longitudinal and lateral traction of passenger-car tires by means of imposing a steer angle and applying a brake torque.) Although a traction dynamometer suitable for truck tires was designed and built in Germany (15) many years ago, only recently have similar programs been initiated in the U.S.. Examples of recently fabricated truck-tire dynamomenters can be seen in Figs. 5 and 6.

In view of the costs associated with the use of traction dynamometers, test engineers commonly conduct braking and cornering tests with a tire-vehicle system to define the "traction quality" of the installed tires. These procedures do not yield absolute measurements. However, they are generally adequate for comparing and ranking tires on a scale of traction quality (15–17). For purposes of this discussion,

References p. 94

Fig. 5. Firestone Truck-Tire Test Trailer.

Fig. 6. MVMA-HSRI Truck-Tire Test Trailer.

however, only data obtained by means of a traction force and moment dynamometer shall be examined. Specifically, we shall review some of the dry traction findings obtained with the dynamometer that has been described by Dugoff and Brown (12).

INFLUENCE OF ROAD SURFACE AND SPEED ON DRY TRACTION PERFORMANCE

Dry traction measurements were made recently on eleven different tire designs to acquire data for tire modeling validation purposes. The longitudinal and lateral components of tire shear force, under combined longitudinal and lateral slip, were measured at one vertical load, F_z, and two speeds, namely, 20 and 40 miles per hour. These tests were conducted at the Automotive Proving Grounds, Inc., in Pecos, Texas (May 8 to June 17, 1972) on two test surfaces having the following specifications:

1. a coarse aggregate bituminous concrete with slight macro and moderate microtexture; ASTM tire friction coefficients at 40 mph: wet surface, 0.75 (peak) and 0.47 (slide); dry surface, 0.85 (peak) and 0.73 (slide); and

2. a brushed Portland cement concrete with slight macro and moderate microtexture; ASTM tire friction coefficients at 40 mph: wet surface, 0.72 (peak) and 0.46 (slide); dry surface, 0.82 (peak) and 0.68 (slide).

Tables 1 and 2 define the eleven test tires with respect to carcass construction, meridian profile, tread geometry, and contact print geometry. Each of these tires was a commercially available production tire and was selected to provide a systematic variation in aspect ratio and carcass construction while keeping tread patterns and tire cord as much alike as possible. All tires were in load range B (four-ply rating) and all five major U.S. manufacturers are represented in this test sample. Although an attempt was made to select tires with a standard longitudinal groove, highway-type tread pattern, tread pattern variations did exist, with the moderately open (with transverse grooving) tread of the 5-rib radial tire (R70A) constituting the most marked deviation from the continuous rib pattern possessed by the remaining ten tires.

TABLE 1

Carcass Construction and Meridian Profile of Test Tires

Tire Code	Tire Size	Rim Width (in.)	Carcass Construction	Cord	H/W*	Periphery**
B178A	F78-14	5.5	Bias	Polyester	0.80	16.15
B178B	F78-14	5.5	Bias	Polyester	0.80	15.75
B170A	F70-14	5.5	Bias	Polyester	0.74	16.75
B278A	F78-14	5.5	Belted-Bias	Polyester-Glass	0.80	15.50
B278B	F78-14	5.5	Belted-Bias	Polyester-Glass	0.80	15.75
B278C	F78-14	5.5	Belted-Bias	Polyester-Glass	0.80	15.80
B270A	F70-14	5.5	Belted-Bias	Polyester-Glass	0.74	16.00
B270B	F70-14	5.5	Belted-Bias	Polyester-Glass	0.74	16.25
B260A	F60-14	7.0	Belted-Bias	Polyester-Glass	0.59	17.25
R70A	FR70-14	6.0	Radial	Rayon-Rayon	0.71	16.75
R70B	FR70-14	6.0	Radial	Rayon-Rayon	0.71	17.00

*Inflated tire aspect ratio
**Bead-to-bead meridian length (inches, uninflated).

References p. 94

TABLE 2

Tread and Contact Print Geometry

Tire Code	Tire Size	Tread Arc*	Ribs	Groove Depths (in.)		Contact Print **	
				Shoulder	Crown	Void (%)	Length (in.)
B178A	F78-14	6.00	7	0.42	0.36	45.0	6.80
B178B	F78-14	5.75	7	0.41	0.34	36.3	6.58
B170A	F70-14	6.50	7	0.41	0.35	31.4	6.30
B278A	F78-14	5.75	7	0.42	0.36	40.6	7.15
B278B	F78-14	5.70	7	0.42	0.36	39.3	7.08
B278C	F78-14	5.75	7	0.42	0.36	34.4	7.00
B270A	F70-14	6.30	7	0.36	0.32	37.4	6.70
B270B	F70-14	6.50	7	0.40	0.36	32.0	6.20
B260A	F60-14	8.00	9	0.32	0.30	42.7	5.60
R70A	FR70-14	6.50	5	0.40	0.36	47.0	7.12
R70B	FR70-14	6.50	7	0.40	0.34	37.0	7.08

*Shoulder-to-shoulder transverse arc length (inches, uninflated)
**At test load (1100 lb.) and 24 psi. The % void includes kerfing.

An examination of the traction fields generated by each of these tires on two different dry surfaces, at two different speeds, clearly indicates, as expected, that tires differ in their peak force capability. Perhaps of more significance is the finding that tires exhibit a shear force dependency on longitudinal and lateral slip that is sensitive to the surface on which it is tested.

Since the data show that the maximum slip angle of sixteen degrees used in these tests was not large enough to establish the peak lateral force that can be generated, it is necessary that a comparison of peak traction force capability possessed by this test sample be restricted to longitudinal forces. Accordingly, Table 3 tabulates the "peak" and "slide" values of the longitudinal force, F_x, ratioed to the normal force, F_z, exhibited by each of the eleven tires. (Note that only $F_x/F_z|_{peak}$ is tabulated for the concrete surface since the curve of F_x versus S_x, as measured on the concrete surface, peaks at very large values of longitudinal slip and frequently reaches a maximum at $S_x = 1.0$.) It will be observed that these results were not obtained at a constant value of normal load and consequently the normalization of the measured longitudinal forces may be contaminated by the existence of a frictional dependency on normal load, F_z. Notwithstanding any biasing of these results caused by testing these tires at normal loads significantly less than the rated load (1280 lbs.), it appears that these tires (1) produce significantly more longitudinal force on these two surfaces than does the ASTM standard tire, and (2) vary significantly in the peak and slide values of the normalized longitudinal forces produced.

As a check on the consistency of these findings, the findings tabulated in Table 3 have been ratioed to each other in the manner indicated in Table 4. For the asphalt

TABLE 3

Normalized Peak Longitudinal Force Measured on Eleven Tires F_x/F_z

Carcass Construction	Aspect Ratio	Normal Force - lbs.	Bituminous Concrete				Portland Cement	
			Peak 20 mph	Slide 20 mph	Peak 40 mph	Slide 40 mph	Peak 20 mph	Peak 40 mph
Bias-Ply	78	800	1.163	0.932	1.008	0.985	1.033	0.959
	78	800	1.01	0.858	0.964	0.883	0.836	0.829
	70	800	1.085	0.872	1.009	0.874	0.971	0.906
Belted-Bias	78	800	1.425	1.087	1.283	1.13	1.185	1.116
	78	800	1.125	0.877	1.027	0.89	0.995	0.895
	78	1000	1.050	0.85	0.985	0.95	0.950	0.87
	70	700	1.130	0.894	1.112	0.897	1.054	0.996
	70	900	1.031	0.791	0.991	0.889*	0.916	0.915
	60	800	1.162	1.00*	1.13	0.935	1.079	0.935
Radial	70	1000	0.996	0.817	1.000	0.933	0.869	0.891
	70	900/800	1.12	0.859	1.099	0.873	1.002	1.017

* - extrapolated data point

TABLE 4

Normalized Peak Longitudinal Force Ratios for Eleven Tires

Carcass Construction	Aspect Ratio	Peak/Slide Asphalt - 20 mph	Peak - 20 mph/ Peak - 40 mph		Peak - Asphalt/ Peak - Concrete	
			Asphalt	Concrete	20 mph	40 mph
Bias	78	1.248	1.155	1.077	1.125	1.05
	78	1.178	1.049	1.01	1.209	1.163
	70	1.243	1.075	1.073	1.118	1.113
Belted-Bias	78	1.31	1.11	1.062	1.202	1.151
	78	1.282	1.095	1.111	1.13	1.148
	78	1.235	1.066	1.092	1.105	1.132
	70	1.265	1.016	1.059	1.071	1.117
	70	1.303	1.041	1.001	1.126	1.083
	60	1.284	1.029	1.153	1.078	1.208
Radial	70	1.22	0.996	0.975	1.147	1.122
	70	1.305	1.020	0.987	1.117	1.08
	Average	1.26			1.13	1.124

References p. 94

surface, we find that, at 20 mph, the ratio of $F_x/F_z|_{peak}$ to $F_x/F_z|_{slide}$ is remarkably consistent over the test sample, with $F_x/F_z|_{peak}$ averaging twenty-six percent higher than $F_x/F_z|_{slide}$. With the exception of one tire, the $F_x/F_z|_{peak}$ produced at 20 mph is consistently higher than the $F_x/F_z|_{peak}$ produced at 40 mph. One sees also that the $F_x/F_z|_{peak}$ obtained on asphalt is, on the average, twelve to thirteen percent higher than the $F_x/F_z|_{peak}$ obtained on concrete, both at 20 and 40 mph. This finding contrasts with the surface discrimination yielded by the ASTM tire which, at 40 mph, yields a ratio of $F_x/F_z|_{peak}$ on asphalt to concrete equal to 1.037 and a ratio of $F_x/F_z|_{slide}$ on asphalt to concrete equal to 1.073. In examining these data, one should be aware that, in general, the data appear to be of higher quality and accuracy when the tests are made at 20 mph. The data also appear to be of higher quality when measurements are made on the asphalt surface.

In addition to the finding that the sample tires differ in their ability to generate maximum values of braking force and that, in general, these maximum forces decreased with test speed and had values that were less on the Portland cement surface than on the bituminous concrete, the data also demonstrate that surface and speed have different influences on the two components of shear force. Consider, for example, the behavior of the B170A tire as described in Tables 1 and 2. This tire, when inflated to 28 psi and tested on dry asphalt at 20 mph with a normal load of 800 lb., exhibits the traction field plotted in Fig. 7. A complex interaction between lateral and longitudinal slip is evident. Consequently, it will be helpful to consider, first, the longitudinal shear force, F_x, and the lateral shear force, F_y, that are produced when the lateral and longitudinal slip, S_y and S_x, are, respectively, equal to zero.

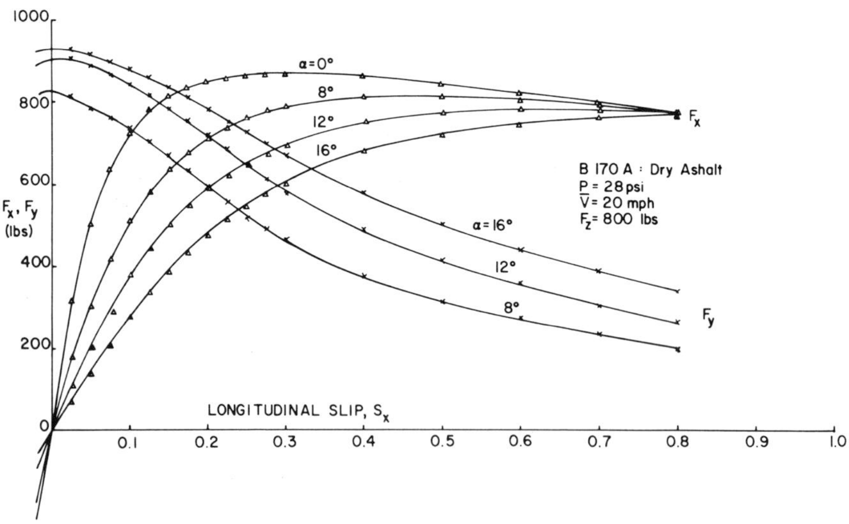

Fig. 7. Tire traction field: F_x and F_y versus S_x.

In Fig. 8, F_x and F_y are plotted as a function of longitudinal or lateral slip, respectively, with lateral slip defined as

$$S_y = \tan \alpha$$

The following observations can be made:

1. The braking (or circumferential) stiffness of this tire is somewhat larger than the cornering stiffness.
2. The tire behaves differently on the asphalt and concrete surfaces. In particular, there is a large change in the character of the F_x versus S_x curve with the peak longitudinal force being substantially reduced on the concrete surface. On the other hand, the F_y versus S_y relationship is approximately the same for the two surfaces used in this study.
3. The asymmetry in the frictional coupling between the tire and the asphalt surface becomes significantly larger when the tire is tested on the concrete surface.
4. At the maximum value of lateral slip used in this study, $S_y = 0.287$ ($\alpha = 16$ deg.), the tire has yet to attain its maximum value of lateral force.

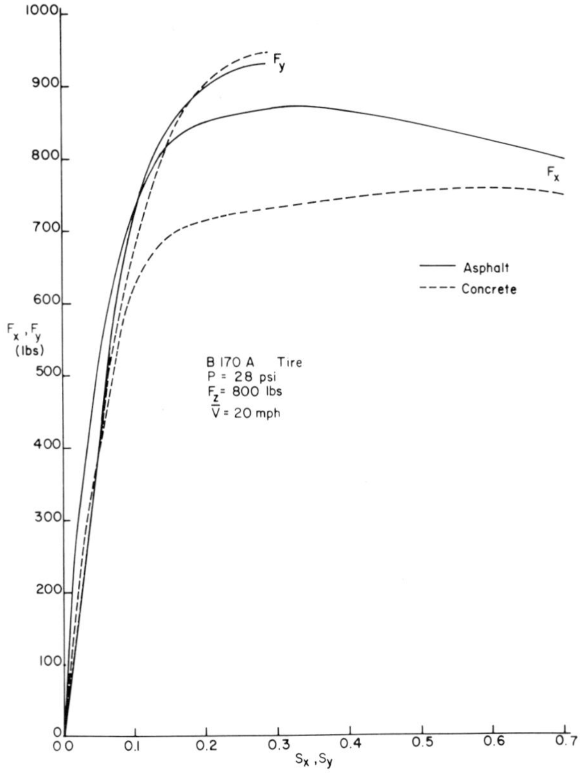

Fig. 8. Longitudinal force versus longitudinal slip and lateral force versus lateral slip: influence of pavement surface.

References p. 94

Fig. 9 presents data comparable to Fig. 8 in order to demonstrate the influence of forward velocity on F_x versus S_x and F_y versus S_y. The following observations can be made:

1. It appears that the braking stiffness of the B170A tire increases with increased velocity, although this finding could easily be an artifact deriving from imprecision in the determination of small values of S_x.
2. Over the range of lateral slip that was imposed on the test tire, an increase in velocity causes a larger decrement in the peak longitudinal force than is the case for peak lateral force.
3. On the asphalt test surface, an increase in speed reduces the value of S_x at which a peak value of longitudinal force is produced by the B170A tire.

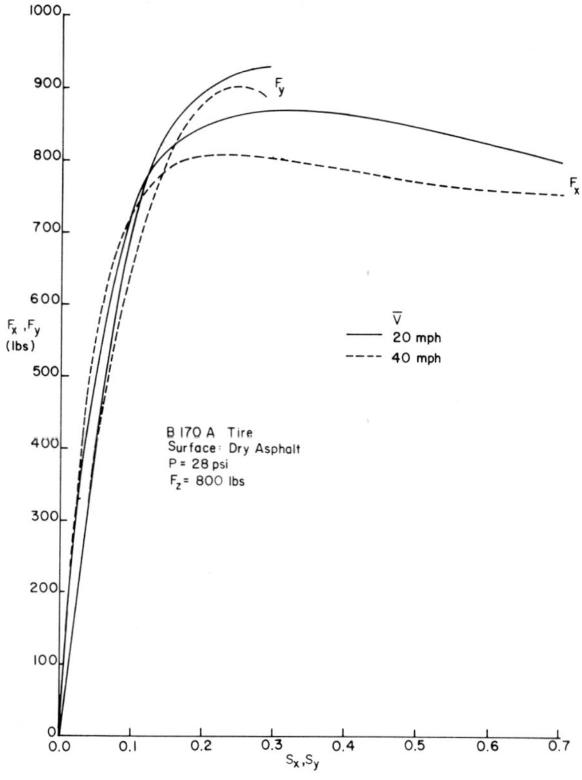

Fig. 9. Longitudinal force versus longitudinal slip and lateral force versus lateral slip: influence of velocity.

Similar findings hold for the other tires in the test sample. For example, Fig. 10 shows a limited portion of the traction field produced by the B178A tire when this tire is tested at 20 and 40 mph on both the asphalt and concrete surfaces. Again, it should be noted (see Fig. 11) that the peak lateral shear force capability of this tire remains to be defined and that the longitudinal force characteristics are markedly surface dependent whereas the lateral force characteristics are relatively independent of the test surface when V = 20 mph. If we plot F_y versus F_x at a constant slip angle of twelve degrees ($S_y = 0.213$), we see, in Fig. 12, that the total shear force produced on concrete is significantly less than that produced on asphalt when the speed of the tire is 40 mph.

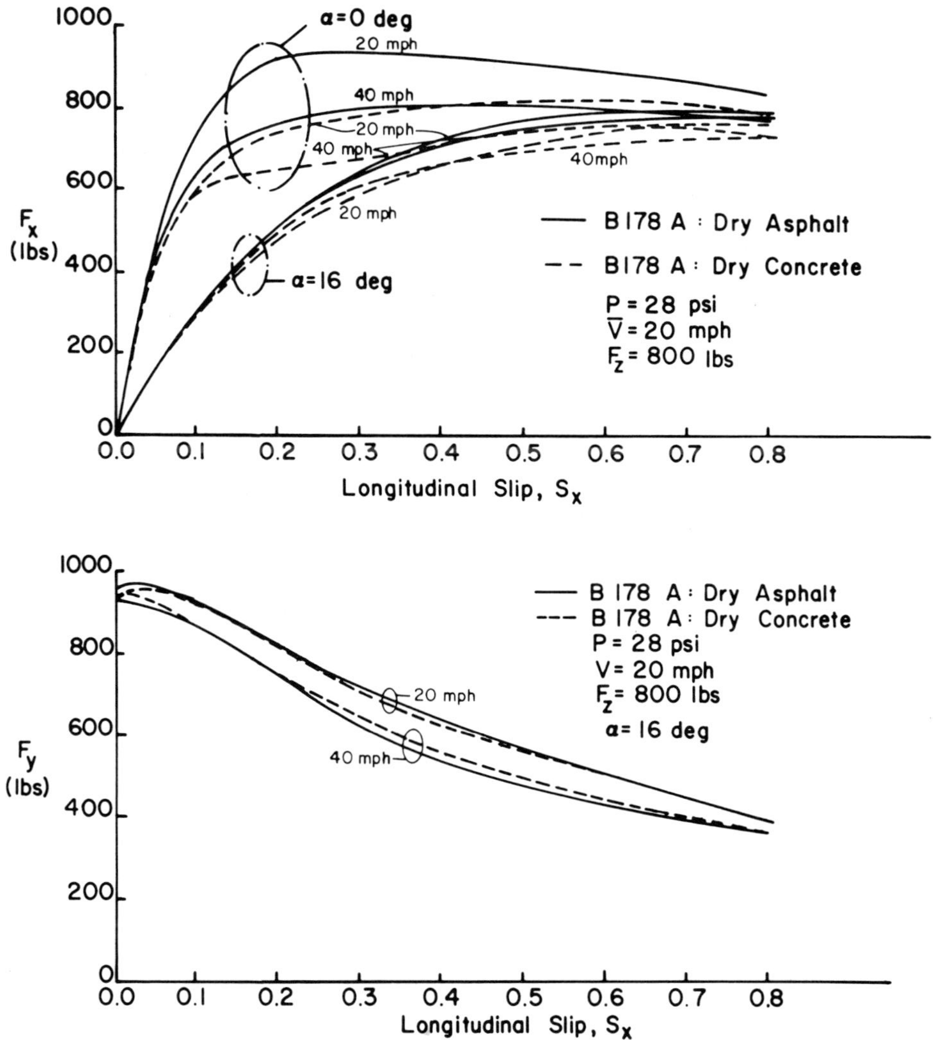

Fig. 10. Tire traction field as influenced by surface and speed.

References p. 94

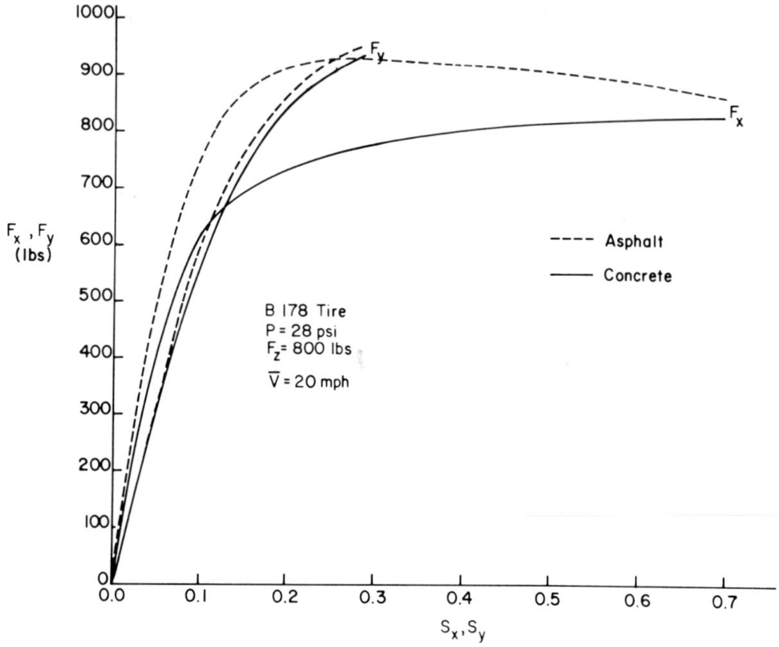

Fig. 11. Longitudinal force versus longitudinal slip and lateral force versus lateral slip: influence of pavement surface.

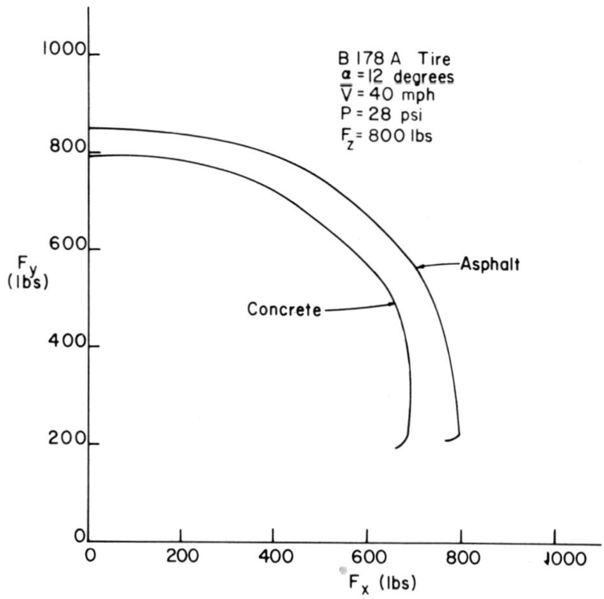

Fig. 12. Lateral force versus longitudinal force at a fixed valve of lateral slip.

INFLUENCE OF TIRE DESIGN VARIABLES ON DRY TRACTION PERFORMANCE

To demonstrate the extent to which design variables influence the traction field of the pneumatic tire, Figs. 13, 14, 15, and 16 have been prepared. Fig. 13 shows the difference in shear-force performance obtained by testing two tires of comparable carcass construction and aspect ratio that were produced by two different manufacturers. Fig. 14 provides a similar comparison for a belted-bias tire. It can be observed that the longitudinal shear force performance exhibited by these two test tires differs to a greater extent on the asphalt surface than on the concrete surface. The influence of carcass construction, with aspect ratio held fixed, and the influence of aspect ratio, with carcass construction held fixed, are shown, respectively, in Figs. 15 and 16. It may be presumed that the performance differences noted in these figures derive, in part, from variations in (1) tread compound, (2) normal pressure distribution, and (3) longitudinal and lateral compliance of the respective tires. It appears that the development of a complete and definitive understanding of the manner in which design variables influence dry traction behavior will require that substantial systematic testing be performed with specially constructed tires and that the analysis and modeling of the mechanics of shear-force generation be advanced beyond its current state.

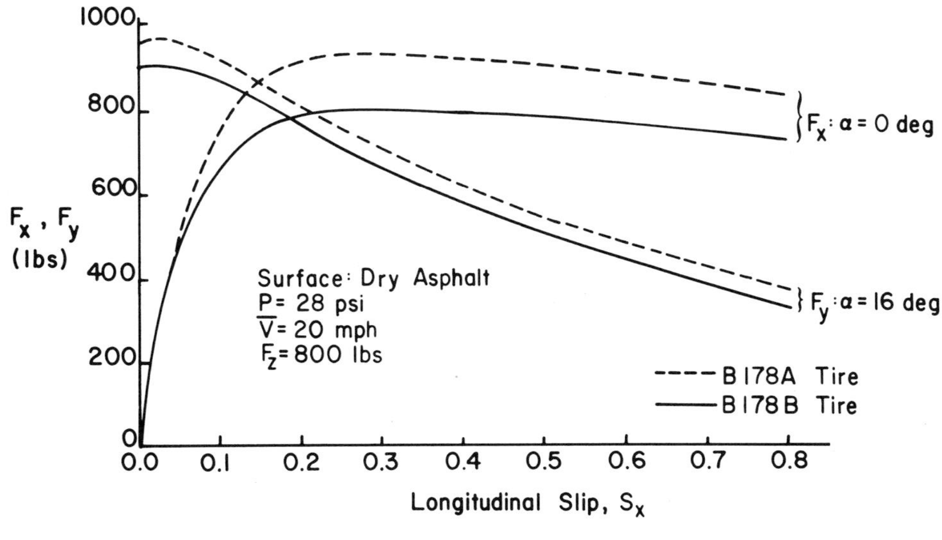

Fig. 13. Traction fields produced by tires of a different manufacturer.

References p. 94

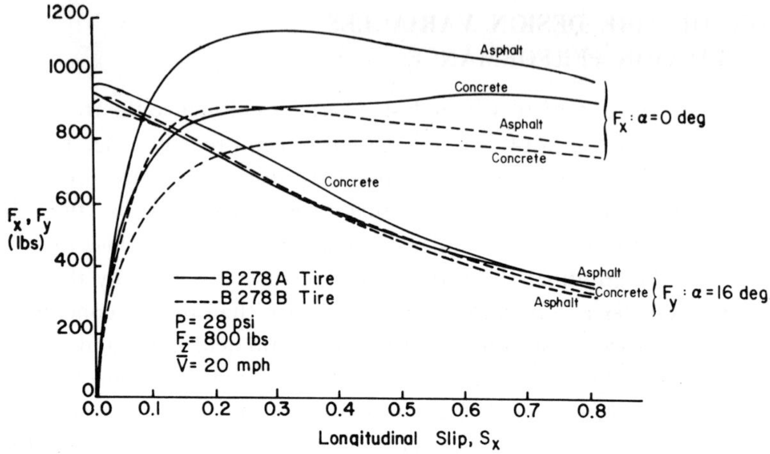

Fig. 14. Comparison of the traction field produced by two tires (of different manufacturer) on two test surfaces.

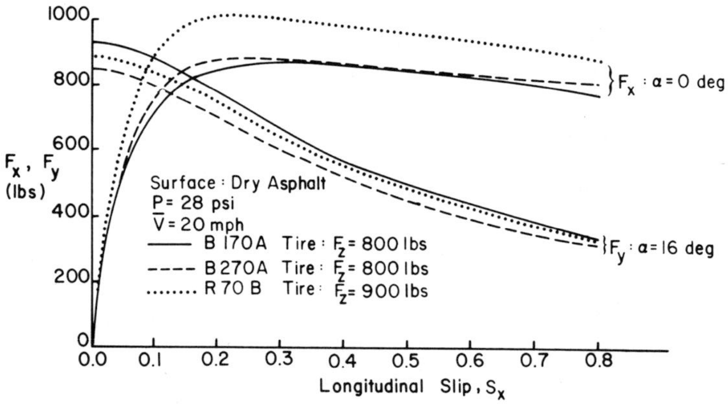

Fig. 15. Influence of carcass construction on the traction field of tires with aspect ratio of .7.

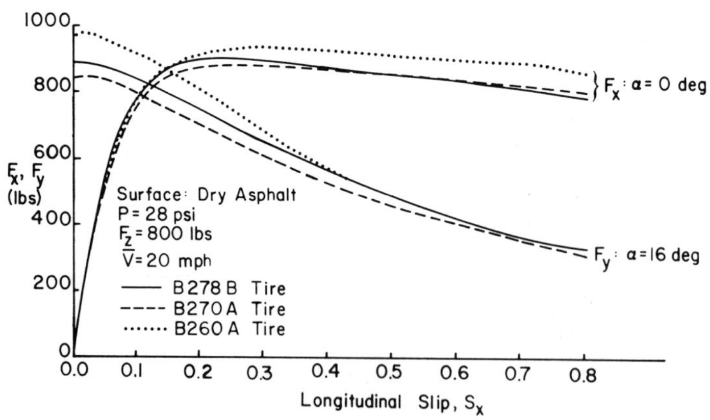

Fig. 16. Influence of aspect ratio on the traction field of belted-bias tires.

THE INFLUENCE OF COMBINED SLIP ON RESULTANT SHEAR FORCE

The traction field plotted for the B170A tire in Fig. 7 can alternatively be plotted in the manner diagrammed in Fig. 17. (Note that an attempt to plot F_y versus F_x for constant values of S_x would result in a less regular set of curves as a result of the greater imprecision encountered in measuring S_x in comparison to S_y.) Fig. 17 constitutes the typical finding that has led some vehicle mechanicians to postulate the theory of the friction ellipse (or circle) to describe the limiting shear force capability of the pneumatic tire. Clearly, the actual behavior of the resultant shear force vector, as defined by Figs. 7 or 17, is more complicated than the assumption that the limit resultant shear force is dependent upon the normal force and an asymmetric friction coefficient.

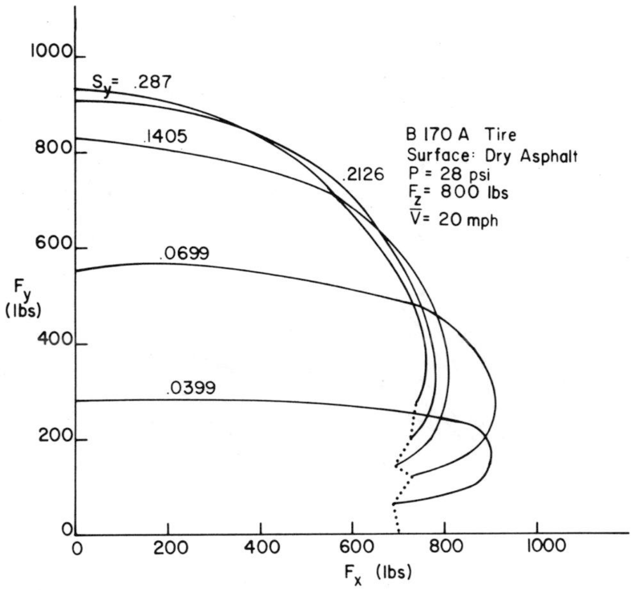

Fig. 17. Lateral force versus longitudinal force at constant values of lateral slip.

On examining Fig. 17 closely, it appears that the locus of F_y versus F_x for $S_x = 1$ and $0 < S_y < 0.287$ behaves in a rather peculiar manner. The lack of any systematic trend in the trajectory of this locus suggests either that the measurements of F_x and F_y are not accurate at the point of wheel lockup or that measurement artifacts arise from other sources. If, however, the experimental data are plotted as shown in Fig. 18, one finds that there is some consistency in the manner in which F_y varies with S_y for constant values of S_x. From simple physical considerations, we would expect F_y to vary linearly with S_y ($\tan \alpha$), when $S_x = 1.0$. Such behavior can be noted at the larger values of lateral slip, S_y, but, for the present, further study is required to determine whether the shape of the $S_x = 1.0$ curve at low values of S_y is an artifact or

References p. 94

real. It is encouraging to note, however, that, notwithstanding the crudity of the assumptions being made by those investigators who are attempting to model the shear-force generation process (1, 2, 3, 4, 5), remarkable success has been obtained in predicting the general shape of the results diagrammed in Fig. 17.

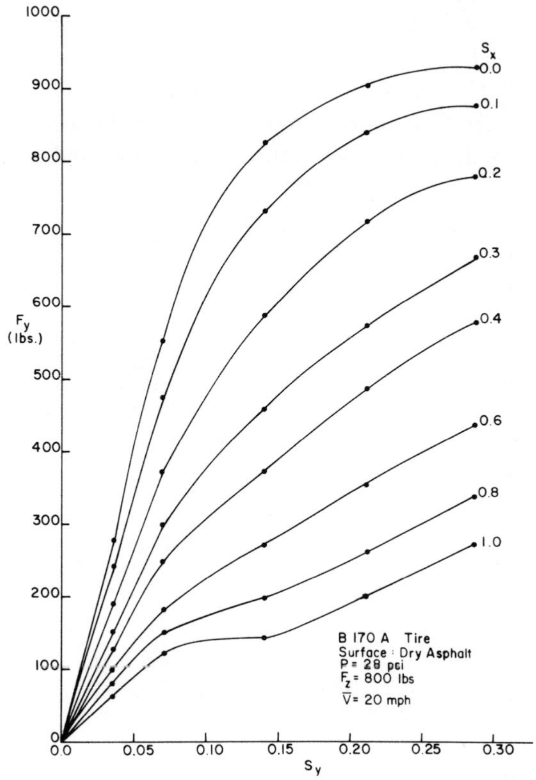

Fig. 18. Lateral force versus lateral slip at constant values of longitudinal slip.

INFLUENCE OF TREAD PROFILE ON PEAK LATERAL FORCE CAPABILITY

The traction findings discussed above were obtained by testing one or two tires of a given set of four over a range of speeds and slip (S_x and S_y) on two pavements, both in a wet and dry condition. It was intended that the tread wear resulting from the test activity would be constrained to be sufficiently small that the measured data would be viewed as results produced by a tire with negligible tread wear.

Whether the dry traction data are, in fact, unbiased by the effects of tread wear is not absolutely clear in view of a recent set of experimental findings that indicate a remarkable sensitivity to tread shoulder wear (18). The discovery of this sensitivity was unanticipated; and in view of the relevance of this phenomenon to traction measurement, in general, and to the prediction of tire-vehicle system maneuvering performance, it seems most desirable that these findings be reviewed in completing this discussion of dry traction.

After some systematic increases in peak lateral acceleration of a motor vehicle were detected in a vehicle test program as test time accrued on a given vehicle-tire combination, continuing and increasing tread wear was indicated as the likely cause of the observed phenomenon. This hypothesis was examined by designing a test program in which maximum or peak side force was monitored as a function of time during which a test tire was exposed to a slip angle of twenty degrees. As test time was accrued on a given tire, tread profile measurements were made to establish the exact nature of the wear that was taking place.

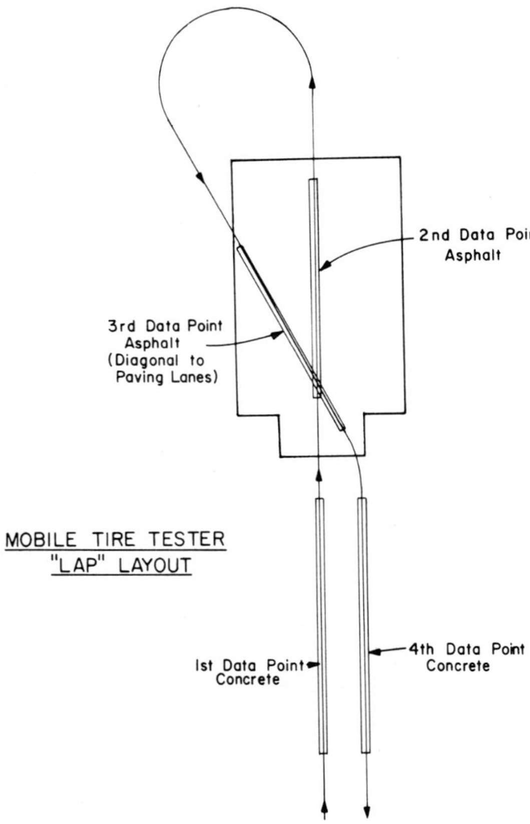

Fig. 19. Diagram of tire test area.

References p. 94

A new tire, broken in by exposing it to 100 miles of normal use in a passenger vehicle, was subjected to a test sequence in which the dynamometer (commonly referred to as the HSRI Mobile Tire Tester), operated at 40 mph, lowered the test tire onto the pavement at a slip angle of twenty degrees. The test tire was loaded with a large normal load representative of the load carried by the heavily loaded tire in an extreme maneuver. Data sets were gathered in groups of four runs each, accumulated as shown in Fig. 19. This figure diagrams the path of the tire tester and the order in which measurements were made on a concrete and asphalt surface. Note that the very first measurement is obtained on concrete, followed by two runs on asphalt, and a fourth measurement on concrete, these four data points constituting one "lap."

Measurements of side force, normalized by the normal force, F_z were obtained over ten laps. Fig. 20 presents the test results, together with the tread profiles that were measured at the conclusion of each lap by means of a contour copying device. An examination of this figure reveals the marked rise in peak side force that accompanies wear that is occurring primarily at the shoulder of the tread. In fact, the increase of side force occurring with initial wear of the shoulder is so rapid that the side force can be seen to increase with time over the nine-second test interval. (See Fig. 21, the raw data record obtained when the new tire is first run over the concrete test surface and then over the asphalt surface.)

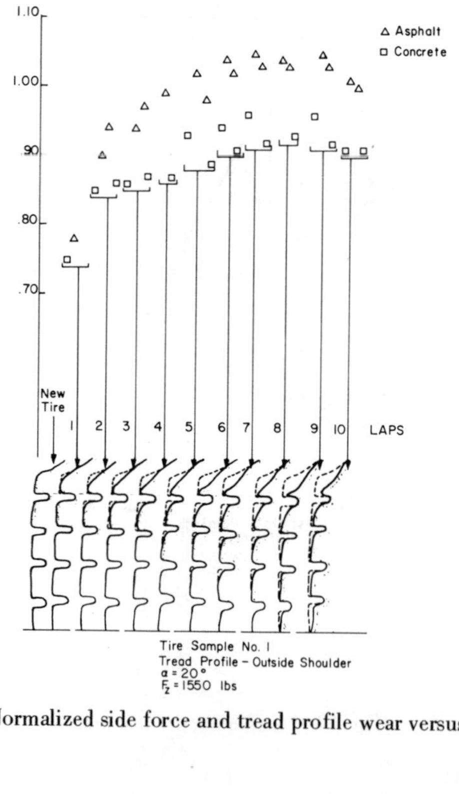

Fig. 20. Normalized side force and tread profile wear versus test laps.

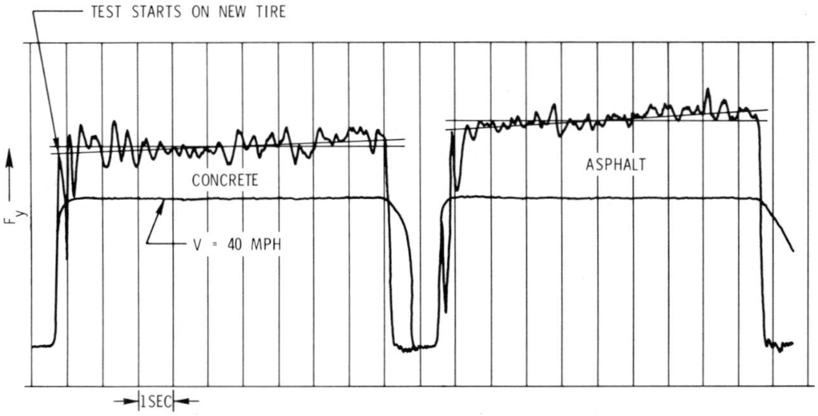

Fig. 21. Raw data record of lateral force measurement.

This experimental procedure was performed on a total of twelve different tires (18). A wide variation in peak side-force sensitivity to shoulder wear was noted. The findings indicate that tires with belted-bias construction are particularly sensitive to tread shoulder wear. Certain tires, however, were found to exhibit peak side-force characteristics that remained constant, irrespective of the shoulder wear that is produced by severe sideslipping on a dry surface. It seems reasonable to hypothesize that the observed phenomenon is associated with the form of the normal pressure distribution, $\sigma(x,y)$, and a marked variation in this distribution with small changes in tread shoulder profile. The observed phenomenon obviously alters the limit maneuvering characteristics of a tire-vehicle system as testing to determine these characteristics proceeds. It is not known whether differences in peak lateral traction (as derived from different states of shoulder wear) are significant in normal use, that is in the case that such tires (namely, tread-wear sensitive tires, with differing amounts of shoulder wear), are present on the multiple wheels of a motor vehicle.

CONCLUDING REMARKS

A considerable amount of testing has been performed to demonstrate that the wet traction performance of pneumatic tires is sensitive to the geometry of the pavement in addition to being dependent on tire properties such as tread compound, tread pattern, tread depth, etc. Very little testing has been performed to measure dry traction performance. Although some data have been gathered to define the traction field produced by a given tire on a given dry pavement, rarely has the same tire been tested thoroughly on other dry surfaces. As indicated earlier, the primary motivation for collecting dry traction data derives from the researcher who is engaged in developing tire traction models and from the vehicle dynamicist who desires to predict how a given tire-vehicle system will maneuver on a given surface in response to operator control.

References p. 94

REFERENCES

1. H. Dugoff, et al., "An Analysis of Tire Traction Properties and Their Influence on Vehicle Dynamic Performance," SAE Transactions, Vol. 79, Paper No. 700377, New York, Society of Automotive Engineers, 1970.
2. P. S. Fancher, et al., "Tire Traction Grading Procedures as Derived from the Maneuvering Characteristics of a Tire-Vehicle System," Vol. II, Final Report, NBS Contract No. 1-35715, Highway Safety Research Institute, Univ. of Michigan, Ann Arbor, June 13, 1972.
3. D. I. Livingston and J. E. Brown, "Physics of the Slipping Wheel, II. Slip Under Both Tractive and Lateral Force," Rubber Chemistry and Technology, Vol. 43, No. 2, pp. 241-261, 1970.
4. H. Sakai, "Theoretical Study of the Effect of Tractive and Braking Forces on Cornering Characteristics of Tires," Bulletin of the Society of Automotive Engineers of Japan, No. 3, March 1971.
5. H. B. Pacejka and P. S. Fancher, "Hybrid Simulation of Shear Force Development of a Tire Experiencing Longitudinal and Lateral Slip," Proceedings of the XIV International Automobile Technical Congress, FISITA, London, pp. 3/78-3/85, June 1972.
6. S. K. Clark, ed., "Mechanics of Pneumatic Tires," NBS Monograph 122, November 1971: A United States Department of Commerce Publication: U.S. Government Printing Office, Washington, D.C.
7. N. Seitz, "Experimentelle Untersuchungen uber Seitenkrafte an rollenden Luftreifen." Automobil-Industrie, pp. 71-77, April 1969.
8. D. L. Nordeen and A. D. Cortese, "Force and Moment Characteristics of Rolling Tires." SAE Transactions, Vol. 72, pp. 325-336, 1964.
9. G. Krempel, "Untersuchungen an Kraftfahrzeugreifen." Automobiltechnische Zeitschrift, Vol. 69, Nos. 1 and 8, pp. 1-8, 262-268, 1967.
10. W. Gengenbach, "Das Verhalten von Kraftfahrzeugreifen auf trachener und insbesondere nasser Fahrbahn." Dissertation submitted to the Universitat (TH) Karlsruhe, 1965.
11. K. E. Holmes and R. D. Stone, "Tire Forces as Functions of Cornering and Braking Slip on Wet Road Surfaces." Proceedings, The Instiitution of Mechanical Engineers, Vol. 183, Part 3H, pp. 35-55, 1968-69.
12. H. Dugoff and B. J. Brown, "Measurement of Tire Shear Forces," SAE Transactions, Vol. 70, Paper No. 700092, New York: Society of Automotive Engineers, 1970.
13. W. E. Meyer and M. O. Schrock, "Review of Test Methods for Tire Friction Characteristics." Journal of Materials, Vol. 4, No. 1, pp. 44-61, March 1969.
14. C. G. Giles, et al., "A Test Vehicle for Studying the Skid Resisting Properties of Tires and Road Surfaces During Controlled Cornering and Braking," Road Research Laboratory Report LR 99, 1967, Crowthorne, England.
15. G. Freudenstein, "Luftreifen bei Schrag und Kurvenlauf. Experimentelle und Theoretische Untersuchungen an Lkw - Reifen." Deutsche Kraftfahrtforschung und Strassenverkehrstechnik, No. 152, 1961.
16. W. Bergman, et al., "Tire Traction Measurement on the Road and in the Laboratory," SAE Transactions, Vol. 80, Paper No. 710630, New York: Society of Automotive Engineers, 1971.
17. A. H. Neill, Jr. and P. L. Boyd, "Research on Wet Tire Traction," Tire Science and Technology, Vol. 1, No. 2, pp. 172-189, May 1973.
18. R. D. Ervin, et al., "Vehicle Handling Performance." (Vol. 1 of 2), Final Report, DOT Contract HS-031-1-159, Highway Safety Research Institute, Univ. of Michigan, Ann Arbor, November 1972.

DISCUSSION

Thieme

Thank you, Dr. Segel.

D. Livingston

I am afraid that I missed your point, Leonard, about why it was necessary to conduct traction experiments on dry, uncontaminated surfaces. Could you explain that further?

Segel

I am sorry you missed that, Dan, because I thought I tried to make the point very clear.

Livingston

Well, it's because I have a preconceived notion that you can measure traction-slip curves just as well on contaminated surfaces.

Segel

I started my paper by noting that the peak forces generated by rubber tires on dry surfaces are very large in comparison to the forces produced on lubricated surfaces. Note that I used the word "lubricated" to cover wet surfaces and other forms of contamination. Further, I observed that the motivation for studying dry traction does not derive from a concern as to whether the tire can generate forces that are high enough to maneuver the vehicle under some emergency condition. Rather, I am talking about the prediction of vehicle performance during hard braking and hard cornering. Accordingly, the data that I presented were not peak forces or slide forces; they were traction fields. These data indicate the variation of lateral and longitudinal force as a function of combined lateral and longitudinal slip. In effect, these data define the mechanical characteristics of a tire, which characteristics are needed if one wants to predict the performance of a tire-vehicle system in response to steering, braking, and the like.

Livingston

But it always seemed to me that by proper parameterization you could isolate the coefficient of friction limit from a force-slip curve and obtain the mechanical properties of a tire.

Segel

Perhaps. Perhaps someone would care to comment on that because I am not sure I am getting the thrust of your remarks.

Livingston

Well, usually, one can describe a force-slip curve, at least approximately, in terms of two quantities; that is, a cornering stiffness, which embodies the mechanical parameters of the tire, plus a peak value which is connected with the coefficient of friction.

We may be in disagreement on this point since I do not believe that the traction field or force – slip curve should be said to depend on the tire mechanical properties alone unless one is also viewing the surface coefficient of friction as a mechanical property.

We have analyzed such curves in this manner repeatedly, and there is considerable confirming literature to the effect that varying the surface material, lubricating, or contaminating the surface – that is, changing it in any way – does not affect the initial slope of the force versus slip relation but rather the peak value. All interpretations of such curves are also to the effect that the initial slope is completely independent of the surface coefficient of friction and connected only with the mechanical properties. Thus, I would look for the mechanical properties of a tire in the initial slope, e.g., the cornering stiffness. It is in this context that I failed to understand the necessity of a dry, uncontaminated surface for the purpose of deriving mechanical properties.

Segel

The written paper makes reference to all the ongoing efforts to model the tire and predict its traction field as a function of certain parameters like cornering stiffness and braking stiffness and your own efforts in that direction have been duly recognized in the written version. However, in this oral presentation. I elected to restrict my remarks to the empirical findings.

Livingston

I now have something of a comment. I would like to suggest that your F_y versus F_x curves curve back as F_y decreases towards the F_x axis for the same reason that there is a peak in the force-slip curve, which is most likely due to the frictional heating.

Segel

The observation is made in the written paper that the efforts that have been made, to date, to model the traction fields that I have presented, do, in fact, predict the coming back of the curve as one goes to large F_X. I find it very encouraging that one can predict this particular phenomenon.

J. Tanner

I have two questions. Were the braking and cornering forces that you measured alined with the velocity vector or the wheel plane, and did you say that you were measuring side forces in a locked-wheel mode that were greater than zero?

Segel

Standing forces?

Tanner

I was interested in knowing if your tests showed the tire developing cornering forces during locked-wheel skids.

Segel

With respect to the first part of your question, the forces are measured with respect to tire axes as per the SAE definition. The procedure that is employed in the dynamometer used to obtain these measurements involves, incidentally, a force balance that is a direct copy of the balance that was developed at Langley many years ago and used by Walter Horne. This dynamometer is characterized by a means of generating torques around the spin axis to control the spin rate of the tire. The operating mode is as follows: the tire is spun at an angular velocity corresponding to the angular velocity of a free rolling tire; the tire is lowered onto the pavement; and then the angular velocity is decreased in a linear fashion until the lock-up condition is reached; the spin velocity is then increased until the free rolling condition is again attained. The tire remains in a locked condition only long enough to get a force measurement. Of course, this procedure is done very slowly in order to assure that the data are not confounded by transient phenomena.

O. Nordström

I would only comment that we have made some dry friction measurements rather similar to what was presented here, and I can tell that we also found a rather considerable increase in friction which could be explained by shoulder wear of the tire. We tried to reach a steady-state level by running the tire in at a constant slip of 20% and no side slip angle. Once we reached this level where we got the same results several times, we measured the side force – braking force interaction characteristics of the tire. When we compared the maximum longitudinal force at 5° slip angle at the beginning and at the end of such a series of measurements we found however an increase in this value of about 10%. This could be an effect of the shoulder wear as the overall tread depth change was small.

Thieme

Do you grind your tires?

Nordström

No, it is a field test. We were wearing them in by running them actually on the road.

J. D. Walter

I think this phenomenon that you have commented on, the increase in side force with wear, has been known for some time. I found a little surprising your statement that you had a tire that did not show this characteristic with tread wear. Could you comment on what type of tire this was?

Segel

I regret that I did not bring another figure along that shows the findings obtained for these 12 tires. The reason I did not, was that it was a slide that identifies the manufacturer of the tires and I wanted to remain anonymous as far as brands are concerned. It is my recollection that radial tires did extremely well in this regard, namely, they did not show an increase in *peak* side force capability as shoulder wear progressed. Now I am only talking about a small amount of wear at the shoulder. I am not talking about a general decrease in tread depth, or what one gentlemen called a reduction in 100% "skid depth" to a lesser "skid depth." This finding, incidentally, was not reported in the open literature but revealed in a research program report submitted to the U. S. Department of Transportation. These results were noted by researchers at the Calspan Corp. and tests were run on their tire facility confirming this finding. After the DOT report was released to all interested parties, many vehicle test people have come forward and stated that they have noted that the maximum cornering capability of a vehicle on dry pavement tends to increase steadily as they test it. All of the data that I am aware of tend to support one another.

Thieme

Gentlemen, I would like to complement Dr. Segel on the presentation of his interesting paper on tire traction on dry surfaces which was also given from the viewpoint of the vehicle suspension engineer. We received today a working definition of tire traction, and he told us about the mechanism of sheer force generation. Particularly interesting were his findings about sensitivity of the peak lateral force to a small amount of shoulder wear. I will not go too much into details, but the paper expresses, I think, the results of test techniques developed by his institute and by the author and has proved to give us valuable information. I therefore propose a vote of thanks to Dr. Segel.

TIRE TRACTION ON SNOW-COVERED PAVEMENTS

ALAN L. BROWNE

General Motors Research Laboratories, Warren, Michigan

ABSTRACT

A generalized theory for the traction of a pneumatic tire on a snow-covered pavement is presented.

A general description of how a tire provides traction on snow is given with several aspects of the complex interaction being presented in detail. Snow is then classified on the basis of physical properties that are of importance in snow traction. Finally, a general mathematical theory for the traction of a pneumatic tire on snow is developed.

I. INTRODUCTION

Snow traction and wet traction are two very important tire traction problems. However, while intensive study over the past several years has helped us learn much about the wet traction problem, snow traction has commanded the attention of few investigators and, thus, today remains very poorly understood.

Recent work in wet traction has revealed that most data obtained in the past cannot be used by today's investigators (1, 2). This is true because many factors that have since been found to influence wet traction were not understood and, therefore, not measured as the data was being obtained. We surmised and later have confirmed that this is the case with snow traction as well. Thus, we considered it necessary to embark on a study of the fundamental mechanisms through which snow traction is developed before beginning any large scale snow traction test program.

References pp. 134–135

The particular study to be reported on today was performed in order to improve the fundamental understanding of the dynamic processes through which a tire provides traction on a snow-covered pavement. One hoped-for result of the study was guidelines for improving the traction performance of snow tires. However, an equally important goal was to provide information for the design of a more effective all weather tire.

Our study of snow traction can be divided into four major topic areas. These are the role of the tire in snow traction, a discussion of the snow properties that are important to snow traction, the formulation of a generalized mathematical theory of snow traction, and a description of those factors needed for successful use of this theory in predicting snow traction.

Let us now treat each of these topics separately.

II. THE ROLE OF THE TIRE IN SNOW TRACTION

Consider first the role of the tire in snow traction. Those features of the tire that are important to the development of snow traction are the tire carcass construction and the tire tread pattern.

Two other important tire related features will be discussed. The first is the lead edge effect or the retarding force produced in the sloping lead edge of the footprint region due to tire penetration of the snow cover. This is strongly related to carcass construction.

The second is the expulsion of snow from the tread pattern and off the tire surface. This is strongly influenced by the tire tread design.

II.1. Tire Carcass Effects on Snow Traction – Let us begin our discussion of tire effects on snow traction with a description of the role of the tire carcass. In previous studies of traction on soil and snow (3, 4), the traditional approach was to neglect the tread pattern and consider a smooth surfaced tire. This emphasized the effects resulting from differences in carcass construction. However, early studies, both experimental and analytical, used a rigid model for the tire body (5, 6). Recent experimental work has been performed to measure the tire deformation, but little advancement in analytical modeling has been made (7, 8, 9). It must be remembered that all of these early tests were performed on snow and soil layers of infinite depth. The present study involves operation on snow-covered pavements where carcass deformation can be much more significant.

The tire carcass construction can influence the traction of a pneumatic tire on a snow-covered pavement in several different ways: through the footprint pressure distribution and varied sinkage, through the dimensions of the contact patch, and through the self-cleaning and squirm action of the tread.

Load and inflation pressure greatly influence the effect that the first two of these factors have on the snow traction of the tire.

II.1.1. Footprint Pressure Distribution and Snow Traction – Let us consider the effects of pressure distribution. Because of the nature of their construction, tires do not establish a uniform normal pressure distribution in their contact patch. A typical pressure distribution for a bias ply tire is shown in Fig. 17 of (10). That for a belted bias ply tire is more uniform. However, the least variation is present in the pressure level in the contact patch of a radial tire. The flat surface that a tire has in its contact patch on a rigid pavement is the equilibrium shape that a tire assumes under such a nonuniform external loading. If a tire were to sink with a rigid flat surface into snow, the pressure distribution would be uniform in the contact patch. This obviously cannot be, and the tire will sink differentially throughout the contact region until the pressure distribution created by nonuniform sinkage is in equilibrium with the deformed shape of the tire surface.

For the bias ply tire, the expected tire surface shape for a lateral cross section would be that depicted in Fig. 1. The deformation for belted bias ply and radial ply tires (especially steel belted) should be progressively less. Tire inflation pressure and load can strongly influence this deformation (and contact pressure distribution) by changing the rigidity of the tire structure (and, thus, the magnitude of deformation in response to a certain pressure loading) and by altering the mean (and details of the) pressure distribution in the contact patch.

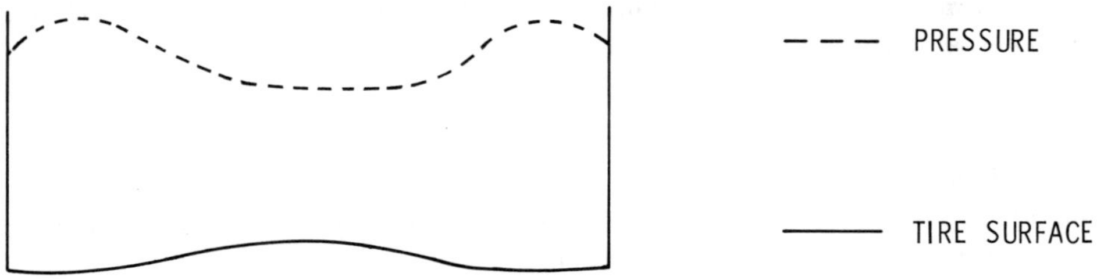

Fig. 1. Tire deformation – lateral cross section.

What does this imply in terms of tire traction capability? Because of nonuniform compaction, the shear strength of snow will vary much more in the contact of the bias ply than the radial tire. This could place either tire better than the other, depending on the details of the pressure distribution. However, the higher pressure levels existing under the shoulders of the bias ply could cause it to penetrate and interlock with some dense packed snowcovers that would not be possible for the radial tire to penetrate.

References pp. 134–135

II.1.2. Footprint Dimensions and Snow Traction – The dimensions of the contact patch are another important feature of the tire construction influencing tire traction on snow.

The radial tire, because of the nature of its construction, may have a larger contact patch than the bias ply tire especially during pure rolling. For hard compacted snows where no tread penetration exists, this means more rubber in contact with the snow and greater traction. For lightly compacted snows, this implies a greater area of snow in shear in the footprint region and greater traction under moderate powered slip or braking conditions. Thus, the radial tire construction may gain an advantage in snow traction performance for these reasons.

II.1.3. Snow Retention and Squirm Characteristics of the Tread – Finally, let us review how carcass construction can affect the tire snow traction capability by altering snow retention and squirm characteristics of the tread. For good snow traction, the tire must exhibit good tread self cleaning properties. Larger grooves clean better. The radial construction allows larger grooving at reduced wear compared to bias tires. Thus, there are major potential gains in self cleaning ability and snow traction capability for the radial tire.

Tread squirm or lateral movement of the tread elements in the footprint region is harmful to snow traction in that snow ridges are sheared laterally, and this process does not contribute to the desired longitudinal traction forces. Radial tires, with their reduced squirm should show increased snow traction.

In sum, from a snow traction standpoint under most snow conditions, radial tires should offer better performance.

II.2. Lead Edge Effects on Snow Traction – Lead edge effects can play an important role in tire traction performance. Since they are carcass-construction related, it is appropriate to discuss them here.

By lead edge effects, we mean the change in net traction due to the motion resistance forces that act on the tire in the sloping-lead edge portion of the footprint. These forces result from gross tire sinkage into the underlying snow cover. Fig. 2 illustrates the forces at the snow-tire interface in this region. (Note that the slip surface or surface of snow-tire interaction can be a snow-snow interface if a layer of snow has adhered firmly to the tire tread surface. This results in shear within the body of snow rather than at the tire surface.) The equations in Fig. 2 depict the external forces acting on a unit area of tire surface. Force f_x acts to decelerate the vehicle in the X direction. The difference $n_y - s_y$ equals the portion of the vehicle weight $\triangle w$ which is carried by this surface element.

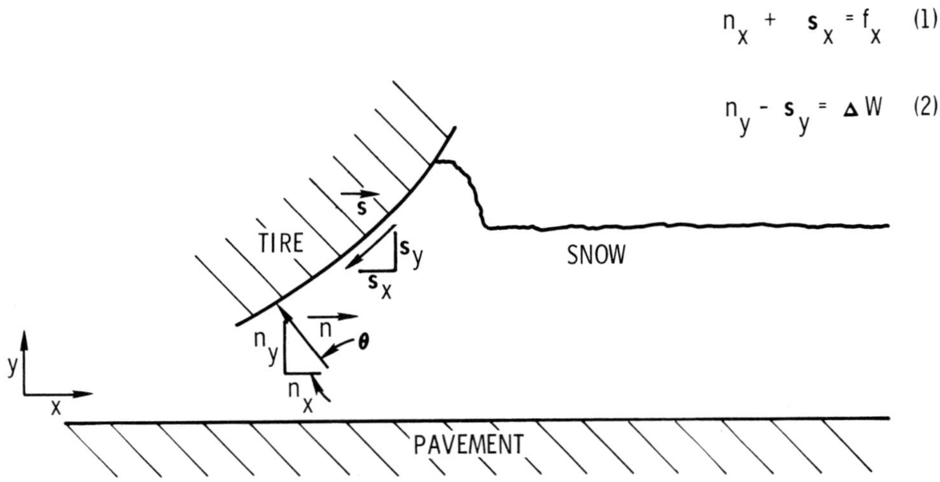

Fig. 2. Lead edge tire – snow interaction.

Force s is a shearing force that is present in the plane of the interface, and s_x is the component of s contributing to the tractive force of the tire. It also assists in the formation of a bow wave of snow at the front of the tire. This bow wave acts to displace some of the snow out of the path of the tire. Force s_y is the component of s contributing to the vertical compaction of the snow as it enters the footprint region.

The force n is perpendicular to the tire-snow interface. In a large portion of the footprint region, n equals n_y and thus, closely approximates the normal load. This normal load is equal to the integral over the footprint area of the pressure distribution under a tire. When operating on compactible snow, there will be a component of n in the x direction in the lead portion of the footprint which acts to decelerate the vehicle. This component – n_x – in conjunction with s_x can provide an appreciable retardation effect on the vehicle. Together (as f_x), they act to produce a compression wave in the snow ahead of the footprint region during forward travel of the vehicle. This results either in snow compaction by horizontal displacement or the creation of a bow wave by shear-plane formation in the leading snow mass. Reference (11) discusses such phenomena observed during the dynamic loading of soils.

It is to be noted, however, that the extent of snow penetration considered in most previous studies (which were concerned with snow trafficability) was on the order of several inches. Such large depths of penetration will not be present for snow traction studies on roads where penetrations of less than an inch are common. Thus, the retardation effect present here will, in most instances, have a much smaller influence than in snow trafficability studies.

References pp. 134–135

Force F_x, which is f_x integrated over the forward curving portion of the tire surface, is proportional to the total depth of tire sinkage. The depth of snow penetration, D_t, the width of the tire, b, and the unconfined compressive strength of the snow σ_c (12) are used to obtain the relationship,

$$F_x = \int_0^b \int_0^{D_t} \sigma_c \cos\theta \, ds. \tag{3}$$

The value of n_y together with the total depth of and pressure-sinkage characteristic for the snow determine the depth of penetration, D_t, during free rolling or steady state locked-wheel sliding.

II.3. Tread Pattern Effects on Snow Traction – Let us continue our treatment of tire effects on snow traction by discussing the role of the tire tread pattern. Bekker (13) and others report that, in the case of snow trafficability, the tire tread pattern – especially insofar as design features are concerned – has very little effect on snow traction. However, large amounts of experimental data do exist showing that the tread pattern does affect the level of traction on snow-covered pavements. Tests consistently show that snow tires have better traction than O.E. tires when on snow (14, 15). Unfortunately, little other than this general conclusion can be inferred from the published test results which usually make no attempt to differentiate between the performance of snow tires with different tread patterns. In this section, we will analyze the effect of the tire tread pattern on snow traction by detailing the roles played by the different features of the tread pattern.

Traction is provided by forces acting on the tire at the interface regions between it and the snow. These interface regions have been arbitrarily grouped under the following five headings:

1. Outer tread surface
2. Effective longitudinal grooves
3. Effective cross grooves
4. Outer sidewall edge surfaces of the tread
5. Sloping lead edge of the tire–snow contact zone.

The traction or motion-resistance forces due to gross tire sinkage into the snow cover which are grouped under heading 5 have been previously discussed. We will, thus, focus our attention on the first four tire-snow interface zones.

II.3.1. Traction Mechanisms at the Outer Tread Surface – The first interface is at the outer tread surface in the footprint region. Fig. 3 illustrates such portions of the tire tread. Since the area of the outer tread surface is the footprint area minus the void area of the tread pattern, we see that the area of this interface region would be a maximum on a smooth surfaced tire.

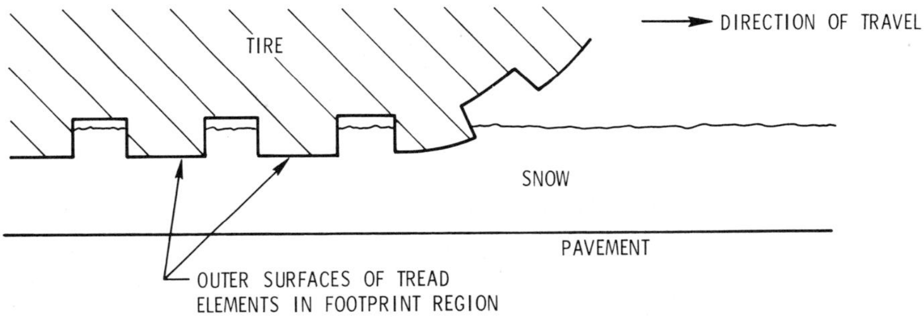

Fig. 3. Outer tread surfaces.

The traction forces developed here have different mechanisms depending on whether the tire is operating with or without slip.

Consider first the case of a tire rolling without slip. Adhesion (or cohesion if the tread surface is snow coated) and mechanical interlock between the top surfaces of the tread elements in the contact patch and the underlying snow cover provide part of the traction for a purely rolling tire. Mechanical interlock can be subdivided into two categories: micro-mechanical interlock by the draping of the rubber over the snow surface texture or macro-mechanical interlock due to the presence of chains, studs, tractionizing materials either added to the tire tread surface or present within the tire tread materials, and sipes or knife cuts.

For a tire operating with slip, one component of traction is due to the "sliding friction" at the interface. This sliding friction which results from relative motion of the surfaces (tire-snow or snow-snow) at the interface is comprised of three parts. The first is an interface force due to the presence of adhesional and cohesional bonds. These are continuously being quickly formed and destroyed which process can result in the mechanical rearrangement of particles in the underlying snow mass. The second component of sliding friction is due to hysteresis in the rubber resulting from the small scale surface texture on the rubber or snow. The third component is a viscous shear force resulting from the shear of the fluid film at the tire-snow interface,if one exists. The other major surface component of snow traction during slip is due to ploughing elements on the tire tread surface. The ploughing component can be in the form of viscoelastic-plastic trenching of the underlying snow layer or a scraping-progressive fracturing of a surface layer of snow off the underlying snow cover during wheel slip by a tire surface irregularity. Examples of irregularities are chains, studs, tractionizing materials in the tread, and edges of sipes and knife cuts.

References pp. 134–135

If there is no tread penetration into the snow cover, this will be the only traction producing interface. In fact, for traction to be increased rather than reduced by the presence of a tread pattern on the tire, the tread must penetrate the snow cover.

II.3.2. Traction Mechanisms for the Effective Longitudinal Grooves – The second interface region important in the development of snow traction is associated with the effective longitudinal grooves in the tire footprint zone. Effective longitudinal grooves are those void areas that can be sighted through longitudinally from the front to the rear of the footprint. This part of the footprint area would not be penetrated by a nongroove area of the tire tread surface if the tire were spun while the vehicle remained stationary. Such areas lie between the dotted lines in Fig. 4.

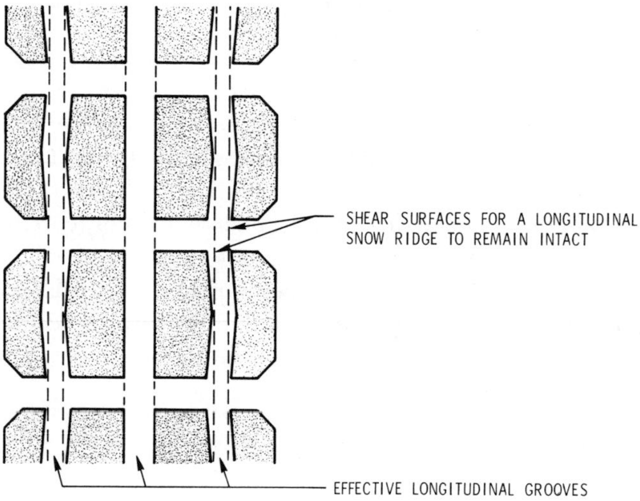

Fig. 4. Tread pattern with effective longitudinal grooves.

The traction forces developed here have different mechanisms depending on whether the tire is operating with or without slip.

For the case of a pure rolling tire, the maximum traction obtained without slip that is associated with an effective longitudinal groove will be equal to the smaller of the following two quantities:

a. the shear strength of the bonds at the snow-snow and snow-rubber interfaces that exist at the side and top boundaries of the longitudinal snow ridge filling the effective longitudinal groove. It is along these surfaces that fracture would have to occur if the tire were to slip and spin with the snow ridge remaining attached to the underlying snow cover.

b. the shear strength of the longitudinal snow ridge at its base, the ridge being broken from the underlying snow cover by the action of the tire.

Considering the geometry of present day snow tire tread patterns, failure invariably will occur in the base plane of the snow ridge.

The source of traction during slip depends on the nature of the failure mechanism for the longitudinal snow ridge when slip first occurred. If bond failure occurred such that the snow ridge remained attached to the underlying snow cover, then traction will be produced by the relative slip between snow-rubber and possibly snow-snow interfaces. However, if failure occurred at the base of the snow ridge such that snow was retained in the longitudinal groove in the tire, then traction will be produced during slip by the sliding at a snow-snow interface.

II.3.3. Traction Mechanisms for the Effective Cross Grooves – That groove area in the footprint region not lying within effective longitudinal grooves is classified as effective cross grooving. Snow wedges in these areas will be sheared at their base during tire slip (Fig. 5).

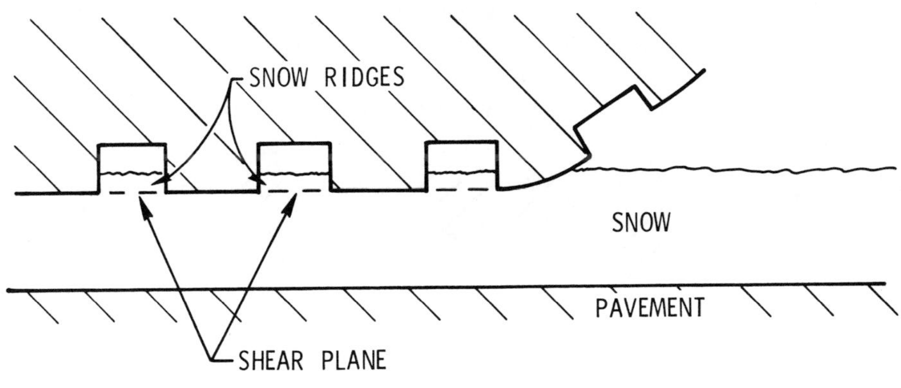

Fig. 5. Snow ridges penetrating the tread pattern.

As is the case with the other tread features, the mechanisms through which traction is developed are different for nonslipping and slipping tires.

For the case of a pure rolling tire, the traction available from an effective cross groove is based on the shear strength (at the dotted cross section plane of the tire tread surface) of the snow ridge that extends into the groove.

When the tire is operating with slip, an appreciable traction force can be produced by the ploughing-progressive compaction-shearing action of the trailing edge of a groove (in the direction of slip) as it plows through the snow wedges penetrating the

References pp. 134–135

tread surface and scrapes off surface layers from the underlying snow cover. Only when the grooves become packed with highly compacted snow do they cease to function in this manner. However, even in this case there can be traction resulting from the slip of compacted snow retained in the tread over the remaining layer of snow under the tire in the footprint region.

Because of their importance, these traction mechanisms involving the effective cross grooving will now be looked at a little more closely.

Limiting traction provided by the effective cross grooves during pure rolling was stated as being equal to the shear strength in the plane of the tread surface of the snow ridges within them. The shear strength of a snow ridge is proportional to its degree of compaction, to the pressure normal to the plane of shear and to its cross sectional area at the plane of failure (12). Thus, tire sinkage beyond tread depth results in greater potential traction from the compaction and pressurization of the snow within the grooves. When tread sinkage is less than nonfilled groove depth, the degree of compaction of the snow in the groove which was present before the tread was impressed on this region determines this shear strength.

Alteration of the groove geometry such as by putting sloping sidewalls on the effective cross grooves (Fig. 6) or by making them shallower can help increase the compaction-pressurization and, thus, increase the shear strength of the snow ridges. Such changes can, however, adversely affect traction in other ways.

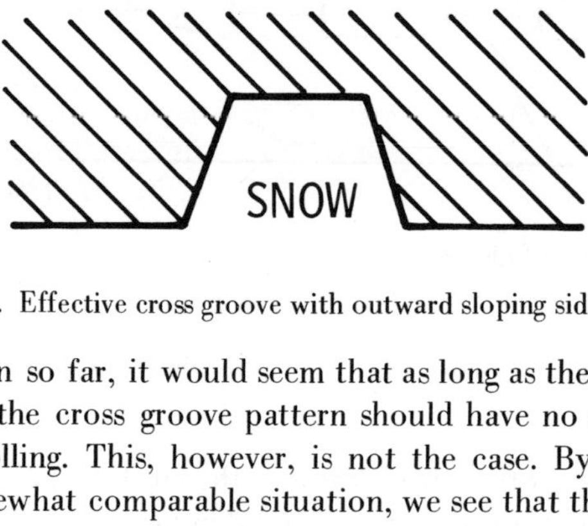

Fig. 6. Effective cross groove with outward sloping sidewalls.

From the discussion so far, it would seem that as long as the effective cross groove area were the same, the cross groove pattern should have no effect on the traction levels during pure rolling. This, however, is not the case. By the study of groove action in soils, a somewhat comparable situation, we see that the traction level is also dependent on the groove spacing and the groove width. The greater the groove width to spacing ratio, the greater will be the contact pressures on the rib surfaces and the greater the chance of tread penetration, even in highly compacted snow; also the greater will be the area of snow in shear. There is also an optimum groove width for maximizing the traction due to this mechanism under pure rolling conditions (Fig. 7). This width is that in which the preferred shear surface of the snow in the effective cross groove just spans the width of the groove. However, this is not the optimum width when operating with slip.

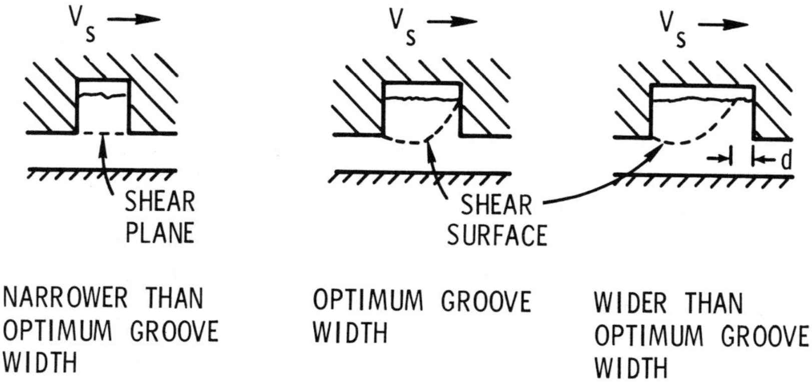

Fig. 7. Optimizing the width of effective cross grooves for a pure rolling tire.

Consider now the traction mechanisms associated with the effective cross grooves when they are operating with slip. Neglecting tread sinkage, for a wide groove, during slip, a progressive fracturing of the snow wedge occurs which supplies a tractive force to the wheel which varies with the amount of slip as illustrated in Fig. 8. Thus, the groove width should be sufficient so that the fracture surface just penetrates the remaining portion of the snow wedge as the groove exits from the rear of the footprint region; that is, so that

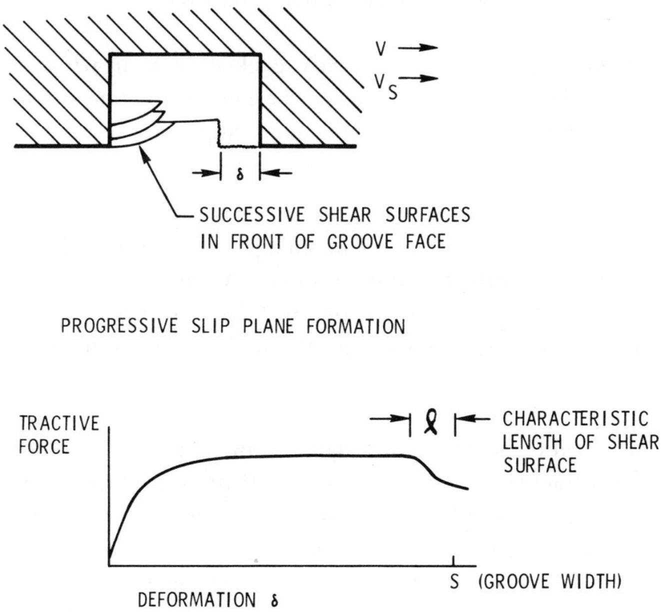

Fig. 8. Traction force curve for the groove.

$$\delta_{exit} = S - \ell. \quad (4)$$

A brief comment should be made of the tread sinkage phenomenon which modifies the above conclusion about optimum width. Consider, for simplicity, effective cross grooves whose widths are less than the characteristic failure length of the snow. Failure first occurs (Fig. 9) in the snow wedges in the cross grooves along the dotted lines shown in A. Slip then occurs with the cross ribs sinking into the less firmly compacted areas that previously lay under grooves B. Slip-sinkage may help increase this penetration (16, 17). Compaction and shear occur in front of the rib as it proceeds through this former groove position.

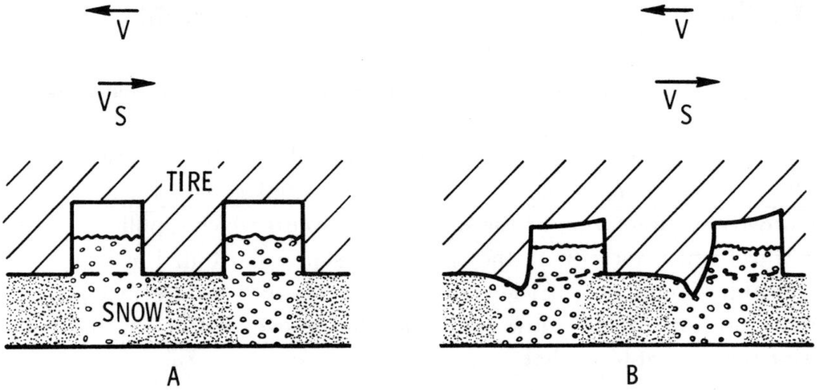

Fig. 9. Digging action of the tread.

The rib next encounters the more firmly packed snow in what was formerly an under-rib position which it will then shear. The number of times that this process is repeated depends on numerous factors, the most important being the slip velocity of the wheel. Through this digging action the tread can sometimes establish contact with the pavement surface.

Let us now include the effect of tread sinkage into the underlying snow layer in our model for slip of the effective cross grooves. As mentioned earlier, the wider the groove is compared to the rib, the greater will be this sinkage. When such sinkage occurs, there will still be a tractive force on the groove faces even when

$$\delta > S - \ell. \quad (5)$$

as the groove face will be ploughing through a snow ridge that previously lay under an adjoining rib. This tractive force may exceed that present during the initial phase of slip since the shear strength of this rib compressed snow is greater. A limiting factor in this analysis is that the groove may become choked with snow. The tire will then begin to act as if it had a smooth surface. Another aspect critical to the ploughing phenomenon is the penetrability of the underlying snow layer. On relatively thin snow layers, great density changes must occur for significant penetration depths to be possible. Also, little penetration can be expected for dense snow. However, the dynamic impact of the tread elements as they collide with the snow cover upon entering the footprint region does permit greater trend element penetration than would be predicted on the basis of static loading. Such penetration is related to the ram hardness of the snow.

A summary of the way in which traction is provided by effective cross grooves during slip can be made on the basis of the sign of the slip. When positive slip exists (the wheel spinning faster than would be the case during free rolling), the trailing edges of the cross grooves (in the direction of spin) act as trenching machines carrying snow through the footprint region. When a tire is operating in the range of wheel slip between pure rolling and fully locked-wheel skid, the grooves act as scraping edges until they are filled with snow, at which point the tire behaves as if it had a smooth surface in its interaction with the underlying snow. During locked-wheel skidding, after an initial transient period in which the grooves are becoming filled, the whole tread region becomes packed with snow, and the tire behaves as if it had a smooth surface.

In general, during shearing of the snow by the groove edges, work is done upon the snow and the energy of the vehicle is decreased, that is, a positive increment is added to the tractive force.

In conclusion, on the basis of this discussion, we can surmise that these contributions to the tractive force will increase as the number of shearing edges and groove width and depth are increased.

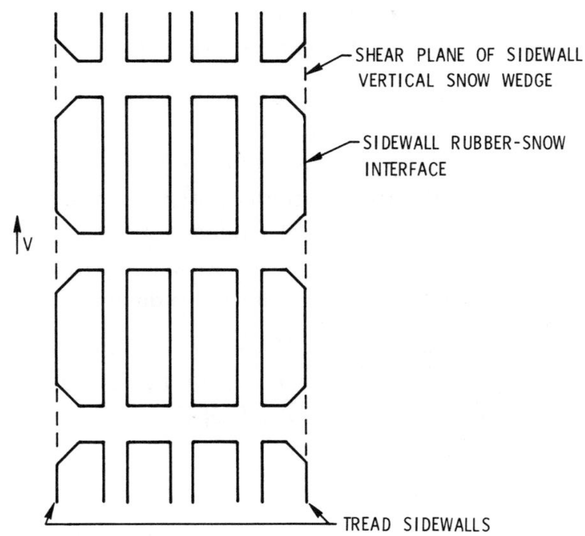

Fig. 10. Tread sidewall perimeter.

References pp. 134–135

II.3.4. Traction Mechanisms at the Outer Sidewall Edge of the Tread – The remaining interface region between tire and snow at which traction forces are provided to the vehicle is the sidewall perimeter of the tire footprint region (Fig. 10). For the tire to slip with respect to the ground, the snow ridges in the openings and indentations in the tread pattern in the tread sidewall perimeter must be sheared at the side edges of the tread.

For pure rolling, the shear strength at the outer vertical faces of snow wedges that are formed in the sidewall indentations at the tread perimeter contributes to the available traction. Cross grooves that fan out at the tire perimeter, or indentations in the side rib at the perimeter, offer opportunities for large increases in shear surface area for only small increases in tread void area. Another increment of tractive force during pure rolling is provided by the adhesive and micro-mechanical bonds between the rubber and snow at the outer surfaces of the sidewall of the tread.

During wheel slip, there is traction provided by the rubber sidewall rib extensions ploughing through the potentially long snow wedges lying between them. There is also a contribution in the form of friction due to the relative motion of the snow and tread sidewall faces of the spinning tire. Since there is negligible normal pressure at this interface, all other contributors to this component of frictional force during slip such as hysteresis loss in the rubber are quite small. There is an additional contribution from shear at vertical snow-snow interfaces in the plane of the sidewall as the tread slips and digs downward into the snow.

II.4. Snow Expulsion: Effects on Tire Traction – There is no point in putting a tread pattern on the tire surface for improving snow traction if the snow cannot be automatically expelled from this tread pattern during each rotation of the wheel.

Thus, in concluding this discussion of the role of the tire tread pattern in the production of snow traction, it is very important for us to consider the problem of snow expulsion from the tire. Five major factors influence the expulsion of snow from either the void areas in the tread pattern or off the surface of a pneumatic tire: tread pattern movement, adhesional and frictional forces between snow and tire, wheel spin velocity, abrupt changes in tread surface curvature at the lead and trailing edges of the footprint, and groove and tread element geometry.

II.4.1. Snow Expulsion: Tread Pattern Movement – The effect of tread pattern movement is different for each of four modes of tire operation: constant speed straight ahead driving, acceleration without wheel slip, braking without wheel slip, and forward travel with wheel slip.

Consider first constant speed straight ahead driving. Three general types of tread movement are noted. The first is tread squirm in the footprint region. This is usually defined as the movement of individual tread elements in a lateral direction. The second is a circumferential contraction of the tread surface in the forward portion of the contact patch. The third is a circumferential expansion of the tread surface in the rear portion of the contact patch.

Tread squirm can result in the fracture or shearing of snow ridges. This reduces the available traction. Belted and radial tires, because of their greater lateral stability (than bias tires), have very little tread squirm.

The circumferential movement of the tread can also result in the fracture of snow ridges. This can occur in the lead portion if the cross grooves are too narrow (18), but generally, this occurs in the rear portion of the footprint as the tread pattern expands and its velocity increases. This action is extremely conducive to the fracture of lateral ridges.

The circumferential and lateral movements assist the expulsion of fractured snow ridges from the tread pattern. Because of the longitudinal compression in the forward portion of the footprint region and the tread squirm throughout the footprint, snow wedges no longer completely fill the grooves. Adhesional and frictional forces at the interfaces are reduced since the snow is no longer packed in the grooves. Longitudinal expansion of the tread at the rear then flips the snow wedges from the tread pattern.

During acceleration (without wheel slip), the longitudinal contraction of the tread is increased throughout the entire footprint region. The contraction gradients (amount of contraction vs. position in the footprint) at the lead and trailing edges are increased. Lateral tread distortion also increases as the tire tends to distort in the footprint (as well as act as if it were trying to wrap itself around the rim).

During braking without wheel slip, the tread pattern undergoes less contraction at the lead edge of the footprint. At the trailing edge, there is likewise less extension. Thus, the results are modified in these respects from those for the case of constant speed driving.

The final mode of vehicle operation considered is forward travel with wheel slip. Under this condition, grooves, in general, became more tightly packed with snow.

II.4.2. Snow Expulsion: Adhesional Bonds and Frictional Forces – The second factor influencing the expulsion of snow from the tire tread is the adhesional and frictional forces between snow and tire which cause snow retention in the tread pattern and on the tread surface.

Adhesional forces that resist snow expulsion from the grooves act at the interfaces present between snow and tire along the sides and base of the grooves in the tread pattern and on the tire tread surface. The strength of the adhesional bonds is dependent on the intimacy of the contacts. This in turn, is dependent on how tightly the snow is packed into the groove. Also, the strength of the bonds is dependent on the moisture content of the snow at the snow-rubber interfaces and the wettability of the tire tread material. Heat transfer occurs from tread to snow, slowly producing a

References pp. 134–135

water film at the interface and weakening the shear strength of the interface and the strength of the interface bonds (19).

Adhesional forces also promote snow retention on the tire tread surface. Frictional forces that resist snow expulsion from the grooves act at the snow-rubber interface. They result from the mechanical interlock between rubber and snow. The magnitude of these forces is dependent on how tightly the snow is packed within the grooves.

As the separated snow peak starts to slide out of a groove, it may develop a partial vacuum beneath it at the base of the groove. Air pressure, thus, can act to retain snow in a groove.

II.4.3. Snow Expulsion: Wheel Spin Velocity – Wheel spin velocity is a third important factor influencing snow expulsion from a groove or off the tread surface. Centrifugal force acts to expel the snow from the grooves and off the tire surface. If ω is the rotational velocity of the wheel, M the mass of the snow, and r the radius of the wheel, then as a first approximation, the snow will be expelled from the groove when

$$Mr\omega^2 \geq \sum_{i=1}^{3} F_i . \qquad (6)$$

The F_i are the previously mentioned forces promoting snow retention. For snow expulsion from the tire tread surface, this equation takes the form

$$Mr\omega^2 \geq F_4 \qquad (7)$$

where F_4 is the adhesional bond strength at the tire-snow interface.

II.4.4. Snow Expulsion: Changes in Tread Surface Curvature – Another major factor to be considered in our study of snow expulsion is the effect of abrupt changes in tread surface curvature at the lead and trailing portions of the footprint region (Fig. 11). The flexing of the tread rubber as it passes through these regions acts to destroy the adhesional bonds and to separate the snow from the tire tread surface as well as from the side edges of the grooves.

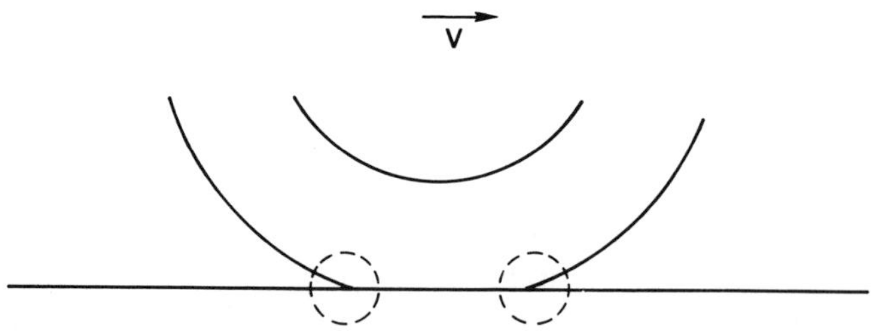

SNOW EXPULSION

Fig. 11. Tread surface curvature at the front and back of the footprint.

II.4.5. Snow Expulsion: Groove and Tread Element Geometry – Groove and tread element geometry is the final factor of importance in determining the snow retention characteristics of a tire.

The centrifugal force which acts to expel the snow from the tread is in the radially outward direction. The component of force at the snow-rubber interface in the radially inward direction acts on the snow to resist this expulsion. For grooves with walls that are radial planes, which is the normal tread configuration, this inward acting sidewall force is maximized. Grooves with outward sloping sidewalls clean more readily than normal grooves since only tension failure of the adhesional bonds is required. For grooves that have wider bottoms than openings the shear strength of the snow becomes important. The rotational velocity must be great enough to force the snow wedge against the constraining sections of the groove sidewall with a force sufficient to shear the snow.

Groove size is also another important consideration. As the groove dimensions increase, the centrifugal force increases more rapidly (for regular shaped grooves) than the restraining force. Thus, large grooves should exhibit better self-cleaning properties.

Groove size has other effects on snow expulsion. The greater the extent of a groove in the longitudinal direction the greater will be the bending that a snow wedge occupying it will have to undergo at the trailing edge of the footprint region. This will place greater tension on those interface bonds holding the snow in the groove and promote their failure and, thus, snow expulsion. Finally, the greater the extent of grooving, the more unstable the tread pattern will be. The resulting increased movement of the tread during its passage through the footprint will promote snow expulsion from the grooved portions of the tread. The adverse effects of tread squirm on snow traction have previously been mentioned.

III. SNOW PROPERTIES IMPORTANT TO TIRE TRACTION

An aspect of major importance in the formulation of a generalized model for snow traction is the determination of those snow properties that are important to the development of traction on snow. However, knowledge of the important snow properties is by itself insufficient. Other steps necessary to complete this model are first, to determine the explicit forms of the functions relating these important snow properties to measurable quantities in the snow and tire and second, to use all of this information to write a traction statement for the tire.

Before making an effort in this direction, it is important to note certain facts. First, snow is an extremely varied substance. Even in a single snowfall it is far from being homogeneous. Many factors produce variations in snow properties; in a dynamic environment such as that experienced in its interaction with the tire, the properties of

References pp. 134–135

the snow are constantly changing. And second, the interaction between snow and tire is, at best, poorly understood. For example, no one has been able to determine which reaction forces are being produced by condensation of the snow – a change in volume without a change in the relative orientation of the snow particles – and which forces are being produced by shear of the snow due to the relative movement of adjacent snow particles (20).

Both these factors should be kept in mind during the following discussion.

III.1. Snow Classification – One way to approach these problems is to begin by classifying the many types of snow on the basis of visual or measurable properties that have been found to have some correlation with snow traction. Density was found to be such a property.

Kinosita and Akitaya (19) made a classification of snow chiefly on the basis of its density (Fig. 12). Of the many types of snow that they described, we will consider those in the range from new to highly compacted snow. Slush (flowable snow) will only be briefly discussed as its treatment is more appropriately a natural extension of the previous work done by the author on the problem of hydroplaning. Traction on granular adhesionless snow is similar to that on sand; this topic has been extensively researched and will not be treated here.

TYPE OF SNOW	DENSITY
NEW SNOW	.1 g/cm^3
POWDER SNOW	.3 g/cm^3
COMPACTED SNOW	.45 TO .75 g/cm^3
ICE	.95 g/cm^3

Fig. 12. Snow classification as to density.

III.1.1. New and Powder Snow – New snow is that snow which has recently fallen and has not been acted upon mechanically, such as by the passage of vehicles, or thermally, such as by large temperature changes or by radiant energy from the sun.

Powder snow is new snow that has been acted upon by mild mechanical forces, such as wind, or by cyclic temperature variations. These produce a more uniform particle size and geometry in which little bonding has occurred between particles.

The density of powder snow is approximately three times that of new snow (see Fig. 12). When the density of snow has been increased by bonding between the particles, the snow will be said to be compacted (or packed). When the density of snow has been increased with little bonding, the snow will be said to be compressed. New snow is very susceptible to compression (if the moisture content is low). Powder snow can usually undergo density increases only through compaction.

Compressible new snow (little compaction will occur if the moisture content is less than 1% and the temperature is below −5°C) collapses considerably on loading. Thus, for traction on new snow, the critical factors are: the depth of the new snow and whether it is compactible or only compressible, the pavement surface texture, and the tire tread pattern.

Powder snow can be densified much less than new snow. Powder snow is compactible by sintering at low (less than 1%) free water content. Much higher pressures must be applied to dry-powder snow for it to be compacted to a certain level of interparticle cohesion than to powder snow with a measurable free water content (greater than 1%) (21).

Powder snow that has not been driven on previously behaves like new snow during the initial pass of a tire except that its density is much greater. Once powder snow has been subjected to automotive traffic, it becomes compacted snow.

III.1.2. Compacted Snow – Snow will pack (be compacted) only if energy is added to the system in the form of work done upon it by external forces or in the form of heat exchange with the environment. Packing is characterized by an increase both in the snow density and in the strength of interparticle bonds. The degree of susceptibility of snow to compaction is critical to the development of a significant level of traction due to the presence of the tread pattern. The shear strength and compressibility of the snow are related to its density and interparticle bond strength both of which are functions of its degree of compaction.

III.1.3. Slush or Flowable Snow – Snow is termed flowable when it behaves like a fluid and deforms under loading to the shape of its container – this being the region between the tire tread surface and the pavement surface. Density, cohesiveness, and moisture content of snow as well as the magnitude of the applied normal pressure determine whether it is flowable. Snow with a high moisture content – 30% or greater (slush) – is flowable and can be considered to be a viscous fluid and treated as such in traction calculations (19, 22).

III.2. Snow Cover Thickness Effects on Snow Traction – Snow locomotion studies in the past have concentrated on the development of criteria for determining the trafficability of snow layers of infinite depth, that is of sufficient depth so that their finite thickness had no influence on the measured results (23, 24, 25, 26). The present study is concerned with pneumatic tired vehicles on snow covered roads where snow layers can be quite thin. The depth of the snow cover under these circumstances can affect the magnitude of the available traction forces.

Snow depth (as well as density, temperature, moisture content, and grain structure) helps determine the depth of tire penetration beneath an undisturbed snow surface which is much less here than for infinite depth snows. Beyond this, there is also experimental evidence suggesting that for thin layers of snow such, as are being considered here, the layer thickness has a significant effect on snow strength. It has been found that as the sample thickness is decreased, the maximum shear stress is increased, and the displacement at which this maximum occurs is decreased (27). The snow layer thickness may well influence the shape and magnitude of the shear strength-displacement characteristics for snow (28).

III.3. Snow Properties Important to Tire Traction – Even though we are not familiar with the intimate details of the traction producing interactions between tire and snow, our general study of tread pattern effects on snow traction has pointed out those snow properties that should most influence snow traction. These are:

$\tau_{SS}(\rho,p,t,M(t),G_S(A_G))$ — the shear strength of snow

$\sigma_C(d_T,d_S,\rho,G_S(A_G),t,M(t),V,V_S)$ — the compression force-displacement characteristic for snow

and

$\sigma_{SS}(d_S,\rho,p,G_S(A_G),t,M(t),V_S,\frac{G_W}{R_W})$ — the shear stress-displacement characteristic for snow.

Additional important expressions, the forms and values for which must be determined for each particular snow and tire combination, are:

$\mu_I(S_S(\rho,G_S(A_G)),S_T(K(W_R)),H_R,\tau_{SS})$ — the coefficient of friction for incipient sliding due to micro-mechanical interlock

$\mu_{SR}(S_S(\rho,G_S(A_G)),S_T(K_S(W_R)),p_I,V_S,H_R,H,R_C,M(t),t)$ — the coefficient of sliding friction at a rubber-snow interface

$\mu_{SS}(\rho,p,V_S,t,M(t),G_S(A_G))$ — the coefficient of sliding friction at a snow-snow interface

and

$\tau_{SR}(\rho,p,t,M(t),G_S(A_G),R_C,H_R,T)$ — the shear strength of the adhesional bonds at a snow-rubber interface.

The following are definitions for the quantities appearing above:

A_G Age of snow, a main determinator of grain geometry and the strength of intergrannular bonds.
d_S Depth of the snow cover prior to the passage of the tire.
d_T Depth of the tread element penetration into the underlying snow cover (Ref. (3) p. 35).
G_S Shape factor for snow; a measure of its graininess.
G_W Width of an effective cross groove.
H Hysteresis loss per deformation cycle per unit volume for the tire tread rubber.

H_R Tread rubber hardness.

K A number which represents the number per unit area, size, and effectiveness (all of which are related to tire wear) of points of mechanical interlock in a tread surface element, the most important of these being the knife cuts or sipes and bits of tractionizing material embedded or mixed in the tread.

K_S A number relating the number per unit area (this being the number of sipes and traction elements extending from the rubber matrix), size, and effectiveness of scraping edges on a tread element surface.

$M(t)$ Moisture content of the snow.

p Pressure.

p_1 Average normal pressure under a tread element.

R_C Composition of tread rubber as it relates to adhesional bond strength between water or ice and rubber.

R_W Length of a rib element in the circumferential direction.

S_S Snow surface texture factor.

S_T Tire tread surface texture factor.

t Temperature.

T Time for bond formation between rubber and snow.

V Vehicle velocity.

V_S Slip Velocity.

W_R Measure of wear as related to the effectiveness of a feature of the tire tread.

ρ Snow density.

A lot of qualitative information exists linking these snow properties and "convenience" expressions to measurable snow, tire, and environmental quantities (12, 29, 30). However, little work has been published containing the exact nature of the linking functional relationships. Those few that were available to us are listed in the appendices of (31).

IV. A GENERAL MATHEMATICAL THEORY FOR SNOW TRACTION

Being now well grounded in traction mechanisms and snow properties, we can develop a general mathematical formulation for the traction of a pneumatic tire on a snow covered pavement.

Mathematical statements will be made for the traction mechanisms associated with each of the five tire-snow interface regions discussed previously. These then will be added to give an expression for the total snow traction capability of a tire-snow pair. The analysis began with rather lengthy mathematical formulations, appropriate for computer solution. The final model presented here has been greatly simplified so that the effect of changes in features of the tread pattern, snow properties, and operating conditions on the level of snow traction can be readily seen. This is accomplished to a major degree by the use of delta functions that allow us to separate out the effect of the many different factors.

IV.1. Traction Forces Provided by the Outer Tread Surface in the Tire Footprint Region – Denote by T_1 the traction forces provided by the outer tread surface in the tire footprint region excluding lead edge effects. Then:

$$T_1 \cong \delta_R \sum_{i=1}^{N_1} (\delta_1 T^i_{1S} + \delta_2 (T^i_{1AR} + T^i_{1M})) + \delta_S \sum_{i=1}^{N} (\delta_1 T^i_{1SS} + \delta_2 T^i_{1R}). \quad (8)$$

Displaying the functional dependence of these quantities on the snow, tire and dynamic properties of the system, we obtain the following form:

$$\begin{aligned} T_1 \cong\ & \delta_R \sum_{i=1}^{N_1} [\delta_1 \tau^i_{SS} A^i_1 + \delta_2 \tau^i_{SR} A^i_1 + \mu^i_I p^i_1 A^i_1] \\ & + \delta_S \sum_{i=1}^{N_1} [\delta_1 f^i_{SS}(\rho_1, p_1, V_S, t, M(t), G_S(A_G)) A^i_1 \\ & + \delta_2 f^i_{SR}(S_S, S_T, p_1, V_S, H_R, t, H, M(t), R_C)\, A^i_1] \end{aligned} \quad (9)$$

or

$$T_1 \cong \delta_R \sum_{i=1}^{N_1} [\delta_1 \tau^i_{SS} + \delta_2 (\tau^i_{SR} + \mu^i_I p^i_1)] A^i_1 + \delta_S \sum_{i=1}^{N_1} [\delta_1 \mu^i_{SS} + \delta_2 \mu^i_{SR}] p^i_1 A^i_1 . \quad (10)$$

If we assume that the snow has uniform properties before it enters the footprint region and that τ_{SS} is linearly related to ρ over the range of p_1 under the tread elements in the footprint region, this equation will take the following simplified form:

$$T_1 \cong [\delta_R \{\alpha_0 \tau_{SS} + \alpha_1 (\tau_{SR} + \mu_I p_1)\} + \delta_S \{\alpha_0 \mu_{SS} + \alpha_1 \mu_{SR}\} p_1] A_1 . \quad (11)$$

The new symbols appearing in the preceding equations are defined as follows:

A_1 Projection on the pavement surface of the area of contact between tread surface elements and the underlying snow in the footprint region.

f_{SR} Frictional force (psi) generated by the relative motion of surfaces at a snow-rubber interface which will be approximated by $\mu_{SR} p_1$.

f_{SS} Frictional force (psi) generated by the relative motion of surfaces at a snow-snow interface which will be approximated by $\mu_{SS} p_1$.

N_1 Number of distinct surface elements (or arbitrary divisions).

T_{1AR} Contribution of snow-rubber interface adhesional bond shear to the maximum tractive force.

T_{1M} Contribution of mechanical interlock due to the surface texture (including sipes and studs if any are present) to the maximum tractive force.

T_{1R} Contribution of relative snow-rubber surface movement in the footprint region to the traction force during slip.

T_{1S} Contribution of cohesional bond shear (snow-snow) to the maximum tractive force.

T_{1SS} Contribution of sliding friction between a snow layer coating the tread element and the underlying snow cover.

α_0, α_1 Fractions of the tread surface for which shear occurs within the snow layer and at the rubber-snow interface, respectively.

δ_s This symbol has the value 1 if there is tire slip; otherwise, it is 0.

δ_R This symbol has the value 1 if the mode of operation is pure rolling; otherwise, when there is tire slip it is 0.

δ_1 This symbol is 1 if $T^i_{1S} < T^i_{1AR} + T^i_{1M}$ which is true when there is shear-plane formation in the snow layer; otherwise, it is 0.

δ_2 This symbol is 1 when δ_1 has the value 0 which is true when bond rupture occurs at the snow-rubber interface; otherwise it is 0.

$(\ )^i$ The value of the quantity within ( ) which is present at the ith distinct surface tread element (or arbitrarily specified tread element region if there are no distinct blocks in the tread pattern) in the footprint region.

IV.2. Traction Forces Provided by the Effective Longitudinal Groove Area in the Tire Footprint Region – Let T_2 denote the traction force associated with the effective straight longitudinal groove area of the footprint region. Then:

$$
\begin{aligned}
T_2 \cong\ & \delta_R \Big[\sum_{j=1}^{N_{2S}} \Big\{ \delta_P \delta^j_2 \sum_{i=1}^{N^j_{2R}} T^{i,j}_{2A} + \sum_{i=1}^{N^j_{2S1}} T^{i,j}_{2S1}) + (1-\delta^j_2)\,(\delta_P (1-\delta^j_F) T^j_{2S} + \delta^j_F T^j_{2SN}) \ \Big] \\
& + \delta_S \Big[\sum_{j=1}^{N_{2S}} \Big\{ \delta_P \delta^j_2 \Big(\sum_{i=1}^{N^j_{2R}} T^{i,j}_{2R} + \sum_{i=1}^{N^j_{2S1}} T^{i,j}_{2SS1} \Big) \\
& + (1-\delta^j_2)\,(\delta_P (1-\delta^j_F) T^j_{2SS} + \delta^j_F T^j_{2SSN}) \Big\} \Big].
\end{aligned}
\tag{12}
$$

Including the functional dependence of the above quantities on snow, tire, and dynamic parameters of the system, T_2 can be rewritten in the following manner:

$$T_2 \cong \delta_R [\sum_{j=1}^{N_{2S}} \delta_P \delta_2^j (\sum_{i=1}^{N_{2R}^j} (\tau_{SR}^{i,j} + \mu_I^{i,j} p_2^{i,j}) A_{2R}^{i,j} + \sum_{i=1}^{N_{2S1}} \tau_{SS}^{i,j} A_{2S1}^{i,j})$$

$$+ (1-\delta_2^j)(\delta_P(1-\delta_F^j)\tau_{SS}^j A_{2S}^j \tag{13}$$

$$+ \delta_F^j \tau_{SSN}^j A_{2S}^j)\Big\} + \delta_S [\sum_{j=1}^{N_{2S}} \Big\{ \delta_P \delta_2^j (\sum_{i=1}^{N_{2R}^j} f_{SR}^{i,j} A_{2R}^{i,j} + \sum_{i=1}^{N_{2S1}^j} f_{SS}^{i,j} A_{2S1}^{i,j})$$

$$+ (1-\delta_2^j)(\delta_P(1-\delta_F^j) f_{SS}^j A_{2S}^j + \delta_F^j f_{SSN}^j S_{2S}^j)\Big\}]$$

or

$$T_2 \cong \delta_R [\sum_{j=1}^{N_{2S}} \Big\{ \delta_P \delta_2^j (\sum_{i=1}^{N_{2R}^j} (\tau_{SR}^{i,j} + \mu_I^{i,j} p_2^{i,j}) A_{2R}^{i,j}$$

$$+ (1-\delta_2^j)(\delta_P(1-\delta_F^j)\tau_{SS}^j + \delta_F^j \tau_{SSN}^j) A_{2S}^j)\ ]$$

$$+ \delta_S [\sum_{j=1}^{N_{2S}} \Big\{ \delta_P \delta_2^j (\sum_{i=1}^{N_{2R}^j} f_{SR}^{i,j} A_{2R}^{i,j} + \sum_{i=1}^{N_{2S1}^j} f_{SS}^{i,j} A_{2S1}^{i,j}) \tag{14}$$

$$+ (1-\delta_2^j)(\delta_P(1-\delta_F^j) f_{SS}^j + \delta_F^j f_{SSN}^j) A_{2S}^j \Big\}].$$

Considering present tread geometries[1] and neglecting terms of minor importance, the expression for T_2 reduces to:

$$T_2 \cong \delta_R [\sum_{j=1}^{N_{2S}} \Big\{ \delta_P (1-\delta_F^j)\tau_{SS}^j + \delta_F^j \tau_{SSN}^j \Big\} A_{2S}^j]$$

$$+ \delta_S [\sum_{j=1}^{N_{2S}} \delta_P (1-\delta_F^j)(f_{SS}^j + \delta_F^j f_{SSN}^j)\ A_{2S}^j]. \tag{15}$$

[1]*Only with wide longitudinal grooves and small depths of penetration is there a possibility of tire slip with longitudinal snow ridges remaining intact.*

Assuming uniform conditions in the snow in the longitudinal grooves of the tire in the footprint region, this expression becomes:

$$T_2 \cong [\delta_R\{\delta_P(1\text{-}\beta)\tau_{SS} + \beta\tau_{SSN}\} + \delta_S\{(1\text{-}\beta)\delta_P\mu_{SS} + \beta\mu_{SS}\}p_5]A_{2S}. \tag{16}$$

The new symbols appearing in the above equations have the following definitions:

$A^{i,j}_{2R}$ Area of a distinct snow-rubber interface (projected on the plane of the pavement surface).

A^{j}_{2S} Area of the base cross section of a segment of an effective longitudinal groove (projected on the plane of the pavement surface).

$A^{i,j}_{2S1}$ Area of a distinct snow-snow interface (projected on the plane of the pavement surface).

f_{SSN} Frictional force (psi) generated by the relative motion of snow surfaces at the surface of severed snow wedges which will be approximated by $\mu_{SS}p_5$.

N_{2R} Number of distinct rubber-snow interfaces along the side and top boundaries of a segment of an effective longitudinal groove, this being dependent on the depth of tread penetration.

N_{2S} Total number of segments into which the effective longitudinal grooves have been arbitrarily divided.

N_{2S1} Number of distinct snow-snow interfaces along the side and top boundaries of a segment of an effective longitudinal groove. (The top boundary is included since the groove may already have been partly packed with snow from a previous revolution.)

P_2 Normal pressure at a snow-rubber interface.

P_S Normal pressure (to the base plane) in a longitudinal snow wedge.

T_{2A} Traction contribution of the shear strength of the rubber-snow bonds along the boundaries of an effective longitudinal groove.

T_{2R} Traction force resulting from relative motion of the tread rubber and the snow.

T_{2S} Base shear strength of a snow ridge in an effective longitudinal groove.

T_{2S1} Traction contribution of the shear strength of the snow at snow-snow interfaces along the boundaries of an effective longitudinal groove.

T_{2SN} Shear strength of a new bond formed between the snow previously entrapped in an effective longitudinal groove and the new underlying snow layer.

T_{2SS} Traction during wheel slip resulting from relative motion between a severed effective longitudinal groove snow wedge and its base.

T_{2SS1} Traction force at a snow-snow interface resulting from relative motion between the snow retained in the effective cross grooving in the tread and an intact longitudinal snow wedge.

T_{2SSN} Traction during wheel slip resulting from relative motion between snow previously trapped in an effective longitudinal groove and the underlying snow layer.

β Fraction of the effective longitudinal groove area entering the footprint packed with snow.

δ_2 This coefficient is 0 when the snow wedge in the longitudinal groove shears at its base so that it is transported by the tread, this occurring when

$$\sum_{i=1}^{N^j_{2R}} T^{i,j}_{2A} + \sum_{i=1}^{N^j_{2S1}} T^{i,j}_{2AS1} \rangle (1-\delta^j_F) T^j_{2AS} + \delta^j_F T^j_{2AN} \tag{17}$$

It is 1 when the tire slips such that the snow ridge remains intact and attached to the underlying snow cover. The opposite condition to that of (17) is then true.

δ_F This symbol has the value 1 if a groove segment is clogged with previously severed snow such that there is no penetration of new snow into it and the value 0 if there is new snow penetration into the groove segment.

δ_P This symbol has the value 1 if there is tire penetration into the underlying snow layer; otherwise, it is 0.

γ_{SSN} Shear strength (psi) of the new snow-snow bonds formed between the snow in a snow filled groove segment (into which there is no new snow penetration) and the underlying snow layer.

$(\ )^{i,j}$ ith interface of the jth effective longitudinal groove segment.

IV.3. Traction Forces Provided by the Effective Cross Groove Areas in the Tire Tread Pattern — Let T_3 denote the traction forces due to mechanical interlock of the effective cross grooves (those groove areas outside of effective longitudinal grooves) with the snow as well as those forces resulting from the retention of severed snow ridges in this part of the tread element pattern.

The following mathematical expression for T_3 is proposed:

$$T_3 \cong \delta_R \sum_{i=1}^{N_3} (\delta^i_F T^i_{3SN} + \delta_P (1-\delta^i_F)\, T^i_{3S})$$
$$+ \delta_S \sum_{i=1}^{N_3} [(\delta^i_F + (1-\delta^i_F)\delta_P \delta_L) T^i_{3SS} + \delta\ (1-\delta_L)\,(1-\delta^i_F) T^i_{3P}]. \tag{18}$$

When the functional dependence of these quantities on the snow and tire properties and on the dynamic environment is included, the above expression becomes:

$$T_3 \cong \delta_R \sum_{i=1}^{N_3} [\delta_F^i \tau_{SSN}^i A_3^i + \delta_P(1-\delta_F^i)\tau_{SS}^i A_3^i] + \delta_S \sum_{i=1}^{N_3} [(\delta_F^i + (1-\delta_F^i)\delta_P\delta_L) f_{SS}^i A_3^i$$
$$+ \delta_P(1-\delta_L)(1-\delta_P^i) F_S^i (V_S, \rho, p_3, M(t), t, G_S(A_G), \ell, d_T, W_R, H_R)] \tag{19}$$

or

$$T_3 \cong \delta_R \sum_{i=1}^{N_3} [\delta_F^i \tau_{SSN}^i A_3^i + \delta_P(1-\delta_F^i)\tau_{SS}^i A_3^i] + \delta_S \sum_{i=1}^{N_3} [(\delta_F^i$$
$$+ (1-\delta_F^i)\delta_P\delta_L)\mu_{SS} p_3 A_3^i + \delta_P(1-\delta_L)(1-\delta_F^i)\tau_{SS1}^i A_3^i]. \tag{20}$$

The force F_s which acts on the trailing groove edge as it scrapes a layer from the surface of the underlying snow cover, has been approximated by $\tau_{SS1} A_3$, the product of the shear strength of the rib-packed snow and the projected area of the effective cross grooves on the plane of the underlying pavement surface.

An additional assumption which appears here, and which will be made throughout the analysis, is that the effective cross grooves become completely packed with snow throughout the footprint due to the scraping action of the slipping tread only during the special case of locked-wheel operation.

If we now assume that average values of τ_{SSN}, τ_{SS}, p_3, and d_T can be estimated directly from the tire tread dimensions and footprint area, the tire load, and the general state of snow compaction-compression, T_3 can be written in the following simplified form:

$$T_3 \cong \delta_R [\alpha\tau_{SSN} + \delta_P(1-\alpha)\tau_{SS}] + \delta_S \Big\{(\alpha + (1-\alpha)\delta_P\delta_L)\mu_{SS} p_3$$
$$+ (1-\delta_L)(1-\alpha)\delta_P \tau_{SS1}]\Big\} A_3. \tag{21}$$

The new symbols appearing in the preceding equations have the following definitions:

- A_3 Projected area of a groove segment in the tread pattern on the plane of the pavement surface (excluding those segments which lie within effective longitudinal grooves).
- ℓ Lateral extent of an effective cross groove.
- N_3 Number of groove segments not included in effective longitudinal grooves.

References pp. 134–135

p_3 Normal pressure (to the plane of the footprint) in the snow ridges in effective cross grooves.

T_{3P} Contribution to the traction during the tire slip of the ploughing action of the groove edges through the underlying snow cover.

T_{3S} Contribution to the maximum tractive force during pure rolling of the mechanical interlock of the effective cross grooves in the tread pattern with the underlying snow cover.

T_{3SN} Contribution to the maximum tractive force during pure rolling of the new bond strength (between snow filled groove segment and underlying snow layer).

T_{3SS} Contribution to the traction during tire slip of the friction due to the relative motion between the snow entrapped in the effective cross grooves of the tread pattern and the underlying snow base.

α Fraction of the effective cross groove area entering the footprint packed with snow.

δ_L This symbol has the value 1 during locked wheel operation. Otherwise, it is 0.

τ_{SS1} Shear strength of rib packed snow.

$(\ )^i$ Value of the quantity within ( ) which is present in the ith effective cross groove segment.

IV.4. Traction Forces at the Sidewall Perimeter of the Tire Footprint Region – Let T_4 denote the traction forces resulting from tire-snow interaction at the tire sidewall surfaces adjacent to the footprint region (Fig. 13). Expressed mathematically, T_4 will have the following form:

$$\begin{aligned} T_4 \cong\ & \delta_R \delta_P \Big[\sum_{i=1}^{N_4} (T^i_{4SN} + (1-\delta^i_F) T^i_{4S}) + \sum_{i=1}^{N_4} T^i_{4A}\Big] \\ & + \delta_S \delta_P \Big[\sum_{i=1}^{N_4} (T^i_{4SS} + (1-\delta_L)(1-\delta^i_F) T^i_{4P}) + \sum_{i=1}^{N_{4R}} T^i_{4R} + \delta_L T_{4S}\Big]. \end{aligned} \tag{22}$$

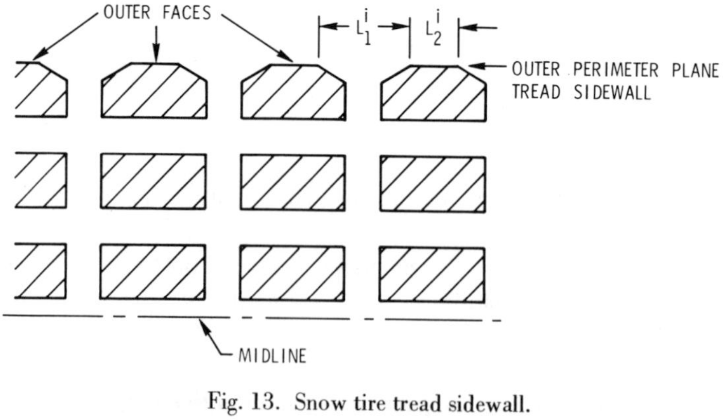

Fig. 13. Snow tire tread sidewall.

Upon inclusion of the functional dependence of these quantities on the snow and tire properties and the dynamic environment, the above expression becomes:

$$T_4 \cong \delta_R \delta_P [\sum_{i=1}^{N_4} \left\{ \tau^i_{SSN}(A^i_4(d)-A^i_{4F}(d_F)) + (1-\delta^i_F)\tau^i_{SS}A^i_{4F}(d_F) \right\}$$

$$+ \sum_{j=1}^{N_{4R}} \left\{ (\tau^i_{SR} + f^j_I(S_S, S_T, H_R, \tau_{SS}))A^j_{4R}(d_T) \right\}]$$

$$+ \delta_S \delta_P [\sum_{i=1}^{N_4} \left\{ f^i_{SS}A^i_4(\delta_T) + (1-\delta_L)(1-\delta^i_F)F^i_S(V_S, \rho, p, M(t), t, G_S(A_G), L_2 \right.$$

$$\left. d_T, W_R, H_R) \right\} + \sum_{j=1}^{N_{4R}} f^j_{SR}A^j_{4R}(D_T) + \delta_L \ell D_T \tau_{SS}] \tag{23}$$

or

$$T_4 \cong \delta_R \delta_P [\sum_{i=1}^{N_4} (\tau^i_{SSN}(d^i-d^i_F)L^i_1 + (1-\delta^i_F)\tau^i_{SS}L^i_1 d^i_F) + \sum_{j=1}^{N_{4R}} (\tau^j_{SR} + f^j_I)L^j_2 d^j_T]$$

$$+ \delta_S \delta_P [\sum_{i=1}^{N_4} \left\{ f^i_{SS}L^i_1\delta^i_T + (1-\delta_L)(1-\delta^i_F)\tau^i_{SS}L^i_1 \frac{\epsilon_T}{2} \right\} \tag{24}$$

$$+ \sum_{j=1}^{N_{4R}} f^j_{SR}A^j_{4R}L^i_2 D^i_T + \delta_L \ell D_T \tau_{SS}].$$

Keeping only those terms which can be significant contributors to snow-tire traction (due to the absence of large normal pressures at the interface) and assuming that the grooves are shorter than the natural fracture length of the snow, then the above expression takes the following form:

$$T_4 \cong \delta_R \delta_P [\sum_{i=1}^{N_4} (1-\delta^i_F)\tau^i_{SS}L^i_1 d^i_F] + \delta_S \delta_P [(1-\delta_L) \sum_{i=1}^{N_4} \tag{25}$$

$$(1-\delta^i_F)\tau^i_{SS}L^i_1 \frac{\epsilon_T}{2} + \delta_L \ell D_T \tau_{SS}].$$

Assuming a homogeneous snow cover and uniform depth of penetration throughout most of the footprint region, T_4 can again be rewritten in a simpler form:

$$T_4 = \delta_R \delta_P \tau_{SS} d_F (1-\gamma) L_1 + \delta_S \delta_P [(1-\delta_L) \frac{\epsilon_T}{2} (1-\gamma) L_1 + \delta_L \ell D_T] \tau_{SS} . \qquad (26)$$

Definitions of those symbols that appear for the first time in the above equations are listed below.

A_4 — Surface area of the snow in a side indentation.

A_{4F} — That portion of the surface area of a side indentation occupied by an unbroken snow wedge, $A^i_{4F} = L^i_1 d^i_F$.

A_{4R} — That portion of the outer face of a rubber element on the sidewall perimeter that is in contact with the outlying snow, $A^i_{4R} = L^i_2 d_T$ for pure rolling; $A^i_{4R} = L^i_2 D^i_T$ for operation with slip.

d — Height of the snow filled portion of the sidewall indentation which satisfies the relationships $d \geq d_T$ (the depth of tread penetration in the new snow layer).

d_F — Penetration depth of an unbroken snow wedge.

D_T — Depth of tire penetration into the snow cover, $D_T \geq d_T$.

f_I — Friction at incipient sliding due to the micro-mechanical interlock (due to shear of the snow by the surface texture of the rubber and snow).

ℓ — Characteristic fracture length of the snow cover.

L_1 — Portion of the two tire tread sidewall perimeter planes that spans sidewall indentations.

N_4 — Number of sidewall indentations in the tread in the perimeter of the snow contact zone.

N_{4R} — Number of tread rubber faces along the sidewall edge of the snow contact zone (this number is always within 1 of N_4).

T_{4A} — Contribution based on the shear strength of the adhesional and the micro-mechanical (surface texture interlock) bonds between the outer faces of the sidewall tread surface and snow to the maximum tractive force during pure rolling.

T_{4P} — Traction contribution of the shear along snow-snow interfaces (on the tread perimeter plane) that occurs because of tread element sinkage during slip.

T_{4R} — Contribution of the forces due to relative motion of the snow and the rubber at the tire tread sidewall to the available traction levels during slip.

T_{4S} — Contributions of the shear strength of the vertical faces of the snow wedges in the indentations in the tread sidewall to the maximum tractive force during pure rolling.

T_{4SN} — Contribution of a new bond strength – between snow retained at the outer face of a side indentation and the new snow at the side edge of the footprint – to the maximum tractive force during pure rolling.

T_{4SS} Contribution of the friction at snow-snow interfaces (between the old snow trapped in the sidewall indentations and the new snow at the tire perimeter) to the available traction levels during slip.
γ Fraction of L_1, corresponding to snow packed indentations.
δ_T Smaller of the two quantities d and D_T.
ϵ_T Additional tread element sinkage that occurs as it slips.
$(\ )^i$ Value of the quantity within ( ) for the ith indentation in the tread sidewall.
$(\ )^j$ Value of the quantity within ( ) associated with outer face of the jth rubber element protruding from the tread sidewall at the footprint perimeter.

The force F_S is provided by shear along the vertical faces of the additional areas of unbroken snow wedges in the sidewall indentations that result from the sinkage of the tread elements during slip.

The maximum contribution to traction from this mechanism will be assumed to be $\frac{\tau_{SS} L_1 \epsilon_T}{2}$ (pp. 83-84 of (31)).

IV.5. Special Traction Forces Present at the Sloping Lead Portion of the Tire-Snow Contact Zone – Let T_5 denote the special interface forces associated with the sloping portions (usually restricted to the lead edge) of the tire-snow contact zone. Expressed mathematically, T_5 has the following form:

$$T_5 = \delta_R \delta_A \delta_P \left\{ \sum_{i=1}^{N_1} T_{5C}^{i,1} + \sum_{j=2}^{3} \sum_{i=1}^{N_j} \delta_F^{i,j} T_{5C}^{i,j} + \sum_{j=2}^{3} \sum_{i=1}^{N_j} (1-\delta_F^{i,j}) \delta_{SS}^{i,j} T_{5C}^{i,j} \right\}$$

$$+ \delta_S \delta_P \left\{ \frac{\delta_A + 1}{2} \left[\sum_{i=1}^{N_1} T_{5C}^{i,1} + \sum_{i=1}^{N_2} (\delta_F^i (1-\delta_L) + \delta_L) T_{5C}^{1,2} + \sum_{i=1}^{N_3} (\delta_F^{i,3} (1-\delta_L) \right. \right. \tag{27}$$

$$\left. + \delta_L \delta_{SS}) T_{5C}^{i,3} + \frac{\delta_A - 1}{2} \left[\sum_{i=1}^{N_1} T_{5C}^{i,1} + \sum_{j=2}^{3} \sum_{i=1}^{N_j} \delta_F^{i,j} T_{5C}^{i,j} + \sum_{i=1}^{N_3} (1-\delta_F^{i,3}) \delta_{SS}^{i,3} \delta_3 T_{5C}^{i,3} \right] \right.$$

Upon including the functional dependence of these quantities on the snow and tire properties and the dynamic environment, the above expression becomes:

$$T_5 = \delta_R \delta_P \delta_A \left\{ \sum_{i=1}^{N_1} \sigma_C^{i,1} \cos\theta^{i,1} a_1^i + \sum_{j=2}^{3} \sum_{i=1}^{N_j} \delta_F^{i,j} \sigma_C^{i,j} \cos\theta^{i,j} a_j^i \right. \tag{28}$$

$$\left. + \sum_{j=2}^{3} \sum_{i=1}^{N_j} (1-\delta_F^{i,j}) \delta_{SS}^{i,j} \sigma_C^{i,j} \cos\theta^{i,j} a_j^i \right\}$$

References pp. 134–135

$$+\delta_S\delta_P\left\{\frac{\delta_A+1}{2}\left\{\sum_{i=1}^{N_1}\sigma_C^{i,1}\cos\theta^{i,1}a_1^i+\sum_{i=1}^{N_2}(\delta_F^i(1-\delta_L)+\delta_L)\sigma_C^{i,2}\cos\theta^{i,2}a_2^i\right.\right.$$

$$+\sum_{i=1}^{N_3}(\delta_F^{i,3}(1-\delta_L)+\delta_L\delta_{SS})\,\sigma_C^{i,3}\cos\theta^{i,3}a_3^i$$

$$+\frac{\delta_A-1}{2}\left[\sum_{i=1}^{N_1}\sigma_C^{i,1}\cos\theta^{i,1}a_1^i+\sum_{j=2}^{3}\sum_{i=1}^{N_j}\delta_F^{i,j}\,\sigma_C^{i,j}\cos\theta^{i,j}a_j^i\right. \tag{28 (Cont'd)}$$

$$\left.\left.+\sum_{i=1}^{N_3}(1-\delta_F^{i,3})\,\delta_{SS}^{i,3})\delta_3^i\sigma_C^{i,3}\cos\theta^{i,3}a_3^i\right]\right\}.$$

We have assumed that: (1) during braking the effective cross grooves are filled soon after entering the footprint region only for the case of nearly locked-wheel operation (the term with coefficient $\frac{\delta_A+1}{2}$); and that (2) when slip occurs during constant speed driving or vehicle acceleration, unfilled effective cross grooves do not fill rapidly enough to provide a significant amount of traction by compression.

Assuming a homogeneous snow cover and effective cross grooves of present day dimensions, we can take the following simplified form for T_5:

$$T_5=\delta_R\delta_P\delta_A\left\{\bar{\sigma}_C\frac{A_1}{A}bD_T+\sigma_C(\alpha\frac{A_2}{A}+\beta\frac{A_{3S}}{A})\,bD_T+\delta_{SS}((1-\alpha)\frac{A_2}{A}\right.$$

$$\left.+(1-\beta)\frac{A_{3S}}{A})\,b\bar{\sigma}_C(D_T-d_G)\right\}$$

$$+\delta_S\delta_P\left\{\frac{\delta_A+1}{2}\left[\bar{\sigma}_C(\frac{A_1}{A}bD_T+(\alpha(1-\delta_L)+\delta_L)\frac{A_3}{A}bD_T\right.\right. \tag{29}$$

$$\left.\left.+(\beta(1-\delta_L)+\delta_L)\frac{A_{3S}}{A}bD_T)\right]+\frac{\delta_A-1}{2}\left[\bar{\sigma}_C\frac{A_1}{A}bD_T+\bar{\sigma}_C(\alpha\frac{A_2}{A}+\beta\frac{A_{3S}}{A})\,bD_T\right]\right\},$$

or

$$T_5 = \delta_R \delta_P \delta_A \Big\{ \bar{\sigma}_C (A_1 + \alpha A_2 + \beta A_{3S}) \frac{b}{A} D_T + \delta_{SS}((1-\alpha) A_2 + (1-\beta) A_{3S}) \frac{b}{A} \bar{\sigma}_C (D_T - d_G) \Big\}$$
$$+ \delta_S \delta_P \Big\{ \frac{\delta_A + 1}{2} [\bar{\sigma}_C (A_1 + (\alpha(1-\delta_L) + \delta_L) A_2 + (\beta(1-\delta_L) + \delta_L) A_{3S}) \frac{b}{A} D_T] + \frac{\delta_A - 1}{2} [\bar{\sigma}_C (A_1 + \alpha A_2 + \beta A_{3S}) \frac{b}{A} D_T] \Big\} \qquad (30)$$

Definitions of those symbols that appear for the first time in the above equations are listed below:

A Footprint area taken in the horizontal plane.

A_1, A_2, A_3 Areas of the portions of the tire footprint region corresponding to the first three features of the tread element geometry.

a_1, a_2, a_3 Areas of the tire tread surface plane associated with the first three distinct tread element features, respectively.

b Width of the tire footprint region.

d_G Unfilled groove depth.

N_1, N_2, N_3 Total number of segments into which the outer tread surface, the effective cross grooves, and the effective longitudinal grooves, respectively, have been divided.

T_{5C} Component of the normal force opposed to the vehicle motion acting on a feature of the tire surface.

δ_A This symbol is 1 during vehicle braking. It is -1 during constant speed driving and vehicle acceleration.

δ_{SS} This symbol is 1 if $d_T \rangle d_G$. Otherwise, it is 0.

θ The acute angle between the normal to the tire surface and the direction of vehicle travel.

$$\bar{\sigma}_C(z) = \frac{\int_0^z \sigma_C dz}{z}$$

$(\ )^{i,j}$ The average value of the quantity widthin ( ) associated with the ith surface segment of the jth feature of the tread geometry.

IV.6. Net Traction on Snow – When the contributions for the five interface regions are added together, we get the following mathematical statement for the net traction force of a pneumatic tire on a snow covered pavement:

$$T = \delta_R [\{\alpha_o \tau_{SS} + \alpha_1 (\tau_{SR} + \mu_I p_1)\} A_1 + \{\delta_P (1-\beta) \tau_{SS} + \beta \tau_{SSN}\} A_{2S}$$
$$+ \{\alpha \tau_{SSN} + \delta_P (1-\alpha) \tau_{SS}\} A_3 + \delta_P \tau_{SS} d_F (1-\gamma) L_1 + \delta_P \delta_A \{\bar{\sigma}_C (A_1 + \beta A_{2S} + \alpha A_3) \frac{b}{A} D_T$$

References pp. 134–135

$$+ \delta_{SS}\bar{\sigma}_C((1\text{-}\beta)A_{2S} + (1\text{-}\alpha)A_3)\frac{b}{A}(D_T\text{-}d_G)\}]$$

$$+ \delta_S[\ \alpha_0\,\mu_{SS} + \alpha_1\,\mu_{SR}\}p_1A_1 + \{\delta_P(1\text{-}\beta) + \beta\}\mu_{SS}p_5A_{2S}$$

$$+ \{(\delta_P\delta_L(1\text{-}\alpha) + \alpha)\mu_{SS}p_3 + \delta_P(1\text{-}\delta_L)\,(1\text{-}\alpha)\tau_{SS1}\}A_3 \tag{31}$$

$$+ \delta_P\tau_{SS}((1\text{-}\delta_L)\,(1\text{-}\gamma)\frac{\epsilon_T}{2}L_1 + \delta_L\,\ell\,D_T)$$

$$+ \delta_P\bar{\sigma}_C\left\{\frac{\delta_A + 1}{2}(A_1 + ((1\text{-}\delta_L)\,\beta + \delta_L)A_{2S} + ((1\text{-}\delta_L)\,\alpha + \delta_L)A_3)\right.$$

$$\left. + \frac{\delta_A - 1}{2}\,(A_1 + \beta A_{2S} + \alpha A_3)\right\}\frac{b}{A}D_t].$$

The terms have been arranged so that the first group of terms with the coefficient δ_R represents the traction during pure rolling, and the second group of terms with the coefficient δ_S gives the traction during operation with slip. As stated before, the many delta functions have been incorporated in the model to allow us to quickly determine which elements in the complex equation (31) are important during the different modes of vehicle operation on the various types of snow covers.

To illustrate the use of the δ functions, assume for example, locked-wheel braking. Here $\delta_R=0$, $\delta_S=1$, $\delta_L=1$ and the traction is represented by the following mathematical form:

$$T = (\alpha_0\mu_{SS} + \alpha_1\mu_{SR})p_1A_1 + \{\ \delta_P(1\text{-}\beta) + \beta\ \ \mu_{SS}p_5A_{2S}$$

$$+ \ \delta_P(1\text{-}\alpha) + \alpha)\}\mu_{SS}p_3A_3 + \delta_P\tau_{SS}((1\text{-}\delta_L)\,(1\text{-}\gamma)\frac{\epsilon_T}{2}L_1 + \delta_L\ell D_T) \tag{32}$$

$$+ \delta_P\bar{\sigma}_C(A_1 + A_{2S} + A_3)b\,\frac{D_T}{A}\ .$$

Furthermore, if the snow cover for this particular case is hard packed permitting no tread penetration, $\delta_P=0$, and the expression becomes:

$$T = \mu_{SR}p_1A_1 + \beta\mu_{SS}P_5A_{2S} + \alpha\mu_{SS}P_3A_3\,. \tag{33}$$

Appendix G of (31) contains a sample calculation using this generalized mathematical theory for snow traction, and the results are found to compare favorably with experimental data.

V. USING THE MODEL FOR SNOW TRACTION

What is needed to implement this generalized model of snow traction?

Two general sets of parameters are needed before the previously given mathematical expression can be utilized for calculating (or predicting) the snow traction of a tire. The first set is composed of those dimensions used in developing the equation which give a general description of the tire tread geometry. The second set of parameters consists of the values of the snow properties appearing in the equations.

While the tire tread dimensions can be obtained with no difficulty, finding good snow data is nearly impossible. The most important reason for this, as stated previously, is that the test procedures used to obtain this data do not take many important factors into consideration. They also are not designed to accurately simulate the mechanics of tire-snow interaction (32); they are designed solely to measure snow "properties".

Because this was the only data that is readily available – that of snow "properties", our mathematical model was derived to utilize it, regardless of its inadequacies. We soon learned, however, that this data is far from comprehensive.

The outgrowth of this unsatisfactory state of affairs is that experimenters must re-evaluate the design and relevance of their test procedures in terms of the usefulness of the data to predicting snow traction.

A redesigned two-element testing program is needed. The overriding principle in both phases is the necessity for close simulation of the tire-snow interaction (25, 33).

One part of the program would involve a study under controlled conditions using individual elements such as tread blocks and studs of the traction producing interactions between these elements and snow. This part of the program would also involve the redesign of bevameter type instruments (23, 26) to obtain pressure-sinkage and shear stress-displacement information under conditions more closely simulating the operation of the tire tread on a snow covered pavement.

The second part would involve testing with tires. These should be carefully controlled tests – such as those done in soil bins for soil trafficability studies – in which single elements of the system are varied so that their effect on tire traction can be identified.

Instrumentation that has been developed in the field of terramechanics for measuring such quantities as pressures on the vertical faces of tread elements (34) should be utilized in both parts of the program.

References pp. 134–135

VI. CONCLUSION

In conclusion, I would like to restate a few major points from this presentation.

I have given a generalized theory for the traction of a pneumatic tire on a roadway with finite depth snow cover. The analytical treatment is designed to allow the rapid determination of the effect of variations in tread geometry, snow properties, and operating conditions on the tractive capability of a tire on snow. However, knowledge of those snow properties which are important in the dynamic interaction of the tire and snow is critical to the usefulness of the present theory. Only a limited amount of such information is currently available. Thus, two recommendations are made for future research: first, experimentation should be conducted with individual elements of the system to carefully determine the nature of the dynamic interactions that occur between tire and snow; and second, a major effort should be made to determine the specific forms of the functional relationships between those snow properties that are found to be important and the measurable tire-snow system parameters.

REFERENCES

1. *A. R. Williams, T. Holmes, and G. Lees, "Toward the Unified Design of Tire and Pavement for the Reduction of Skidding Accidents," SAE Paper No. 720162, January 1972.*
2. *R. R. Hegmon, T. D. Gillespie, and W. E. Meyer, "Measurement Principles Applied to Skid Testing," ASTM STP 530, pp. 78-90, April 1973.*
3. *D. Freitag, A. Green, and N. Murphy, "Normal Stresses at the Tire-Soil Interface in Yielding Soils," Highway Research Record No 74, pp. 1-18, 1965.*
4. *G. Dagan and M. Tulin, "A Study of the Steady Flow of a Rigid-Plastic Clay Beneath a Driven Rigid Wheel," Journal of Terramechanics, Vol. 6, No. 2, pp. 9-27, 1969.*
5. *Z. Janosi, "Theoretical Analysis of the Performance of Tracks and Wheels Operating on Deformable Soils," Transactions of the ASAE, p. 133, 1962.*
6. *J. Wong and A. R. Reece, "Prediction of Rigid Wheel Performance Based on the Analysis of Soil-Wheel Stresses. Part I. Driven Rigid Wheels," Journal of Terramechanics, Vol. 4, No. 1, 1967.*
7. *S. Knight and D. Freitag, "Comments on Mobility Research," SAE Paper No. 782B, 1964.*
8. *G. Krick, "Radial and Shear Stress Distribution Under Rigid Wheels and Pneumatic Tires Operating on Yielding Soils with Consideration of Tire Deformation," Journal of Terramechanics, Vol. 6, No. 3, pp. 73-98, 1969.*
9. *L. L. Karafiath, "Soil-Tire Model for the Analysis of Off-Road Tire Performance," Grumman Aerospace Corporation Research Department Memorandum RM-541, May 1972.*
10. *W. M. Horne and U. T. Joyner, "Pneumatic Tire Hydroplaning and Some Effects on Vehicle Performance," presented at the SAE International Automotive Engineering Congress, Detroit, Mich., January 11-15, 1965.*
11. *C. Gupta and A. Pandya, "Behavior of Soil Under Dynamic Loading: Its Application to Tillage Implements," Transactions of the ASAE, pp. 352-358, 1967.*
12. *M. Mellor, "Properties of Snow," U.S. Army Material Command, Hanover, New Hampshire, 1964.*
13. *M. G. Bekker, Theory of Land Locomotion, The University of Michigan Press, p. 277, 1956.*

14. *R. Moyer, "Braking and Traction Tests on Ice, Snow, and on Bare Pavements," Highway Research Board, Proceedings of the 27th Annual Meeting, Washington, D.C., pp. 340-360, 1947.*
15. *R. Smith and D. Clough, "Effectiveness of Tires Under Winter Driving Conditions," Highway Research Board, Proceedings of the 51st Annual Meeting, January 17-21, 1972.*
16. *M. Hvorslev, "The Basic Sinkage Equations and Bearing Capacity Theories," Technical Report M-70-1, U.S. Army Engineer Waterways Experiment Station, 1970.*
17. *T. Ikeda and S. Persson, "A Track Shoe for Soft Soil," Transactions of the ASAE, pp. 746-749, 753, 1968.*
18. *D. Nordstrom, "Polytrac – A Unique Approach to Engineering Problems," SAE Paper No. 720745, 1972.*
19. *S. Kinosita and E. Akitaya, "Classification of Snow and Ice on Roads," Highway Research Board Special Report No. 115, 1970.*
20. *Anonymous, "Review Section: The Vehicle/Soil Relationship," Automobile Abstracts, pp. 3-4, August 1971.*
21. *P. Schaerer, "Compaction or Removal of Wet Snow by Traffic," Highway Research Board Special Report No. 115, 1970.*
22. *A. L. Browne, H. Cheng and A. Kistler, "Dynamic Hydroplaning of Pneumatic Tires," Wear, 20 pp. 1-28 1972.*
23. *S. Weiss, W. Harrison, L. Abarca and M. Bekker, "Preliminary Study of Snow Values Related to Vehicle Performance," Report No. 2, Land Locomotion Research Lab., 1956.*
24. *A. Assur, "Locomotion Over Soft Soil and Snow," SAE Paper No. 782F, 1964.*
25. *C. Nuttall and R. McGowan, "Scale Models of Vehicles in Soils and Snows," First International Conference on the Mechanics of Vehicle Soil Systems, Turin, Italy, 1961.*
26. *W. Harrison and T. Czako, "Over-Snow Vehicle Performance Studies," U.S. Army Ordinancy Tank-Automotive Command RR-46, 1961.*
27. *F. Zoz and G. Steinbruegge, "Effect of Section Thickness on Shear Characteristics of an Artificial Soil," Transactions of the ASAE, pp. 6-10, 1968.*
28. *S. Cho, H. Schwanghard and H. von Sybel, "The Spacing Effect of Track Shoes on Loose Soils," Journal of Terramechanics, Vol. 6, No. 3, pp. 21-45, 1969.*
29. *M. Mellor, "Snow Mechanics," Applied Mechanics Reviews, Vol. 19, No. 5, pp. 379-389, 1966.*
30. *P. Hobbs, "The Effect of Time on the Physical Properties of Deposited Snow," Journal of Geophysical Research, Vol. 70, No. 16, pp. 3903-3907, 1965.*
31. *A. L. Browne, "Traction of Pneumatic Tires on Snow," GM Research Publication GMR-1346, March 1, 1973.*
32. *K. Wiendieck, "Stress-Displacement Relations and Terrain-Vehicle Mechanics: A Critical Discussion," Journal of Terramechanics, Vol. 5, No. 3, pp. 67-85, 1968.*
33. *M. G. Bekker, Introduction to Terrain Vehicle Systems, The University of Michigan Press, Ann Arbor, 1969.*
34. *T. Liang and C. Yung, "A Microscopic Study of Tractive Performance of a Lugged Tire Operating on Sand," Transactions of the ASAE, Vol. 9, pp. 513-515, 1966.*

DISCUSSION

Thieme

Well gentlemen, having heard this interesting lecture, I presume there will be some questions too. Maybe not so much on the mathematical formulation but on the more practical information about what was going on. Yes, please.

A. G. Clary

Do you believe man-made snow can simulate actual snow for experimental purposes? Have you experimented with artificial materials?

Browne

I have reviewed the attempts to test with snow substitutes that have been reported in the literature. Though I have analyzed the results and have sought some solutions (and consider the use of snow substitutes to simulate certain special conditions quite feasible), I have not had the opportunity to conduct any experiments of my own.

Most people that have worked in the area of soil trafficability realize that even when you are talking about soils, to get any sort of repeatability in traction data you have to restore the soil after every test to the original condition that it had before the test. This, of course, is essential to repeatability in snow traction testing as well. Unfortunately, it is next to impossible unless you have an unlimited area of snow with uniform properties in which to test. There have been some attempts to develop or discover a substance which will compact nicely under the tread like snow and develop shear strength due to this mechanism but which can be easily restored to its original condition for repeated testing. Some of these tests have been fairly successful in producing intial traction data comparable to that obtained on snow. Much attention, however, must still be focused on the need for property regeneration, the means for which must be found before one can conduct successful repetitive testing.

There are materials that have been used fairly successfully as snow substitutes to test snow traction under very limited snow property conditions. For instance, Yuma sand has been used to model very granular snow where there is little compaction and little cohesion between the particles.

Thieme

Maybe in addition you could tell us a little about different types of snow.

Browne

A major part of my study involved trying to develop an understanding of the different types of snow that are normally encountered. In the written text, I have pointed out the considerable effect that the type of snow can have on the available traction. Kinosita and Akitaya determined that one of the major factors affecting snow properties is snow density. As the density increases, the shear strength of the snow increases and, thus, the available traction increases. Snow density is increased in many different ways. For example, density as well as the strength of the cohesional bonds between the snow particles will increase with time even when there are no external forces acting on the snow. In general, snow can be grouped under four different major catagories: new snow, lightly compacted snow that can still be further compacted; firmly compacted snow; and slush. The development of tire traction on these four major classifications of snow is described in the written text.

Thieme

Thank you. Other questions?

H. C. Hodges

In the four catagories that you use to define differences of snow, you mentioned density. Do you find that the temperature in any given period of time has a greater effect?

Browne

Temperature will have a greater effect than density on snow traction if the temperature is close to that which is necessary for the snow to start building up an appreciable moisture content. Here, small changes in temperature would have a large effect on the shear strength of the snow and, thus, on the available tire traction. If you are talking about lower temperatures, then the snow properties (and, thus, tire traction) are fairly insensitive to changes in temperature.

Hodges

Actually, I was referring to the thermal effect on the particulate interlock between the snow members or the snow elements and (consequently) on the surface condition when you run a tire over it. When these (locked elements) are compacted, you have a change in lubricity, if you will, or a change in the frictional property right around the elements themselves. Is that temperature oriented (related), or both?

Browne

Snow properties under those conditions will certainly be affected by the pressure loading especially if the energy added through the compacting action of the tread raises the snow temperature sufficiently to cause melting and the formation of a water film. Such energy can also be provided by heat flow from the tire itself, which, if the snow temperature is close to the melting point, is quite effective in creating a water film right at the tire-snow interface.

Hodges

May I ask one more question? Going back to your film, you showed a tire cleaning as a result of wheel slip. Was this because of a change in temperature and a release of the bond due to temperature; and, is it a fact that when the tire is slipping under those conditions, the tire is developing a lower traction value than when it has the snow bond on it?

Browne

Yes, the OE tire that was shown spinning with a very large slip velocity (which produced a clean surface) was definitely developing a lower traction value. This is because there was a sufficient amount of slip in the interface to scrape the tire surface clean on the underlying snow and, most importantly, to cause substantial heat generation in the tire tread near the surface (due to hysteresis effects) causing the surface snow to melt and shed from the tire. The presence of a fluid film at the interface combined with the lack of any digging action by the tread definitely indicated low values for the available traction.

Hodges

Thank you very much.

M. R. Barone

I seem to recall a product of delta functions. Could you explain physically what this implies?

Browne

The delta functions were simply used for convenience to allow the rapid visualization (and quick determination) of the effects of different operating conditions. For instance, there was a product $\delta_L \delta_P$. If δ_P was one, it meant tread penetration into the underlying snow cover. If there was no penetration, δ_P was zero. Let us say that δ_P is equal to one. We now must look at δ_L. δ_L is one if there is locked-wheel operation, otherwise it is zero. If it is zero, then regardless of what δ_P is, the product of δ_L with δ_P would be zero, indicating that the term with the coefficient $\delta_L \delta_P$ did not contribute to the traction for this particular set of operating conditions. However, if both δ_L and δ_P are one, this indicates that, under locked-wheel operation with tread penetration, terms with this coefficient contribute to the traction production by the tire. To reiterate, the functions are used simply as a sorting mechanism.

Barone

It does not really imply any deeper relationship?

Browne

No, it is just a way in which somebody with no familiarity with that equation could, by merely defining his operating condition, quickly obtain a simplified expression which would point out to him the factors important to snow traction under those conditions.

S. K. Clark

Do you visualize any values for the delta functions other than zero or one?

Browne

In traction studies in soil, researchers correctly employ a μ_S, a coefficient of friction during slip, which is a function of the amount of slip. Here I have chosen to differentiate between operation with slip and the pure rolling condition. That division of operating conditions into these two categories is quite artificial and was for convenience in presentation only, though it, in itself, introduces no inaccurracies into the formulation. Differences in traction that result from variations in the amount of slip can be obtained by introducing relationships (that are a function of slip) for the different coefficients of friction. As an example of a case in which the amount of slip is important, at a point in my talk, I said, that the traction provided by wide effective cross grooves will be maximized if the shear of the snow wedge is just completed as the trailing edge exits from the rear of the footprint. However, if you have a lot higher slip, you may have greater traction due to the larger number of cross grooves passing through the footprint region. Thus, in truth, everything is slip dependent.

Thieme

Thank you. As chairman of this meeting, I would like to congratulate Dr. Browne on his excellent paper on a subject in which he seems to have become an expert. I know that he has put a large amount of work into it. His presentation this morning should enable the audience to understand things many of which may have been more or less a mystery and may have been beyond full comprehension previously because it is a very difficult topic to be treated. Also, Dr. Browne succeeded to get information into formulae that were very difficult. But he presented a generalized theory for the traction of a pneumatic tire on a snow-covered pavement, and he really made clear to us why this symposium is called the Physics of Tire Traction. A major thing about the physical and mechanical properties of the different types of snow was the heat exchange altering these properties of snow as discussed this morning. So furthermore, he presented his paper in an excellent manner, and this is a great service to us all. And I feel sure that congress members would like me to thank Dr. Browne for the able way in which he dealt with the questions. I therefore propose a vote of thanks to Dr. Browne. Once again I would like to thank all the speakers and all those who have contributed in the discussions this morning. A vote of thanks also goes to Mr. Sherry Tiderington who put the slides before us at exactly the right moment.

SESSION II

FUNDAMENTAL ASPECTS OF RUBBER FRICTION

Session Chairman
D. F. MOORE

National University of Ireland
Dublin, Ireland

INTRODUCTORY REMARKS (D. F. Moore)

While I am privileged to act as your chairman during this session, Session II, on the fundamental aspects of rubber friction, we are indeed fortunate to have as speakers, four of the most eminent and active researchers in the field of tire technology. Two of these are from the United Kingdom, one from the German Federal Republic, and one from the U.S.A., and they are all renowned for their basic contributions to the fundamental understanding of the nature of the tire-road frictional contact. Now whereas the emphasis in Session I has been on general topics such as traction under dry, wet, and snowy conditions, the subjects discussed in this session are of a more fundamental and basic nature. Here the physical mechanism of rubber friction will be examined in the light of new or unusual experimental evidence and an attempt will be made to seek some form of agreement and conclusions in areas where perhaps doubt or uncertainty previously existed. We must at least be clear on the roles played by adhesion and hysteresis both with and without the presence of lubricants, and we must accurately assess the influence of material properties, slip speed, temperature, and surface texture in the light of this new experimental evidence. We cannot overemphasize the significance of a fundamental and complete understanding of rubber friction in terms of eventual optimized tire design and ultimately vehicle safety. Those of us who are active in research and development as applied to rubber-like materials are well aware of the truly interdisciplinary nature of tire-to-ground traction. Physics, chemistry, metallurgy, dynamics, tribology, thermodynamics, heat transfer, elasticity, viscoelasticity, rheology, elastohydrodynamics, and lubrication technology play complex and intertwined roles in determining the magnitude of the frictional coupling that ultimately exists in the contact patch. It has been claimed that at least 50 variables are operative during the simple free rolling of a pneumatic tire. Such complexity necessitates the adoption of a simplistic and intuitive approach in seeking a further understanding of contact friction. All of our speakers today have followed this method, consciously or unconsciously, and have hereby provided remarkable insight into the nature and causes of rubber traction. Such advances are then used by the tire designer who synthesizes, compromises, and improvises until a satisfactory optimized design emerges. But here we emphasize the role played by fundamental and original research and its meaningful interpretation.

THE SPEED AND TEMPERATURE DEPENDENCE OF RUBBER FRICTION AND ITS BEARING ON THE SKID RESISTANCE OF TIRES

K. A. GROSCH

UNIROYAL European Tire Development Center, Aachen, Germany

ABSTRACT

The observation that the ranking of tire tread compounds in skid tests is largely independent of the testing conditions, such as speed, type of surface, whether the surface is dry or wet and whether locked-wheel braking or cornering tests are carried out is explained by means of the temperature and speed dependence of rubber friction.

It is shown that speed and temperature are in every case related by the WLF equation, provided the sliding speeds are sufficiently low for the temperature rise in the contact area either to be negligible, or to be known. Master curves, therefore, can be obtained under all relevant conditions, and these describe completely the friction between the rubber and the surface under investigation. The shape of the master curve depends on the type of rubber, on the black filler content, on the type of track surface, and on the extent of lubrication.

The range of combined temperature – speed variable, log $a_T v$, achievable in tire skid experiments, is small because of the opposing nature of speed and temperature effects. The crossing of the friction – log $a_T v$ curves of different polymers on any one surface is unlikely. If it does occur, as in the case of the comparison of oil extended rubbers with unextended rubbers of the same polymer, it shows itself also in a reversal of the ranking in a skid trial at high speeds.

Because the temperature range is limited on ice, the part of the master curve corresponding to high $a_T v$ values comes into play. Since in this region the ranking of polymers, particularly NR and SBR, is reversed, a similar reversal in ranking is also observed in tire skids.

References pp. 162–163

The likely improvement which a particular compound can make to the skid resistance of tires can, therefore, be estimated correctly from simple laboratory comparisons.

INTRODUCTION

A number of extensive skid trials carried out on different types of road surfaces over a wide range of speeds, using different tread profiles and different tire constructions, have shown that the ranking of a compound relative to a control is largely independent of all the imposed outside parameters provided that the surface on which this trial is carried out is dry or wet (1, 2, 3). If reversals in ranking are observed, they occur only at high speeds and then only if a highly hysteretic tread compound is compared with a low hysteretic one. In these cases, the high hysteretic compound is better at low speeds, and at high speeds the opposite can be the case.

An entirely different behavior is observed if skid experiments are carried out on ice (4, 5, 6). Here a very significant reversal occurs between SBR and NR based compounds, if experiments are carried out just above and then just below the freezing point of water. This paper attempts to explain this well-known skid behavior of tread compounds in terms of the temperature and speed dependence of rubber friction.

RUBBER FRICTION ON DRY SMOOTH SURFACES

Fig. 1 shows the friction coefficient of an acrylonitrile gum rubber compound (NBR) on a glass surface at different speeds and temperatures (7). In these experiments, the sliding speeds were kept so low that the temperature rise in the interface between the rubber and glass surface could be neglected, i.e. temperature and speed were independent of each other. It is seen that at high temperatures and low sliding speeds the friction coefficient is very low. It rises with increasing speed and with decreasing temperature. At temperatures between +10°C and –5°C, the curves taken over a range of speeds pass through a maximum. This maximum shifts towards lower speeds at lower temperatures. At temperatures below –5°C, the friction coefficient decreases continuously with increasing sliding speed. This speed – temperature dependence is typical of the viscoelastic behavior observed, for instance, when the dynamic modulus is measured at different frequencies and temperatures. Indeed, shifting the friction curves obtained at different temperatures along the logarithmic speed axis by suitable amounts results in a single master curve as shown in Fig. 2 (7). The shift factors are a function of the temperature interval between a reference temperature and the experimental temperature. It has been shown extensively in the literature that the shift factors obtained in such a way are identical with those obtained from other viscoelastic experiments, (7) and they agree with the functional relationship generally known as the WLF equation (8). Thus, the frictional behavior of a simple rubber compound on a smooth surface is entirely described by the master curve of the friction coefficient as a function of the combined speed and temperature variable log a_Tv where a_T is the WLF coefficient and v is the sliding speed. Similar curves have been obtained for all the polymers used in tire technology. They all show a similar behavior in that the friction coefficient is low at low log a_Tv values, passes through a maximum and again drops to low values when log a_Tv is very large. The whole behavior extends over about 10 decades. The only difference between different polymers is in the position which the maximum takes on the log a_Tv axis, particularly if the reference temperature is the same for all compounds. In this case, the position of the maximum is determined by the glass transition temperature of the polymer. The lower the glass transition temperature, the more the friction curve is placed towards higher log a_Tv values. This is demonstrated in Fig. 3 which contains the master curves for the coefficient of friction on glass of five different polymers that are unfilled but cross-linked with sulfur to the same degree. Therefore, the viscoelastic nature of rubber friction becomes apparent not only through the possibility of transforming speed and temperature dependence but also from the position which the master curve takes in relation to its glass transition temperature.

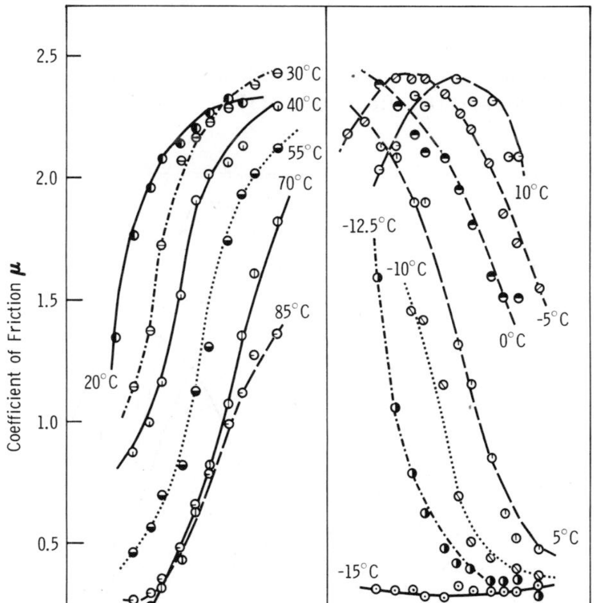

Fig. 1. Coefficient of friction of a pure gum acrylonitrile rubber on glass as a function of the sliding speed at various temperatures (7).

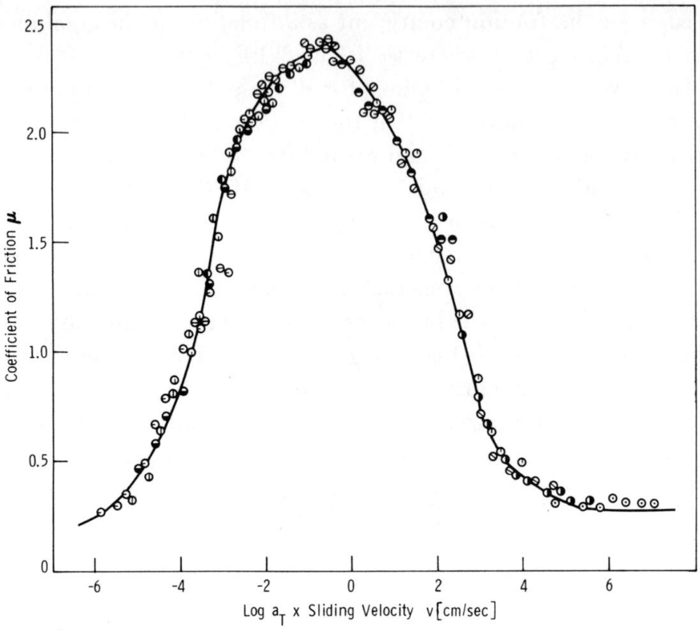

Fig. 2. Coefficient of friction of a pure gum acrylonitrile rubber on glass as a function of the transformed sliding speed log $a_T v$. Data from Fig. 2 of (7).

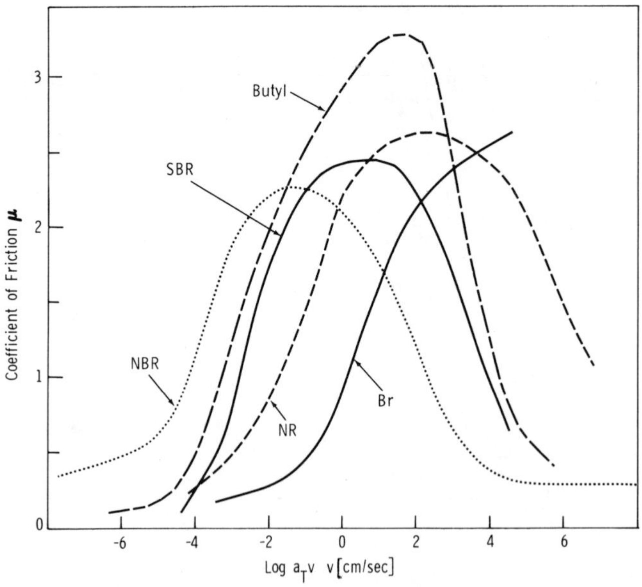

Fig. 3. Coefficient of friction master curves for five gum rubbers with different glass transition temperatures (7).

RUBBER FRICTION ON DRY ROUGH SURFACES

Fig. 4 shows the master curve of the NBR rubber model compound on a carborundum surface. Also, for comparison, the master curve obtained on glass is again shown. The shift factors followed, as in all cases, the well-known WLF equation. It is seen again that the friction curve rises as the $a_T v$ values increase, i.e. as either the speed increases or the temperature decreases. At the point where the friction on glass reaches its maximum, the curve shows a distinct change in rate of increase. However, instead of falling at log $a_T v$ values larger than those at which friction on glass was a maximum, the curve continues to rise, reaching its maximum at very much higher values of $a_T v$. Beyond the maximum, the friction drops very rapidly to very low values. The difference in the shape of the master curves is explained in detail in the literature (7) as being due to two different friction mechanisms. The one dominating on glass is termed adhesional friction and can be shown to be molecular in nature, (7) whilst the additional mechanism which comes into operation on rough tracks is termed deformation friction (9, 10). Large regions of rubber around the asperities of the track are deformed. Energy is lost as the rubber passes through deformation cycles while gliding over the asperities, and this energy appears in the form of an increased frictional force. As in the case on glass, the master curve, although different in shape, is strongly viscoelastic in nature. It describes completely the speed and temperature dependence of the friction coefficient of the rubber on that surface, provided that speed and temperature are independent of each other. The change in the rate of increase of the friction coefficient with increasing log $a_T v$ values at the

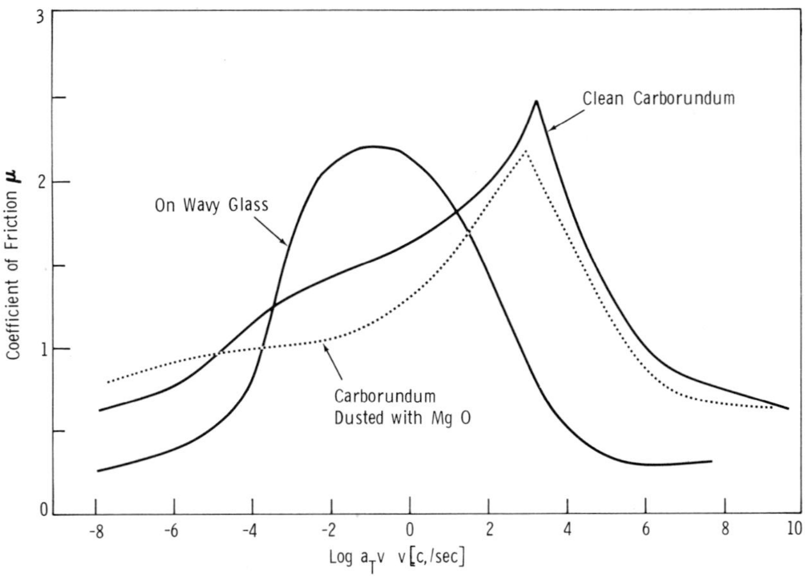

Fig. 4. Coefficient of friction master curve for an acrylonitrile gum rubber on clean and dusted dilicon carbide abrasive respectively with the curve obtained on glass shown for comparison (7).

References pp. 162–163

point at which the friction coefficient reaches its maximum on glass is associated with the fact that adhesion is present on a rough track as well. When a fine powder is applied to the track (which prevents the direct contact between rubber and track), this "hump" (which appears on a master curve obtained with rubber on clean carborundum) is eliminated without noticably affecting the maximum due to deformation friction, as shown in Fig. 4.

THE EFFECT OF CARBON BLACK FILLERS

Fig. 5 shows master curves obtained on three types of surface: glass, carborundum dusted with magnesia oxide powder, and clean carborundum for three rubber compounds, these being NBR gum rubber, the same polymer filled with 20 parts of HAF black, and NBR filled with 50 parts of HAF black (11). It is seen that, on all three surfaces, the effect of the speed and temperature dependence of the friction coefficient is reduced in magnitude as the black content is increased. The position of the curves on the log $a_T v$ axis is virtually unaffected. With 50 parts of black filler added, the curve shows a flat plateau on clean carborundum which starts approximately at the log $a_T v$ values at which the adhesion of the gum rubber was a maximum on glass and tends to $a_T v$ values at which the deformation friction had its maximum on carborundum. The detail of the curve in between these two points has been largely obliterated by the black filler. It appears, therefore, that the friction coefficient of black filled compounds is less sensitive to temperature and speed variations than that of gum rubbers.

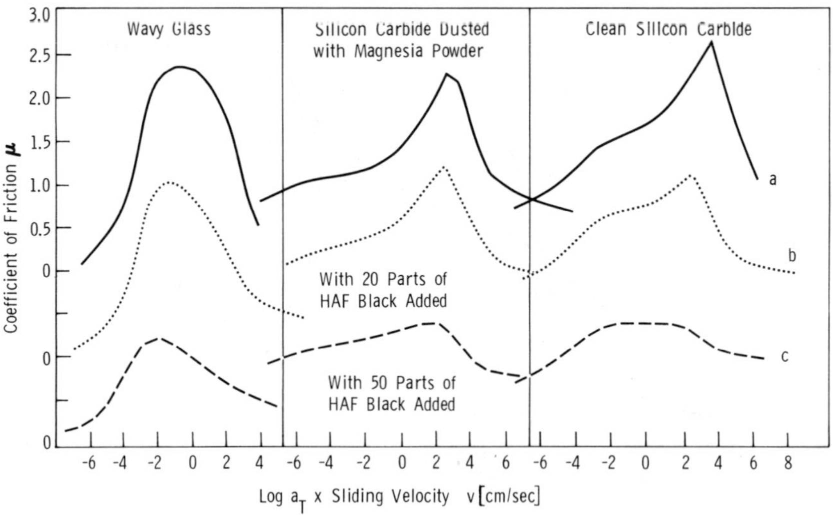

Fig. 5. Master curves of the friction coefficient of an acrylonitrile rubber (a) as gum rubber (b), filled with 20 pphr HAF black and (c) filled with 50 pphr HAF on three different surfaces.

Fig. 6 shows that this is also the case for other types of polymer commonly used in tire technology, except for butyl. In this case, the carbon black has a much smaller damping effect on the speed and temperature dependence of the friction coefficient.

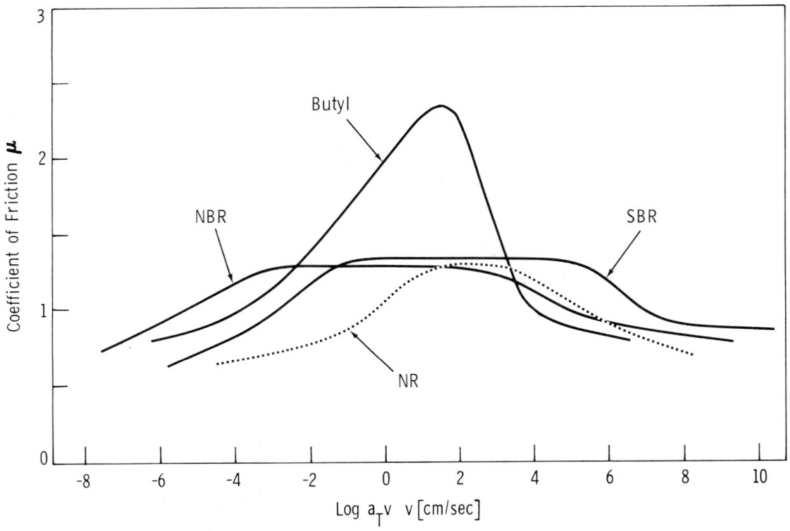

Fig. 6. Master curves of the friction coefficient on carborundum paper for four rubbers of different glass transition temperature, all filled with 50 pphr HAF black (7).

RUBBER FRICTION ON WET ROUGH TRACKS

Fig. 7 shows the master curve of the frictional coefficient as a function of the combined variable log $a_T v$ for the model NBR gum compound on a carborundum surface wetted with distilled water in the one case and distilled water with a small proportion of wetting agent added in the other. The shape of the two curves shows a striking similarity with those obtained on dry clean and dusted carborundum. At low $a_T v$ values, the friction rises with increasing log $a_T v$ values. On carborundum with only distilled water, the friction shows again a change in rate of increase at those $a_T v$ values at which one would expect the adhesion friction on glass to reach its maximum value. At higher $a_T v$ values, the friction continues to increase up to those log $a_T v$ values at which the deformation maximum would be expected to be reached on dry carborundum. At that point the experiment had to be broken off because the distilled water reached its freezing point. Because the maximum experimental speed was also limited in order to avoid any temperature rise in the interface, the obtainable log $a_T v$ values were also restricted. Adding a small proportion of wetting agent had the same effect as that observed when adding magnesia oxide powder to the dry carborundum track; it surpressed the rise in friction near the location of the adhesion friction maximum on glass. Presumably the wetting agent prevents the formation of rubber adhesion bonds on a wet surface in the same way that the magnesia oxide impeded their formation on the dry surface.

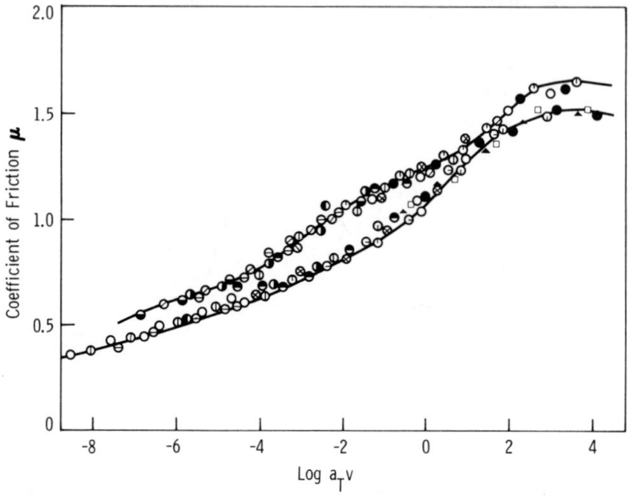

Fig. 7. Master curves of the coefficient of friction obtained for the acrylonitrile gum rubber on a water-lubricated silicon carbide track, with distilled water as a lubricant, upper curve; with 5% detergent added to the water, lower curve. Different symbols denote different temperature.

However, in both cases adhesion must play a part in the total friction mechanism, because the magnitude of the friction coefficient is too large to be explained solely by deformation losses in a deformation cycle normal to the sliding direction. Although rubber adhesion may be surpressed by the presence of the powder or the wetting agent, adhesion between the powder particles sticking to the rubber and those sticking to the carborundum paper or between the molecules of the wetting agent sticking to the rubber and those sticking to the wet carborundum must still be present. This is particularly obvious for very small log $a_T v$ values. In this region, the loss modulus is low and the sliding speed is low, so that internal losses themselves also are low. Even so, on rough surfaces, wet or dry, the friction coefficient is larger for these $a_T v$ values than it is for the sliding friction on dry glass. In this case, even elastically stored energy may contribute to the friction process. The fact that rubber abrades is evidence of this. All the energy stored in the rubber volume which is abraded is, of course, being dissipated and has to be supplied by the frictional force.

FRICTION WITH SINGLE HARD SLIDERS ON LUBRICATED RUBBER

In order to estimate the importance of adhesion in rubber friction on lubricated tracks, experiments were carried out with spherical steel sliders on lubricated rubber tracks over a range of speeds at various temperatures. The sliding speeds were kept low so that no significant temperature rise occurred in the interface. Fig. 8 shows experimental data for a NR + 50 HAF compound. These data have been transformed into a single master curve by shift factors to obtain a best fit. The scatter of the points is much larger than is the case when sliding large rubber test pieces on hard tracks, and the nature of the shift function is, therefore, less certain. However, this large collection of data obtained from tests with the same polymer under diverse testing conditions, including different lubricants, loads, and indentor radii, provides strong evidence that at low $a_T v$ values the transformation follows the WLF equation.

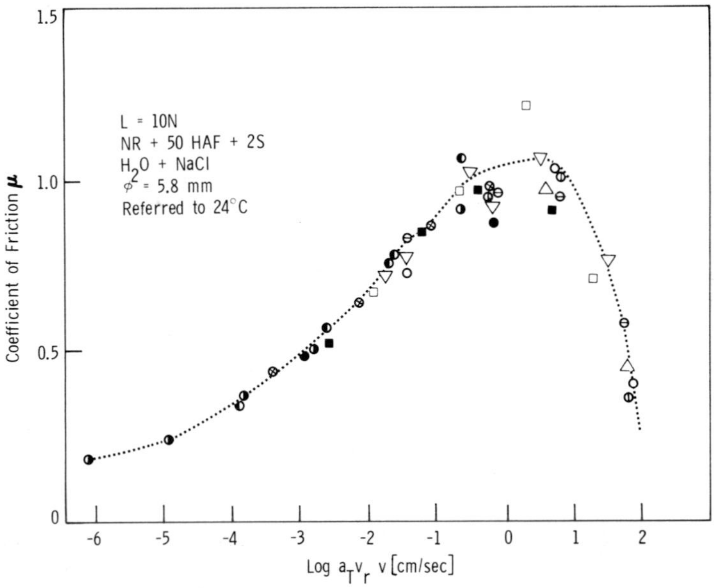

Fig. 8. Master curve of the friction coefficient of an NR + 50 HAF compound obtained by sliding a steel sphere over the rubber track lubricated with H_2O + NaCl. Different symbols refer to different temperatures from +80°C to −15°C.

The shape of the master curve is similar to those obtained on dry smooth surfaces. At low $a_T v$ values the friction coefficient is low, and it rises as log $a_T v$ increases. It reaches a maximum value, however, at a log $a_T v$ value which is entirely different from that obtained on dry glass. The position of the maximum value on the log $a_T v$ axis depends in this case (See Fig. 9.)

a. on whether the compound is a gum rubber or is black filled; (the position of the maximum occurs at a higher $a_T v$ value for the black filled rubber than for the gum)
b. on the viscosity of the lubricant; (the position of the maximum occurs at higher $a_T v$ values for lubricants of lower viscosity)
c. on the normal load acting on the slider; (at the higher load, the maximum occurs at higher $a_T v$ values) and
d. on the radius of the slider; (the position of the maximum occurs at higher $a_T v$ values for the smaller radius.)

References pp. 162–163

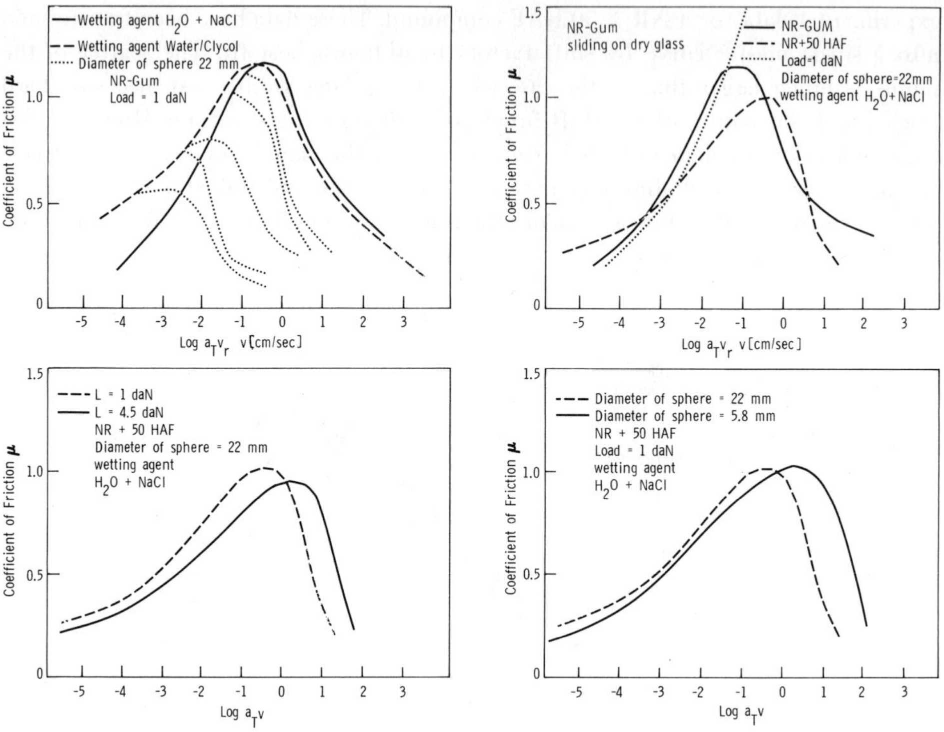

Fig. 9. Friction master curves obtained with a steel slider on a lubricated rubber track, Upper Left: two different lubricants. For the high-viscosity lubricant only the rising part forms a master curve. Upper Right: Comparison of gum NR with black filled NR. Dotted lines shows the master curve of gum NR on dry glass. Lower Left: Two different loads. Lower Right: Two different slider diameters.

All four are consistent with the assumption that the maximum position is determined by the lubricant film thickness. At low $a_T v$ values direct adhesional contact exists between rubber and slider. This is also borne out by a comparison of the friction master curve on dry glass with the corresponding one for a spherical slider on a lubricated rubber track. (Fig. 9, diagram b, hatched curve). At the $a_T v$ value for which the friction becomes a maximum, the lubricant film thickness is such that adhesion between rubber and slider begins to break down. Clearly, increasing the pressure, i.e., increasing the load or decreasing the slider dimensions in the contact area, or decreasing the viscosity of the lubricant requires higher $a_T v$ values in order to achieve the same lubricant film thickness as predicted by classical lubrication theory. For the friction with the higher viscosity lubricant, only the rising part of the individual curves obtained at a particular temperature form part of the master curve. In both its magnitude and the $a_T v$ value at which it occurs, the maximum friction depends on temperature.

RUBBER FRICTION ON ICE

Fig. 10 shows the master curve of the friction coefficient of a NR gum compound on ice. Also shown is the master curve for the friction of NR on dry glass. Because the a_T values cannot be lower than those corresponding to 0°C, only a portion of the total master curve can be obtained (12). Extensive data collected by Southern and Walker (13) also demonstrate clearly that the transformation of speed and temperature for friction of rubber on ice follows the WLF equation.

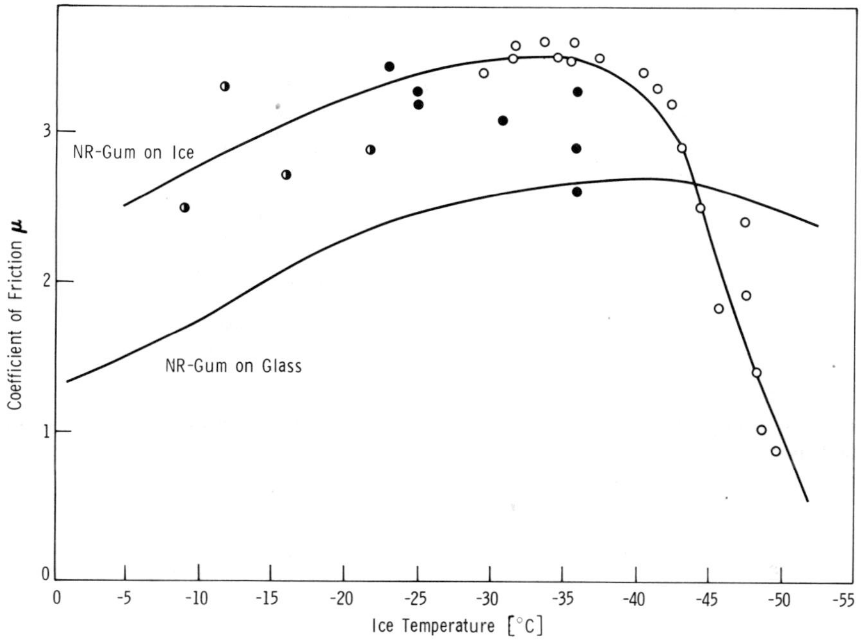

Fig. 10. Master curve of gum NR on ice; also shown for comparison is gum NR on dry glass. Experimental points refer to different ice temperatures.

In Fig. 11 friction coefficients are shown that were obtained on ice with different types of gum rubber at one sliding speed but with different temperatures. By comparison with the master curves on glass (transformed into a temperature variable), it is obvious that the friction on ice is similar in nature to that on glass. If the ice is free from contamination, even higher friction coefficients than on dry glass are possible. The maximum obtainable friction, however, depends on the type of rubber and the temperature of the ice. Between 0°C and about –5°C the friction is lower than would be expected from the master curve for all rubbers on glass. It is possible that some melting occurs in the contact area. At temperatures below about –5°C, the true friction between rubber and solid ice has been reached, and the maximum is determined by the adhesional friction for the corresponding log $a_T v$ values. For NBR, the maximum on glass occurs at much lower $a_T v$ values (higher temperatures) than are possible on ice, and the friction on ice is correspondingly low. The adhesional friction for SBR is also past its maximum on ice. That of NR, however, occurs at a range between 0°C and −20°C for a sliding speed of 1 cm/sec., so that this polymer is particularly suited for winter tire use. BR has its maximum at much lower temperatures, so that its friction coefficient continues to rise with decreasing temperature down to the lowest temperatures obtainable with the experimental apparatus (See Fig. 12).

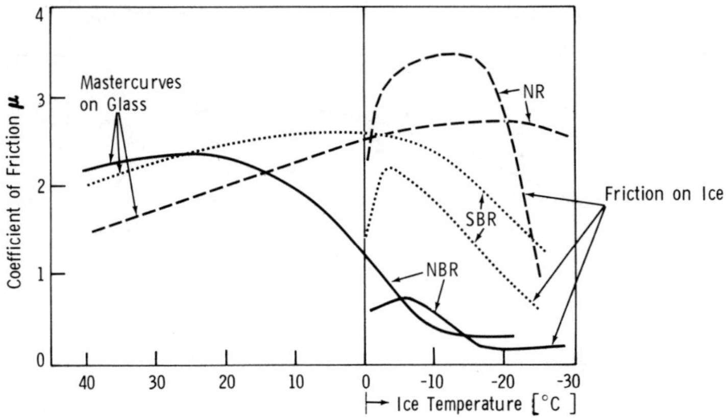

Fig. 11. Friction coefficients on ice at a constant sliding speed of 5 cm/sec for different gum rubbers. For comparison master curves on dry glass are shown (transformed to temperature at constant speed).

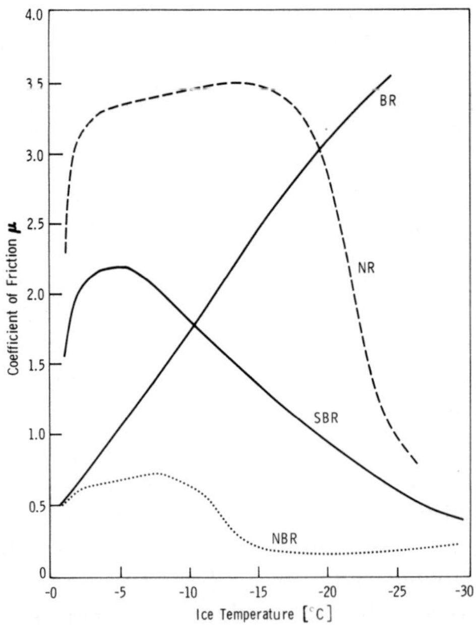

Fig. 12. Coefficient of friction of four different gum rubbers on ice with particular reference to the BR gum compound.

THE EFFECT OF OIL EXTENSION

It is known that oil extension can raise the skid resistance of a tire tread compound on wet surfaces without having detrimental effects on its skid properties on ice (5, 11,14). Fig. 13 shows the master curves on wet carborundum for both an oil extended SBR and a similar unextended SBR and also for both an oil extended NR and a similar unextended NR. In both cases there is a range of a_Tv values over which the oil extended rubber has a higher friction coefficient than the corresponding unextended one. It will be shown in the next section that this range of a_Tv values is of importance in tire skid experiments. As the a_Tv values decrease, the friction of the oil extended compounds decreases more rapidly than that of the unextended rubber, so that there is a cross-over point at lower a_Tv values. The unextended compound then has the superior friction coefficient. This may be interpreted as a reversal in ranking with increasing speed in a tire skid trial, as will be discussed in the next session.

FRICTION AT HIGH SLIDING SPEEDS

Temperature Rise – In practical skid trials, very high temperatures occur in the interface between rubber and track surface, so that temperature and speed can no longer be considered independent variables. Schallamach first measured interfacial temperatures (15) using a compensated thermocouple. The author has recently repeated these measurements using a SBR – BR tread compound, spherical and conical glass sliders into which fine thermocouple wires had been embedded, and sliding speeds on a dry and wet rubber track of up to 150 km/h. The speed unit has been chosen as km/h in order to facilitate a comparison with tire skids. Fig. 14 shows the temperature as a function of the sliding speed on the dry rubber surface for both the sphere and the cone with the same load. It is clear that the temperature is not only load dependent, as shown by Schallamach (15), but for the same load, it also depends on the shape of the slider, so that it presumably is pressure dependent. At very low speeds, the rise is very low, but once a speed has been reached at which the temperature rise has become significant, it rises very nearly with the square root of the sliding speed, i.e. it follows the classical laws of heat conduction of a moving heat source over a semi-infinite surface (16). The dependence on load and indentor shape is the subject of further investigations. Fig. 15 shows that considerable temperature rises can also occur on wet tracks. At moderate sliding speeds, the temperature rises again with the square root of the speed. At a certain speed, which depends on the load and the slider shape and therefore, presumably on the pressure in the contact area, the temperature begins to decrease with further increase in speed. Since it coincides also with a drop in the friction, it appears that lubrication becomes effective. For the spherical slider, no temperature rise with increasing speed could be measured, despite a very high load. In this case the lubrication must have been

References pp. 162–163

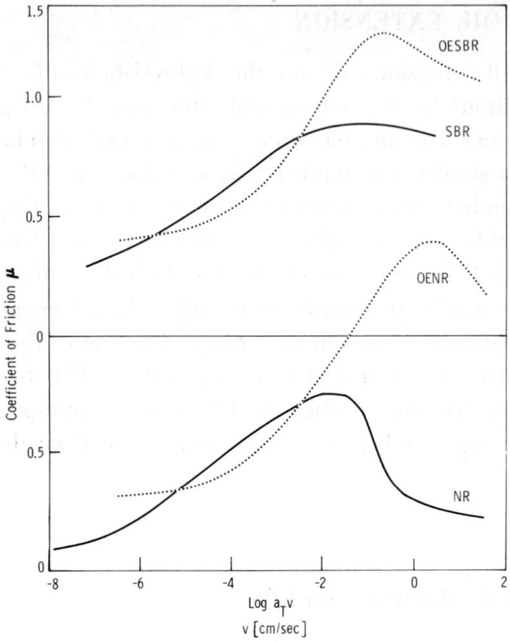

Fig. 13. Master curves of the friction coefficient on wet carborundum for OESBR and SBR (upper part) and OENR and NR (lower part), all compounds having 50 parts of HAF black based on rubber plus oil.

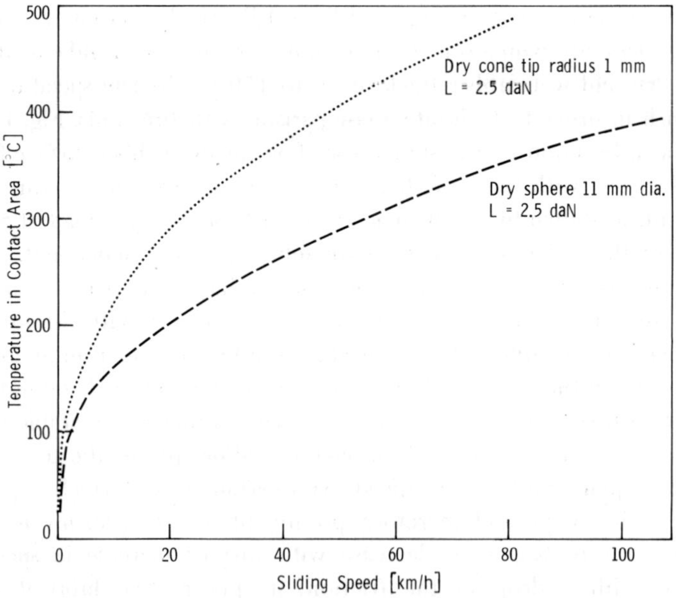

Fig. 14. Temperature in the contact area between a rubber track (60 SBR, 40 BR, 50 HAF) and a slider as function of sliding speed.

substantial at all speeds. The fact that considerable temperature rises may be obtained bears witness to the existence of adhesional friction between rubber and the wet track, since the present temperature effects clearly occurred at the surface of the rubber. This is unlike the destruction of the rubber that was observed by McCarty & Leland (17), and Rieger (18). In these cases the temperature effects occurred underneath the rubber surface through deformation friction.

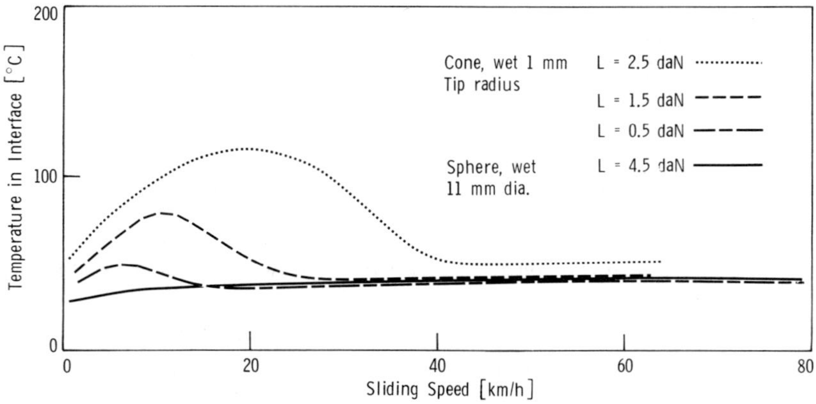

Fig. 15. Temperature in the contact area between a wet rubber track (60 SBR – 40 BR + 50 HAF) and different shaped sliders under various loads as function of sliding speed.

Friction Coefficient at High Speeds – Figs. 16 and 17 show the corresponding friction coefficients as a function of sliding speed, again given in km/h. It is seen that

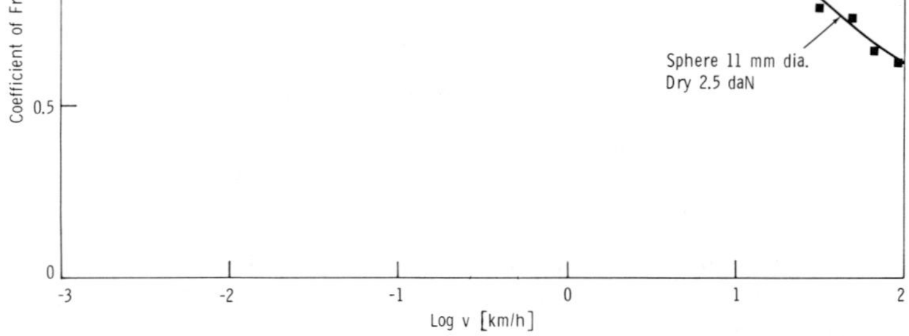

Fig. 16. Coefficient of friction on a dry rubber track for two sliders at the same load (see also Fig. 14).

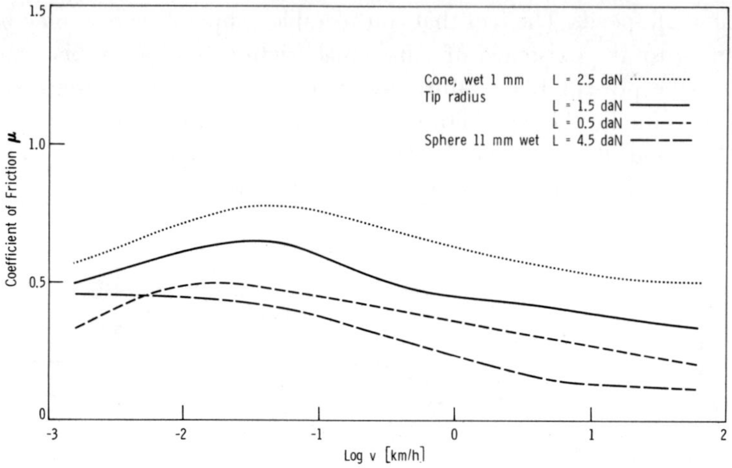

Fig. 17. Coefficient of friction on a wet rubber track for two sliders at different loads (see also Fig. 15).

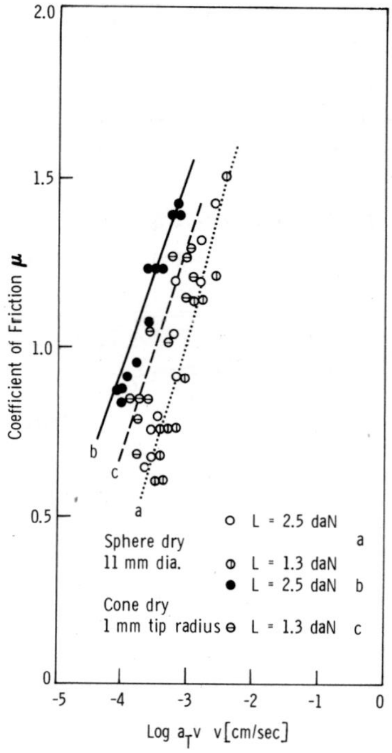

Fig. 18. Master curves obtained from friction on a dry track at high sliding speeds and the resulting temperatures in the contact area (see Figs. 14 to 17). Different master curves are obtained for different loads for the conical slider.

the friction coefficient rises with increasing speed and reaches a maximum value at quite moderate sliding speeds. Position and height of the maximum depend on the load and the shape of the slider and on whether the track was wet or dry. At higher speeds the friction coefficient decreases with increasing speed. Since both temperature and sliding speed are known, it is possible to plot the friction coefficient as a function of log $a_T v$, using the WLF equation to obtain the relevant a_T values. This is shown in Figs. 18 and 19. For the spherical slider on the dry track, the points lie on the rising part of a master curve, and the friction coefficients obtained at different loads fall on the same master curve.

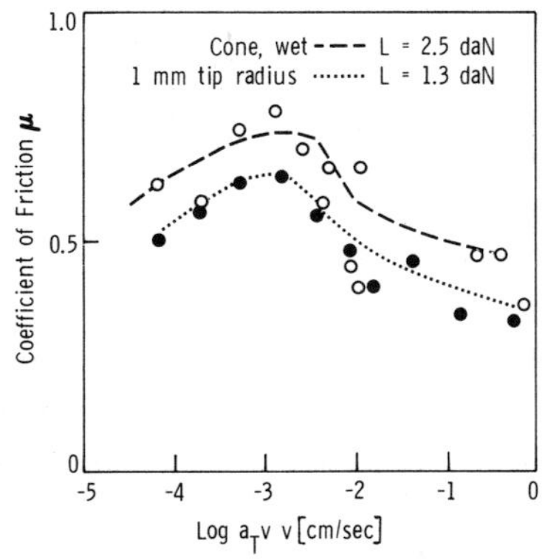

Fig. 19. Master curves on a wet rubber track obtained from friction data at high sliding speeds and the temperature in the contact area.

For the conical slider, the friction coefficient also rises distinctly with increasing $a_T v$ values. However, different master curves are obtained for different loads, the coefficient of friction increasing with the load. Sabey previously showed that the coefficient of friction of conical sliders on rubber tracks increases with the contact pressure, and she attributed these findings to deformation losses (10).

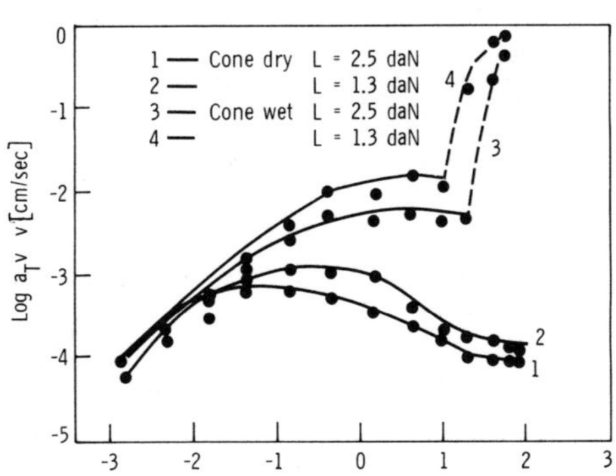

Fig. 20. GTV values pertaining to the speed range in tire skids, obtained from sliding speed and the corresponding temperatures in contact area.

Of particular interest are the log $a_T v$ values which pertain to the wide range of sliding speeds from 10^{-3} to 10^2 km/h. Because the WLF equation is based on a reference temperature which depends on the polymer in question, slightly different ranges of log $a_T v$ values are relevant for different polymers. Fig. 20 shows the $a_T v$ values for the SBR - BR blend used in the high speed friction experiments as a function of the sliding speed. Because the temperatures are lower on the wet surface than on the dry one, higher $a_T v$ values are obtained. At the point at which lubrication sets in, the temperature drops and the sharp rise in $a_T v$ values occurs. However, in this range the friction itself is already very low.

Since the speed dependence of the relative skid rating of compounds (all lubrication effects will be taken to be similar for all compounds) is determined by the $a_T v$ values pertaining to that speed range, reversals in ranking of compounds with speed can be expected to occur only if reversals in the ranking within the $a_T v$ range of the master curve occur. Fig. 21 shows some compound comparisons on dry and wet rough surfaces over the relevant $a_T v$ ranges. The data curves were deduced from the high speed friction experiments with conical sliders. Even if the values of temperature at individual points in the contact region of a carborundum track were considerably different from the values from which the $a_T v$ ranges have been obtained, the influence of this temperature difference on the $a_T v$ range would be quite small. It is seen that the range of $a_T v$ values of practical importance on dry carborundum is quite small, and no reversals between NR and SBR occur, SBR having the higher friction coefficient. On wet surfaces, the range of $a_T v$ values is somewhat wider, but again SBR has the higher friction coefficient over the whole range. Of particular interest are

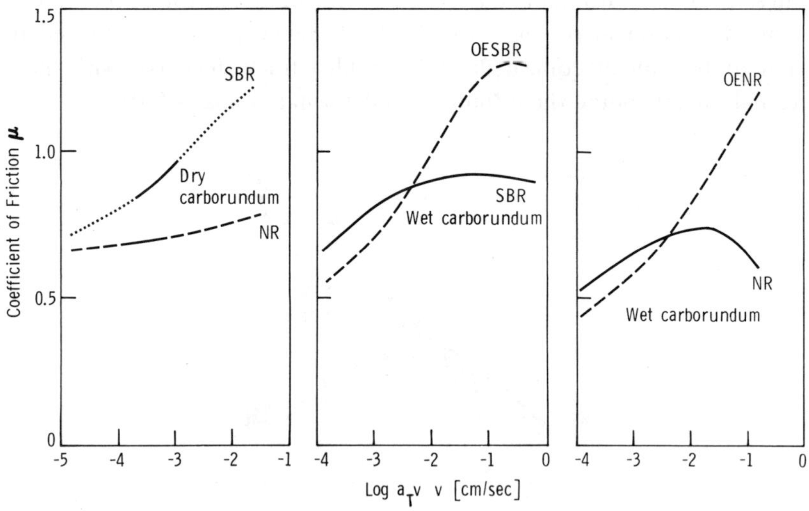

Fig. 21. Parts of the friction curves of NR and SBR tire tread compounds which come into play in tire skids on dry and wet rough surfaces, taken from friction experiments on carborundum (see Figs. 6 and 13).

the comparisons between OESBR and SBR on the one hand, and between OENR and NR on the other. On wet tracks, the master curves cross in the range of the expected practical $a_T v$ values. At high $a_T v$ values, i.e. low sliding speeds, the oil extended rubbers have superior friction, whilst at low $a_T v$ values (high sliding speeds) the reverse is the case. Such reversals in ranking of oil extended rubbers in relation to non-extended ones were found by Grosch and Maycock in skids on wet surfaces (2).

Fig. 22 shows the relevant parts of the friction master curves of NR and SBR gum on glass that apply to high speed sliding on dry and icy surfaces, respectively. For ice the $a_T v$ range has been deduced assuming that the temperature in the interface cannot exceed 0°C. It is seen that SBR is better for the friction on dry smooth surfaces, but the reverse is true for the friction on ice.

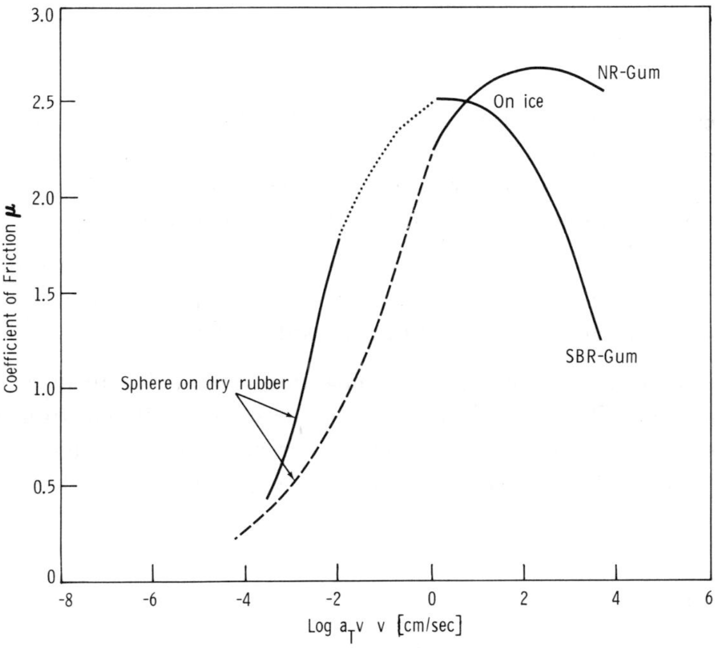

Fig. 22. Parts of the master curves of NR and SBR gum rubbers on dry glass which correspond to tire skid experiments on dry smooth and icy surfaces respectively.

Finally, it becomes obvious, why the British Road Research Laboratory skid tester is capable of ranking tread compounds according to their skid rating in the correct order (19, 20).

It has an average sliding speed of about 11 km/h. At a temperature rise of about 80°C on a wet track this corresponds for NR to an $a_T v$ value of about −2. Hence the skid value of the pendulum skid tester falls within the range of $a_T v$ values obtained also in actual skid trials.

References pp. 162–163

CONCLUSIONS

The temperature and speed dependence of rubber friction at low sliding speeds is reviewed for different polymers and surface conditions including wet and icy surfaces. It is possible in all cases to transform the data into a master curve of the friction coefficient as a function of the combined temperature – speed variable log $a_T v$, where a_T is a temperature dependent variable described by the WLF equation.

It is shown that friction coefficients at high speeds are still part of the master curve; but because temperature and sliding speed have opposing effects, only a small part of the master curve is realizable in practical skid experiments.

Since NR and SBR show no reversals in ranking of the friction coefficient in the relevant $a_T v$ range on either smooth glass or very sharp carborundum either in the wet or dry state, no reversals in compound ranking for these rubbers would be expected in tire skids on different types of surface.

If no reversal of ranking occurs in the relevant part of the master curve, no reversal in ranking between the peak and sliding condition of the tire skid would be expected either, since the peak value corresponds to the higher $a_T v$ value and the sliding one to the lower $a_T v$ value within the small part of the relevant $a_T v$ range.

Cross-overs which occur in the relevant parts of the master curves for oil extended SBR and NR and their corresponding non-extended compounds show themselves as a cross-over in the speed dependence of the tire skid coefficients.

The reversal of the skid ranking when going from a wet surface to an icy one occurs because the temperature conditions on ice are necessarily different from those on a wet surface. Hence, a different part of the master curve comes into play in which NR is near its maximum friction whilst the friction coefficient of SBR is declining.

REFERENCES

1. *G. Maycock, Proceedings of IME, Vol. 180, p. 122, 1965-66.*
2. *K. A. Grosch and G. Maycock, Trans. I.R.I., Vol. 42, T 280, 1966.*
3. *J. K. Meades, Report L.R. 73, Road Research Laboratory, Crowthorne, Berks., England, 1967.*
4. *T. French and R. G. Patton, Fourth Rubber Conference, p. 196, London 1962.*
5. *K. A. Grosch, Rubber Age, Vol. 99 (10), p. 63, 1967.*
6. *K. A. Grosch, A. Schallamach, E. Southern and L. P. McSwift, International Rubber Conference, Moscow, 1969.*
7. *K. A. Grosch, Proceedings at Royal Society, Vol. A 274, p. 21, 1963.*
8. *M. L. Williams, R. E. Landel, and F. D. Ferry, Journal American Chemical Society, Vol. 77, p. 3701.*
9. *F. A. Greenwood and D. Tabor, Proceedings of Physical Society, Vol. 7, p. 989, 1958.*
10. *B. Sabey, Proceedings of Physical Society, Vol. 71, p. 979, 1958.*
11. *K. A. Grosch, ACS Conference, Miami Beach 1971.*

12. *W. Gnorich and K. A. Grosch, Journal of the IRI, Vol. 6, No. 5, p. 192, 1972.*
13. *E. Southern and A. W. Walker, Nature Physical Science, Vol. 237, No. 78, p. 142, 1972.*
14. *K. A. Grosch, Stereo Rubbers, Chapter XIV: "Tires," ed. W.M. Saltman, Marcel Dekker, N.Y., in press at the time of writing.*
15. *A. Schallamach, Transactions of I.R.I., Vol. 32, p. 143, 1956.*
16. *J. C. Jaeger, Proceedings of Royal Society, N.S.W., Vol. 76, p. 203, 1943.*
17. *J. L. McCarty, T.J.W. Leland, Tire Science and Technology, Vol. 1, No. 2, p. 121, 1973.*
18. *H. Rieger, PHD Thesis, Technical University Munich, 1968.*
19. *E. M. Bevilaqua and E. P. Percapio, Rubber Chemistry and Technology, Vol. 41, p. 832, 1968.*
20. *C. G. Giles, B. E. Sabey and C. H. E. Cardew, "Development and Performance of the portable skid resistance tester."*

DISCUSSION

Moore

Thank you very much, Dr. Grosch, for a very interesting talk. Are there any questions?

R. M. Gallaway

The numbers that you used for coefficients of friction were a little puzzling to me. Would you explain those, like 2 and 2-1/2?

Grosch

For gum rubbers on absolutely smooth surfaces, a coefficient of friction as high as 3 can be obtained. In general, if a rubber contains a black filler, as is the case with a tire compound, the maximum that can be obtained is about 1.2 - 1.3. As was said in the paper, black filler reduces the friction coefficient. The printed version of the paper shows a diagram for both black-filled rubbers and gum rubbers. Again on dry carborundum 1.3 is not uncommon, so this is the difference between a tire skid on a contaminated surface, if you like, which was discussed this morning, and that of a lab experiment in which higher friction can be achieved. In addition, one must remember, that at the maximum friction one is in the absolute optimum range of viscoelastic behaviour, and whether this can be achieved in practice in a tire skid is another question.

Gallaway

Does the slip occur at the interface between the rubber and the glass?

Grosch

This is a complete sliding experiment. A block of rubber slides on a surface, procuring 100% slip.

Moore

Are there any other questions?

R. Hegmon

I would like to know how the temperatures were measured.

Grosch

In the experiments in which temperature and speed were kept as separate variables, the whole experiment was carried out in a constant temperature environment and the temperature of the environment was changed; and the experiment is carried out so slowly that temperature equilibrium is established at any time. In a lubricated experiment at low sliding speeds, the lubricant serves as the temperature medium. In the high-speed sliding experiment, the thermocouple is placed directly at the tip of the low heat conduction slider. Dr. Schallamach, when he first did the experiment some years ago – and which we practically only repeated in a way – exercised great care to compensate for heat losses along the thermocouple. We have not done this. We have just measured the time constant of our thermocouple which was in the region of about 1/4 second. So that after two or three seconds we could take measurements which were close to the real interfacial temperature.

B. D. Gujrati

My interest in your work arises because I am also going to report on this subject at this symposium. Dr. Grosch, it seems that you have used a hard sphere on a rubber surface lubricated with water for the high speed test conditions.

Grosch

Yes, water.

Gujrati

In comparison to low speed sliding experiments with rubber on rough surfaces, the hydrodynamic film condition at the sphere-rubber interface at high speeds is quite different. I believe this difference is important in relation to friction and must be emphasized.

Grosch

Yes, I would like to point out, that this is true both for the friction experiments and for the temperature measurements. It shows itself in the temperature experiments, in a decrease in temperature with increasing speed, when lubrication becomes effective. In the frictional experiments, similarly, a decrease in the friction coefficient is observed which is entirely determined by a so-called critical lubrication film thickness. That film thickness can be influenced by pressure, by the shape of the slider, by the viscosity of the lubricant, and by the modulus of the rubber. In the printed version much more detailed data on this aspect are given.

Gujrati

One other question; do you consider the mechanisms in the two cases, rubber blocks sliding on glass surfaces lubricated by water at low speeds to be the same as lubricated sliding of a sphere on a rubber surface?

Grosch

One has either perfect lubrication and then one can speak of hydroplaning which was discussed this morning in which the friction becomes very quickly, very small or one has to a larger or smaller degree vestiges of rubber adhesion. How it acts is certainly subject to investigation. But rubber adhesion occurs even at quite substantial sliding speeds. And then the whole WLF transform can be applied. This is, in fact, the proof that proper adhesion exists even if one might think that lubrication was dominating.

Gujrati

Considering the deformation loss part, how are the two cases different?

Grosch

That I did not follow.

Gujrati

In case of sliding of a rubber block on rough surfaces, beside adhesion there should be what you call the deformation loss. Similarly, there is deformation loss involved in sliding of a sphere on a rubber surface. Are the two cases the same?

Grosch

First of all, I think the deformation peak which we obtained on rough surfaces is influenced by the adhesion itself, either by rubber adhesion or by adhesion of the dust particles which we introduced. The magnitude of the deformation peak is so large that it could not have occurred through normal deformation alone. Tangential stresses produced through vestiges of adhesion must contribute to it. This must occur also during lubricated friction because rubber can be abraded under water. The main deformation peak as shown on the master curves plays probably no part in tire friction because it occurs at $a_T v$ values which are not realizable in practice.

ELEMENTARY EFFECTS IN THE CONTACT AREA OF SLIDING RUBBER

A. SCHALLAMACH

Barnet, Herts., England

ABSTRACT

Visual observation of the contact area of sliding rubber reveals that the motion between the frictional members is often not sliding in the accepted sense. Often motion is by "waves of detachment" passing along the contact in the same direction as the relative displacement of the rubber. These waves are folds in the rubber surface that are almost certainly produced by buckling. Adhesion appears to be complete between the waves. Buckling is attributed to compressive tangential stresses in the contact area which are theoretically predicted and qualitatively confirmed by experiment. The motive force driving the waves is a tangential stress gradient. The mechanism of wave initiation for hard sliders on rubber tracks is different from that for rubber sliders on hard tracks. The measured frictional force reflects predominantly the energy dissipation accompanying the propagation of the waves.

INTRODUCTION

Investigations into the temperature and velocity dependence of rubber friction have established that it is a viscoelastic process but the mechanism of sliding is a matter of speculation (1, 2). This paper reviews conclusions on this subject which have been drawn from visual observations of contact areas between a hard slider and a rubber track and between a rubber slider and a hard track. The experiments, carried out with relatively soft rubbers, have shown that the motion between the frictional members is not necessarily sliding as it is commonly understood. Often motion is brought about by "waves of detachment" (3). These waves are seen under the stereoscopic microscope to be folds in the rubber surface at right angles to the

References p. 177

imposed sliding velocity. They travel in the same direction in which the rubber moves relative to the other frictional member; a physically more significant expression of this finding is that the wave propagation takes place in the direction opposite to that of the frictional force on the rubber.

EXPERIMENTAL RESULTS

The experimental arrangement is sketched in Fig. 1; the directions of imposed sliding velocity v, wave velocity w, and frictional force F on the rubber are indicated.

In order to see the contact area, either the track or the slider must be transparent. A transparent rubber track was made of synthetic polyisoprene; the slider in this case was a hard hemisphere of 19mm radius. The slider on opaque tracks was a meniscus lens of 25mm radius. All rubber tracks were 12.5mm thick. A Perspex plate served as the hard track for the hemispherical rubber sliders of 19mm radius. The contact area, lit by frustrated internal reflection, was either photographed at exposure times of 1/500 sec, or filmed at a frequency of 64 frames per second.

When changing the imposed sliding speed, either the indentation of the rubber track, or the compression of the rubber slider was kept constant. The frictional force was not measured.

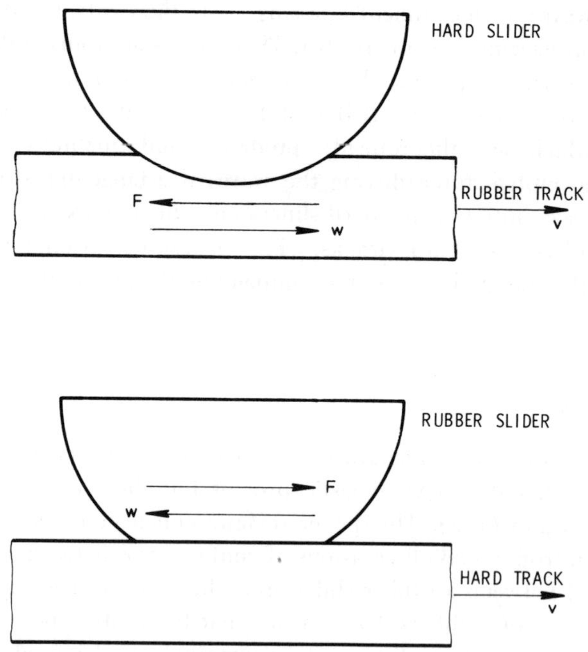

Fig. 1. Sketch of experimental arrangement. v = imposed sliding velocity = track velocity; w = wave velocity; F = frictional force on rubber. After (3).

Fig. 2 reproduces eight alternate frames of a film of the contact area between the hard hemispherical slider and the polyisoprene track. The waves of detachment, seen here as dark lines, start directly at the right edge of the contact and move toward the other side at a speed which considerably exceeds the imposed sliding speed. During their travel, the lines become thinner, break up and loose their indentity. Secondary effects occur at the exit point where parts of the contact become detached from the main body and then suddenly fuse with it again. In between the waves, no relative motion between the slider and the track can be detected so that all motion appears to be connected with the waves of detachment. The mechanism of frictional displacement between rubber and track is, therefore, effected by lifting the rubber locally off the track, by breaking adhesive bonds in the direction normal to the contact.

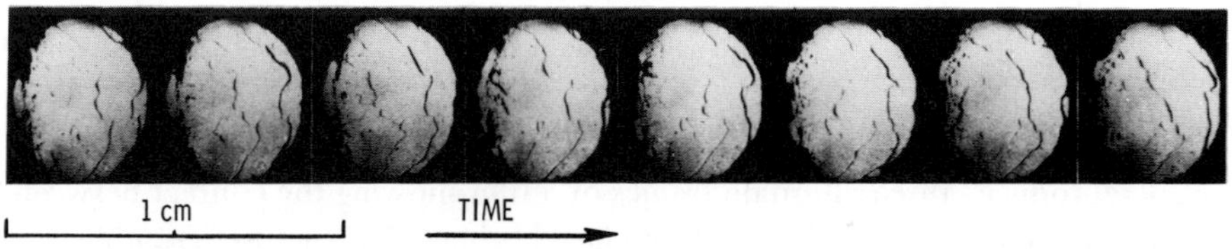

Fig. 2. Eight frames at 1/32 sec intervals of a film of the contact between a hard spherical slider (relative motion to the right) and a polyisoprene track at v = 0.043 cm/sec. From (3).

Fig. 3 is a similar film clip showing the contact between a slider of unfilled natural rubber and a Perspex track. As in Fig. 2, conditions tend to get confused on the left hand side of the contact.

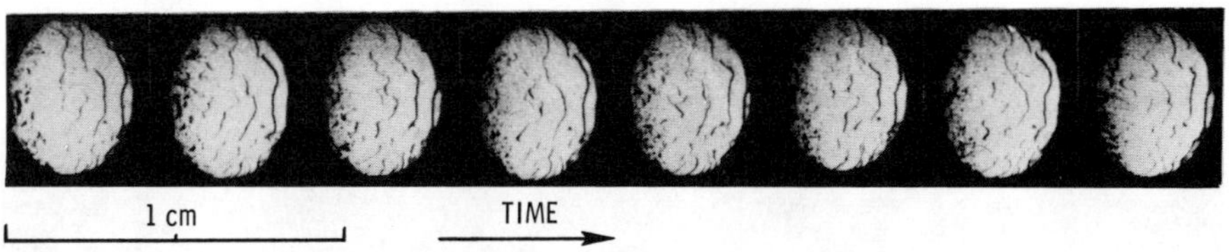

Fig. 3. Eight frames at 1/32 sec intervals of a film of the contact between a spherical natural rubber slider (relative motion to the left) and a Perspex track. v = 0.043 cm/sec. From (3).

The length of the two films from which Figs. 2 and 3 were taken was sufficient to determine with some confidence the wave velocity, w, and the frequency of wave formation, n, as a function of the imposed sliding velocity v. These data, given in Table 1, show a qualitative difference between the effects observed when the track is rubber and when the slider is rubber. With the hard slider on the polyisoprene track, the frequency of wave formation is roughly proportional to the imposed speed, and the wave velocity increases at a slower rate; with a rubber slider on the hard track, the

References p. 177

wave velocity is almost proportional to the imposed speed, and the frequency of wave formation increases at a slower rate. An interpretation of these differences will be advanced presently.

TABLE 1

Frequency of Wave Formation n (sec^{-1}) and Wave Velocity w (cm/sec) at Various Sliding Speeds v (cm/sec). From (3)

	Hard slider on polyisoprene track		Natural rubber slider on Perspex track	
v	n	w	n	w
0.024	5.6	0.86	5.2	0.36
0.043	11.7	1.07	7.1	0.76
0.093	23.6	1.45	9.2	1.34

Fig. 4 reproduces three alternate frames of a film showing the contact between the meniscus lens slider and a track of unfilled butyl rubber. The picture resembles Figs. 2 and 3 but is different in that the waves break up into fine ripples on the left hand side of the contact; a similar effect has also been observed with natural rubber sliders at somewhat higher speeds (3). Fig. 5, which is a single frame of a film of the contact between a butyl slider and a Perspex track, stands in marked contrast to all previously shown photographs; the contact area has been reduced to narrow parallel bands oriented at right angles to the direction of motion and remains unchanged during sliding, no waves of detachment being observable.

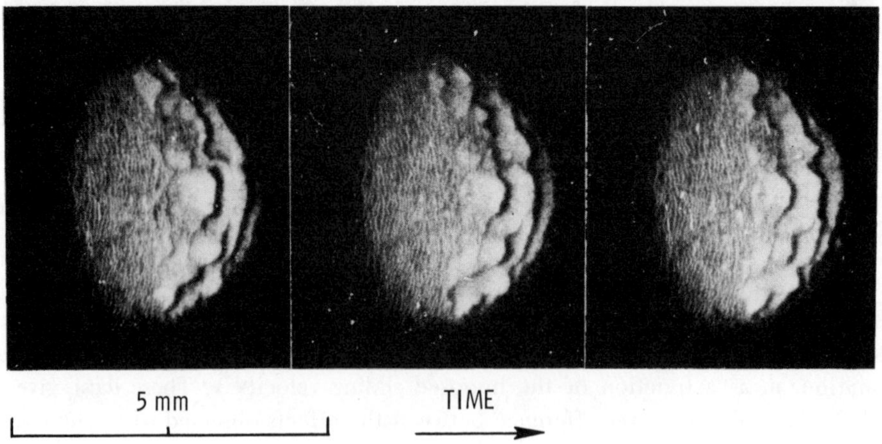

Fig. 4. Three frames at 1/32 sec intervals of a film of the contact between a meniscus lens slider (relative motion to the right) and a butyl rubber track at v = 0.043 cm/sec. From (3).

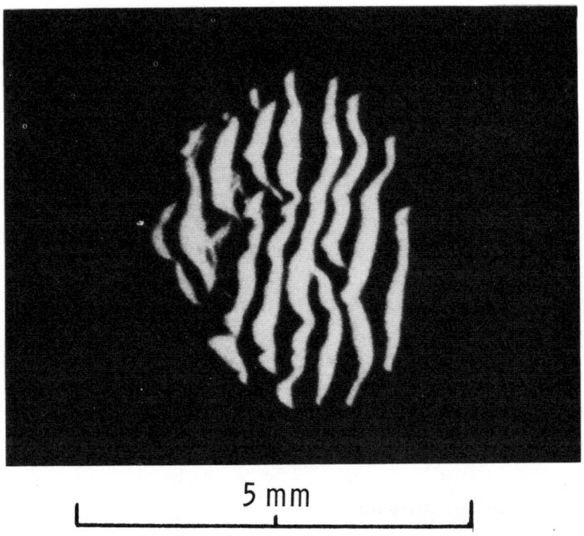

Fig. 5. Single frame of a film of the contact between a spherical butyl rubber slider (relative motion to the left) and a Perspex track at v = 0.043 cm/sec. From (3).

THEORY

The folds in the surface of sliding rubber which constitute the waves of detachment have been attributed to an elastic instability leading to buckling, the instability originating from tangential compressive stresses in the contact area (3). Fig. 5 may be taken as a direct demonstration of this effect. The strips of contact are obviously ridges between which the rubber has buckled inward. A theory has been developed to assess qualitatively the tangential stresses in the contact between a hard slider and a rubber track. The calculations are based on the one-dimensional model of the rubber surface shown in Fig. 6. It is composed of a longitudinal Hookean spring and Hookean cantilevers which support it at regular intervals. Only deflections in the plane of the contact are considered.

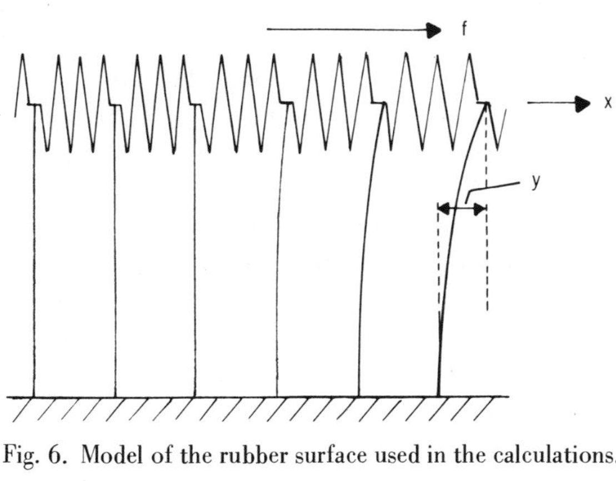

Fig. 6. Model of the rubber surface used in the calculations.

References p. 177

If the cantilever force per unit length of track is ky (y=deflection) and the total spring force is Edy/dx (dy/dx=strain), the force f per unit length of track in the x-direction is given by

$$f = k(y - \sigma^2 d^2y/dx^2) \tag{1}$$

where

$$\sigma = \sqrt{E/k}.$$

Outside the contact, where f=0, the deflection decreases exponentially outward from either end by the factor e^{-1} over the distance σ. Inside the contact, f equals the intensity of the frictional force. Assuming Amontons' law to hold with the friction coefficient μ, f inside the contact is then

$$f = \mu p \tag{2}$$

where p is the normal contact pressure.

The original calculations in (3) were carried out with a parabolic pressure distribution under the slider; the following results were obtained with a 4th-power distribution which is broader and is thought to represent the actual pressure distribution more realistically. Putting the origin in the centre of the contact of length 2a, the pressure is then given by

$$p = p_m (1 - x^4/a^4) \tag{3}$$

where p_m is the maximum pressure in the centre. The solution of the differential Eqn. (1), with f given by Eqns. (2) and (3), contains two integration constants which are derived from the condition of continuous deflection and slope at the ends of the contact. The final result is:

$$y = \mu p/k \, [1 - 24\, \sigma^4/a^4 - 12\, \sigma^2 x^2/a^4 + 4\, \sigma/a(1 + 3\, \sigma/a + 6\, \sigma^2/a^2 + 6\, \sigma^3/a^3)\, e^{-a/\sigma} \cosh x/\sigma)]. \tag{4}$$

The upper graph in Fig. 7 shows the tangential deflection of the rubber surface according to Eqn. (4) when a slider moves over it to the right; σ has been put equal to "a" for the purposes of the calculation.

The change in surface strain that is caused by the varying deflection can be visualized from the lower part of Fig. 7. This depicts a long strip of the rubber surface which initially had equidistant lines before being deformed by the slider. The crowding of the reference lines near the right hand edge of the contact indicates a compressive stress; the stress is tensile at the other edge. The strain distribution is shown in Fig. 8; the strain gradient is almost constant between the stationary values of the strain at either end.

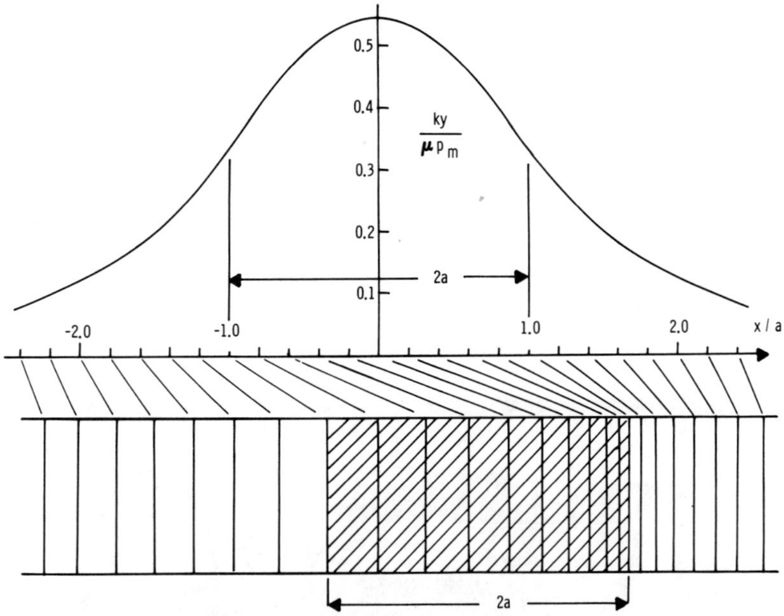

Fig. 7. Tangential displacement of the model rubber surface by a horizontal force with the 4th power distribution of Eqn. (3) inside the contact length 2a.

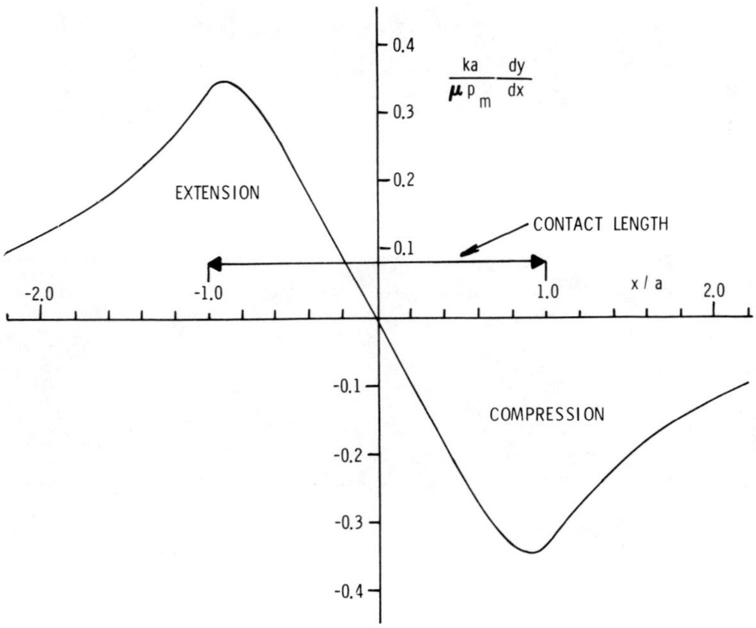

Fig. 8. Distribution of the surface strain on the model rubber surface.

References p. 177

A consequence of the strain distribution along the contact is that the local sliding speed v_s also varies along the contact and is given by

$$v_s = v(1 + dy/dx). \tag{5}$$

Thus, v_s equals v only in the centre of the contact. This effect will, however, be neglected in the later discussion.

The theoretical surface deformation predicted by Eqn. (4) and shown in Fig. 7 may be compared with experimental results. Fig. 9 shows the deflection produced by a spherical slider on a transparent rubber sheet with a 2mm square lattice marked on its surface. The distortion of the lattice in the horizontal direction strongly resembles the

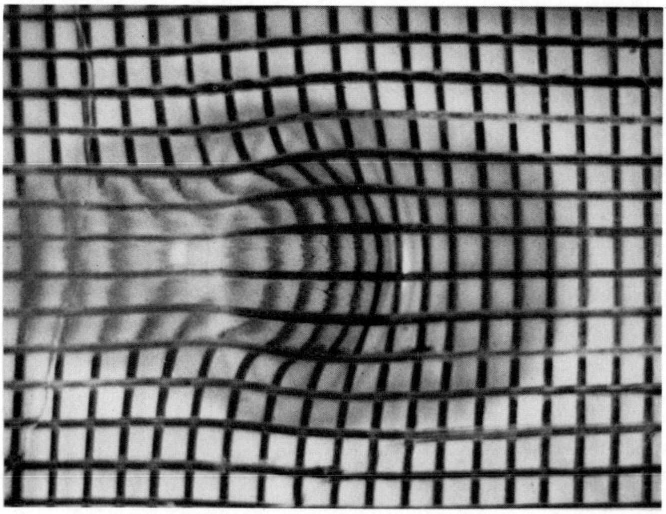

Fig. 9. Distortion of the surface of a transparent rubber slab by a hard spherical slider. From (3).

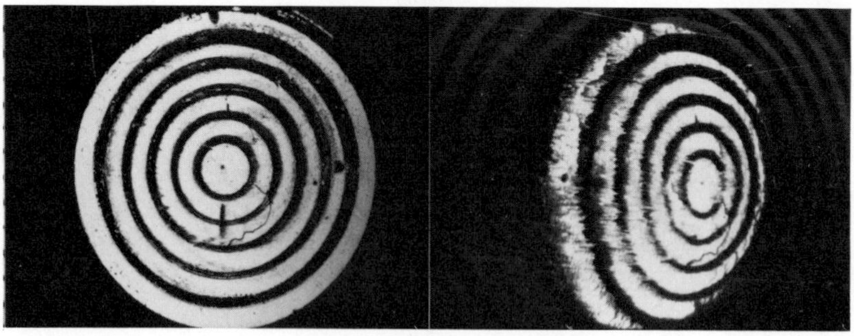

(a) (b)

Fig. 10. Contact between a rubber sphere with concentric surface markings and a Perspex plate when (a), resting and (b), sliding to the left. From (3).

lower part of Fig. 7. The theory also describes qualitatively the deformation of a rubber slider on a hard track. Fig. 10(a) is a picture of a spherical rubber slider at rest; the concentric rings were originally 1mm apart. In Fig. 10(b), the sphere slides toward the left; as in Fig. 9, the spacings between the reference markings have become more closely spaced on the right, and are farther apart on the left. The photograph also shows, by the different number of rings passing through the centre line on the two sides of the apex, that the sphere has rolled over to the left.

In spite of the apparent similarity between the surface deformation of a rubber track and a rubber slider, there is an important difference between them. A rubber slider maintains its gross deformation once a steady state has been reached but the deformation of a rubber track travels with the slider, and fresh rubber is continuously brought into the contact area.

DISCUSSION

Theoretical considerations and experimental facts confirm the presence of compressive stresses in the contact which can be made responsible for buckling, the most immediate illustration of this being shown in Fig. 5. This figure demonstrates, at the same time, that buckling does not necessarily lead to waves of detachment, and that "true" sliding, i.e. a mechanism proceeding on a molecular scale, is an alternative to waves of detachment under certain conditions.

The experimental fact that rubber has a static coefficient of friction proves that adhesive bonds between rubber and track can maintain indefinitely a certain maximum tangential stress in the contact. The theories of dynamic rubber friction reviewed in (2) deal with the breaking and re-making of these bonds, and with the strains in the rubber surface under the influences of the accompanying shearing forces. It is interesting to note that static friction is difficult to explain on this basis. The experimental evidence presented here suggests that, in a good number of cases, limiting friction is reached when the compressive tangential stresses in the contact are great enough to produce buckling and waves of detachment.

The operation of the waves of detachment is best understood by regarding them as waves of stress relaxation. The waves first turn up at the edge of the contact, in the region of greatest compressive stress, and presumably originate from a fold just outside the contact. At its birth, the fold leads to a transient strain relaxation which allows the slider to move forward by a small amount δx. The fold is subsequently driven through the contact by the strain (or stress) gradient of Fig. 8. During its travel, it partially relaxes the contact stress, and the compressive stress in front of the contact can increase again to make formation of a new fold at that point possible. The initiation of the wave depends, however, on whether a hard slider moves over a rubber track, or a rubber slider moves over a hard track.

References p. 177

A hard slider moving at an imposed speed forces the creation of a fold in the newly compressed rubber of the track at the lead edge of the contact. At a frequency of wave formation n, the sliding speed is simply

$$v = n\ \delta x. \tag{6}$$

The predicted linear relation between n and v, which is largely borne out by the first 2 columns of Table 1, was predicted previously. The wave velocity is expected to be proportional to the tangential stress gradient which, in its turn, is proportional to the measured coefficient of friction μ. Grosch's data for similar rubbers (4) show that, in our experimental range, the increase in friction with increasing velocity is roughly the same as the increase in wave velocity (Table 1, column 3). This mechanism of wave initiation by bodily displacement between the frictional members can clearly not operate on rubber sliders because the same rubber remains involved in the sliding process. It is thought that with rubber sliders, wave initiation is not directly dependent upon the imposed velocity but upon the tangential stress build-up, and therefore upon the overall friction coefficient. Comparison of columns 4 and 3 in Table 1 shows, indeed, that the frequency of wave formation on the rubber slider increases with increasing velocity v very nearly at the same rate as does the wave velocity on the rubber track under the hard slider. In order to accommodate the imposed sliding speed, the wave velocity has now an enforced value. The time taken by a wave to traverse the contact of length 2a is 2a/w so that

$$v = (w/2a)\ \delta x. \tag{7}$$

This expected proportionality between sliding speed v and wave velocity w is in good agreement with the data in the 5th column of Table 1.

The propagation of waves of detachment is, like every deformation of rubber, accompanied by mechanical losses because of the hysteresis of rubber. In the present instance, the losses are incurred as the strain configuration around the folds travels over the rubber surface. Although friction comes ultimately from the adhesive bonds, what is measured as frictional energy dissipation is primarily these dynamic losses; in addition, any difference between the energy of adhesion expended in detaching the rubber from the track in the front part of the wave, and the work gained in re-establishing adhesion in its rear part will appear as a component of the frictional force. Whether, in fact, the rubber responds to an imposed sliding speed by developing waves of detachment or by true sliding depends on which of these two is the energetically more advantageous process. It seems that a slider of butyl rubber, which is highly hysteretic, finds it easier to move by true sliding than by waves of detachment. Because the hysteresis of rubber is a viscoelastic phenomenon depending on temperature and rate of deformation, it is quite possible that in friction experiments over extended ranges of temperature and speed the mechanism of sliding changes over from one kind to the other at some critical value of these variables. It is also likely that the type of wave initiation determines the kind of sliding, and that for this reason waves of detachment develop on a butyl track (see Fig. 4) but not on a butyl slider (see Fig. 5).

The discussion of our experimental results is based on relatively few data and is therefore largely conjectural. Whether, in particular, waves of detachment play a role in the friction of harder rubbers, like tyre treads, can only be decided by further investigations.

REFERENCES

1. *K. A. Grosch, Proceedings of the 1974 GMR Symposium on The Physics of Tire Traction, Plenum, 1974.*
2. *A. Schallamach, Rubber Chemistry Technology, 41, p. 209, 1968.*
3. *A. Schallamach, Wear 17, p. 301, 1971.*
4. *K. A. Grosch, Proceedings Royal Society, A274, p. 21, 1963.*

DISCUSSION

Moore

Thank you very much Dr. Schallamach, for a most instructive talk. Are there any questions from the floor?

J. D. Walter

You showed compressive and tensile strains in the contact zone, and indicated they were symmetrical about the center of contact. The strain distribution over the contact zone should add up to equal the net shear force in the contact zone, not zero force. And since one was plus and one was minus, you would have a net zero force if the distribution was symmetrical.

Schallamach

The surface strains must add up to zero because the overall length of the track remains constant. This, however, does not imply that the sum of the surface *tractions* is zero. The tractions producing compression in front of the slider and elongation at the rear must have the same direction. The total frictional force contains, in addition, the component due to the deflection of the cantilevers (see Fig. 6).

Moore

Are there any other questions?

D. B. A. Phillips

I would like to ask whether the wave formation mechanism in these models has anything to do with the squeal that you get in the tires. Is it the same mechanism taking place that causes tire squeal on smooth road surfaces?

Schallamach

It is not known whether waves of detachment form on harder rubbers like tire treads but the surface stresses should be similar to those described here. They might produce squeal when equalizing, without necessarily generating waves of detachment.

Moore

Thank you very much, Dr. Schallamach.

LUBRICATION STUDIES OF SMOOTH RUBBER CONTACTS

A. D. ROBERTS

University of Cambridge, England

ABSTRACT

By producing optically smooth spheres and cylinders of rubber, it has been possible to study the thickness and contour of a liquid film between surfaces by optical interferometry. Normal approach between surfaces clearly shows the entrapment of a bell of liquid and reveals the load bearing capacity of electrical double layer forces at small distances of separation. In the absence of such forces, a liquid film collapses with time leading to areas of adhesion over most of the contact region. When a normally loaded rubber sphere is slid tangentially, the entrapped bell lifts up its leading edge to form a convergent wedge of liquid capable of supporting the normal load. Such action gives the horseshoe shaped contour typical of elastohydrodynamic lubrication. Recently, the effect of relative spin between contact surfaces has been examined. An unrestrained rubber ball rolled with spin between plates in dry contact experiences tangential contact tractions resulting in a creep motion of the ball perpendicular to its direction of rolling, as anticipated by earlier work. In the presence of a lubricant, there is still creep, but in the *opposite* perpendicular direction. Tilt between the contact surfaces generated by elastohydrodynamic action clearly contributes to this expected result, though it does not fully explain it.

References pp. 193–194

INTRODUCTION

A fluid film interposed between sliding rubber surfaces can substantially reduce the coefficient of friction. For example, a smooth rubber sphere drawn over glass in dry contact can show a coefficient of 10 or more, but when lubricated with soap solution this can fall to 0.001 or less. From early studies, it appeared that the effectiveness of lubrication depended upon sliding speed, contact pressure, lubricant bulk viscosity, and surface smoothness (1, 2). It was also noticed that if stationary surfaces were kept in contact under load for some time, the lubricant became squeezed out, and the friction ultimately reached a value nearly the same as that of unlubricated surfaces (2, 3). Over the last decade, there has been considerable interest in elastomer lubrication. Following on from the success with metal surfaces (4), elastohydrodynamic lubrication theory has been extended to the case of soft flexible rubber-like surfaces in which thin-film lubrication is dominated by elastic deformation, and in which fluid viscosity variation may be neglected. Theoretical solutions have been obtained for simple surface geometries in rolling, sliding and normal approach (5-9). One difficulty, however, in testing these solutions has been that of obtaining reliable measurements of film thickness between non-metallic boundaries. Film thickness measurement is desirable not only as a sensitive test of theory but also for design purposes where it is the most useful and sought-after parameter.

In a study made of the extrusion of liquids between highly elastic solids (10), it became desirable to examine in detail the shape of liquid films trapped between rubber and glass surfaces. It was decided to study the contour of the rubber surface relative to the glass by optical interference. Interferometric methods have been used in a number of recent lubrication investigations (11-18), but the main obstacle to using them in lubrication studies in which at least one surface is made of rubber is the quality of this surface. Ordinary rubber surfaces tend to be rough and poorly reflecting which makes interferometry impossible. This difficulty can be avoided either by covering the rubber surface with a thin sheet of smooth plastic aluminized on its outer face (19) or by preparing optically smooth rubber surfaces (20). In this latter case, no reflective metal coatings are required. Most of the work described in this paper has been carried out using the optically smooth rubber surfaces. The apparatus for the lubricant film studies comprised two parts: a mechanical arrangement for bringing rubber and glass surfaces into contact and an optical arrangement (see Fig. 1) for determining interferometrically the optical gap between the lubricated rubber and glass surfaces. The former brings surfaces into contact under a normal dead load by means of a lever arm, and a system of traverses permits alignment and controlled sliding between the surfaces. During sliding the frictional force can be measured by a strain gauge system.

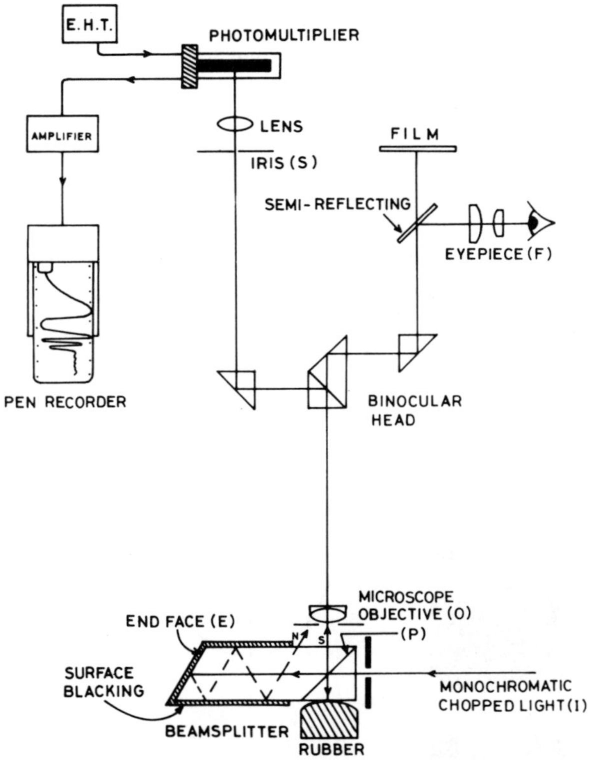

Fig. 1. Optical arrangement of apparatus for determining interferometrically the optical gap between non-metallised rubber and glass surfaces. The glass beamsplitter employed ensures visibility of the fringe pattern for lubricated contact. Interference patterns are observed through a low power microscope, and the absolute separation between surfaces is determined by counting fringes born or lost at the centre of contact as the surfaces separate or come together. Films thinner than 100 nm show no fringes, but a black tint develops as the film approaches zero thickness. The tint intensity recorded photoelectrically may be used to calculate film thickness according to the relation $R = R_0 \sin^2(2\pi nh/\lambda)$ where R is the observed reflected light intensity, R_0 the intensity for an optical gap of $\lambda/4$, n the film refractive index, h the film thickness, and λ the wavelength of the reflected light. The method is sensitive down to about 50Å.

NORMAL APPROACH

By loading a sphere of rubber onto a glass surface in the presence of a viscous silicone fluid, a series of interferograms like those shown in Fig. 2 is obtained (21). A viscous silicone ($1\ m^2\ s^{-1}$) was employed to slow down the approach process to make observations and photography possible. The deduced profiles show that a bell of liquid is trapped within the contact zone and then gradually extruded as surfaces come closer. Towards the end of the process, the pressure excess at the centre diminishes and the depression gradually becomes less pronounced. When the separation is small, asperities break through the liquid film but support only a fraction (no more than 0.1%) of the normal load so that the load borne by the film may be assumed to be equal to the applied load. For film thicknesses less than about 100 nm (1000 Å), the surfaces are almost parallel; but when the thickness is less than 40 nm, the surfaces suddenly spring together at various points leading eventually to intimate contact over the major part of the contact region. The reason for the sudden film collapse is not clear. There will be a small but finite attraction between the surfaces due to Van der Waals forces, but for gaps greater than 10 nm, the attractive force will be rather small compared with the applied load (less than 2%). It may be that contact first occurs at some protusion where the film gap may be less than 10 nm. This may more often occur near an edge of the contact zone; and once a point of contact is established, this pulls the rest of the rubber into contact. A growing point of contact is apparent at the perifery (at four o'clock) to the contact zone in the last interferogram (t = 293,300) of the sequence shown in Fig. 2.

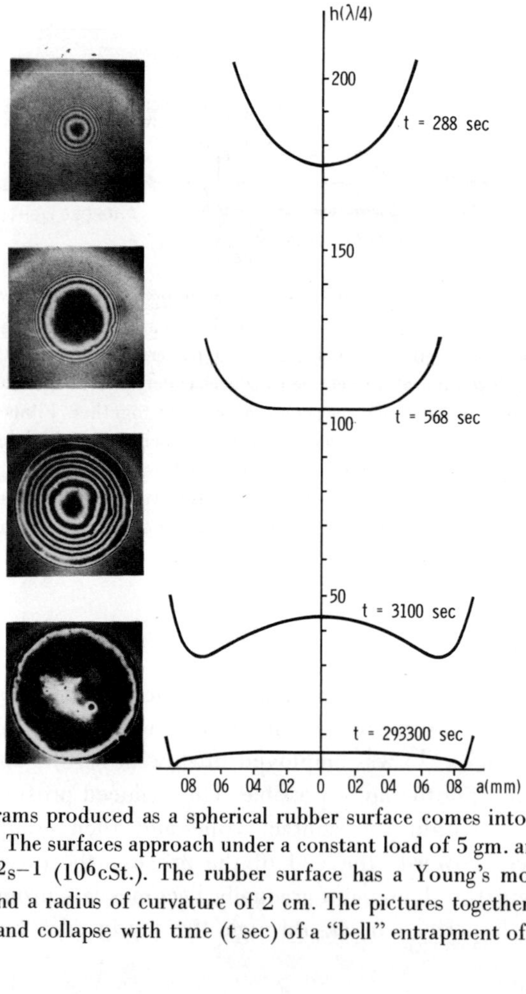

Fig. 2. Interferograms produced as a spherical rubber surface comes into normal contact with a lubricated glass plate. The surfaces approach under a constant load of 5 gm. and squeeze out silicone oil of viscosity 1 m^2s^{-1} (10^6cSt.). The rubber surface has a Young's modulus of 0.6 MNm^{-2} ($6X10^6$dyn.cm^{-2}) and a radius of curvature of 2 cm. The pictures together with deduced profiles show the formation and collapse with time (t sec) of a "bell" entrapment of liquid (λ = 546.1 nm).

With water the squeeze process is very rapid and difficult to observe, but the behavior is qualitatively the same as for silicone fluid. At the stage of film collapse, surfaces seal together exceedingly quickly and trap small islands of water as shown in Fig. 3. After a few hours, these disappear suggesting that although the contact appears to be "dry" a few monolayers of liquid remain between the rubber and glass. The interferometric method (tint intensity) is not sufficiently sensitive to detect, with certainty, the presence of these residual films. This film collapse appears to be characteristic of all liquids which do not contain a sufficient concentration of free ions. If instead of distilled water a dilute solution of sodium dedecyl sulphate (SDS) is used, a drastically different result ensues (see Fig. 3). The surfaces no longer spring together trapping islands of liquid, but instead remain apart at an almost uniform separation of about 20 nm. The film does not collapse with time provided the surfaces are kept moist. Experiments (22) show, that in this case, the thin film is stablized by electrical double layer repulsive forces which support the normal load. The SDS is absorbed on the rubber surface (see Fig. 4) with its negative polar end groups in the water. It is possible that there is similar absorption on the glass, but it is more likely that the glass itself acquires a negative charge by the absorption of water on the Si = 0 groups on its surface to form HO

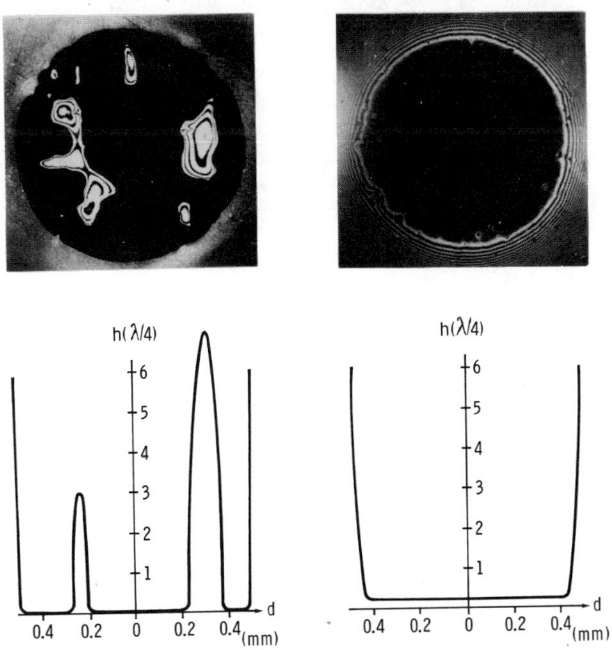

Fig. 3. Interferograms and deduced profiles showing the marked difference in contact between distilled water (left) and water containing 0.01M SDS. Distilled water becomes trapped in pockets but the SDS solution forms a thin equilibrium film of about 200 Å uniform thickness. (λ = 546.1 nm).

References pp. 193–194

Si – 0⁻ plus a free proton lost to solution. The two negatively charged surfaces attract ions from the solution establishing a double layer of charge. The equilibrium thickness of the film is reduced if the contact pressure is increased, and the results shown in Fig. 5 are in close agreement with those obtained for soap films of SDS suspended in air (23). If the ionic strength of the solution is increased with sodium chloride (see Fig. 5), the interacting double-layers of charge becomes increasingly screened and the equilibrium film thickness is reduced. Stable equilibrium films of other soap solutions can equally well be formed between rubber and glass or between two smooth rubber surfaces in point contact (22). Experiments such as these provide a direct means of determining the magnitude of the repulsive forces arising from electrical double layers. They show directly that contact pressures up to about 1 atmosphere (10^6 dyn. cm^{-2}) can be supported. For example, when a 0.01 M SDS solution is squeezed under this pressure the film thickness is about 10 nm. Provided the surfaces are very smooth, the film is not

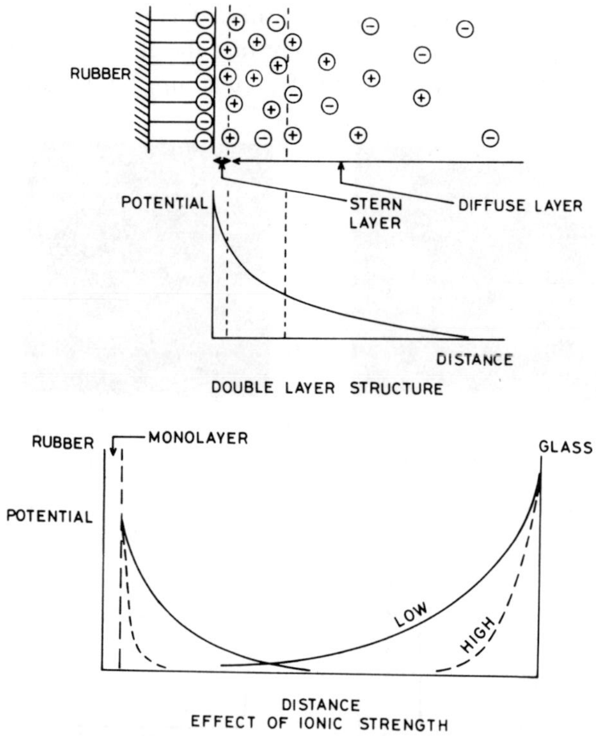

Fig. 4. Assumed charging of rubber and glass surfaces in contact giving rise to electrical double layer repulsive forces. Excess counter-ions near the monolayer surface screen the electrostatic attraction for counter-ions further away with the result that the electric potential drops off rapidly with distance. An increase in electrolyte concentration (ionic strength) magnifies the screening effect and so leads to a more rapid decline in potential and hence a fall in the equilibrium film thickness.

penetrated, and resistance to sliding is due only to the viscosity of the liquid film. The same experiments also offer a method for determing the viscosity of liquid films less than 200 nm thick by studying the rate at which the surfaces approach through the liquid film (21) and by studying their resistance to shear (see below). Charged double-layers, thus, provide a means of lubricating surfaces loaded relatively lightly together. The double-layers do, however, still exist at heavier loads. Other work (24) has shown that at 100 atmospheres the equilibrium film thickness is 1 to 2 nm. For hard metallic surfaces covered with large asperities this seems not to be thick enough to act effectively as a lubricant film; but for flexible surfaces (rubbers, plastics, soft biological tissue), asperities can be much more readily deformed, and double-layers may well play a part in their lubrication.

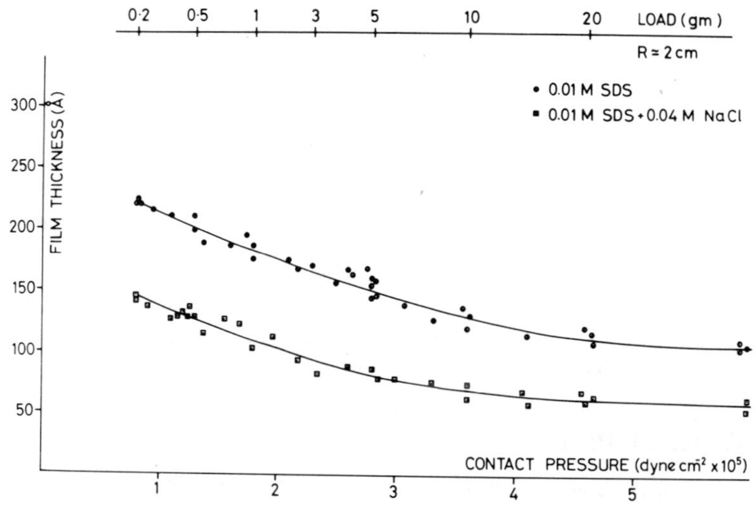

Fig. 5. The equilibrium thickness of SDS films sandwiched between a rubber sphere (R = 2 cm) and glass plate under different contact pressures and for solutions of two ionic strengths. The upper curve is just 0.01M SDS whereas the lower contains the addition of 0.04M NaCl.

In the presence of SDS solution, the contact area between the smooth rubber and glass surfaces under a given load follows closely that predicted by Hertz's equations for the contact of elastic bodies. There is no detectable adhesion. If a liquid with no free ions is used, the liquid eventually becomes completely expelled under pressure, and there is a small but finite adhesion. The most marked effect is observed if dry surfaces are pressed together. The area of contact is found to be larger than that given by the Hertz equation, and there is an appreciable adhesion arising from surface forces which pull the surfaces into increased contact until the stored elastic energy in the rubber just equals the free surface energy released by the creation of the enlarged interface (25). The diameter of the circle of contact observed for a rubber sphere resting on a rubber flat is shown in Fig. 6. The broken line is the theoretical Hertz equation showing zero contact size for zero load. The full line is the experimental result showing a finite contact area for zero load. The contact is reduced to zero only when a negative normal force is applied. If the surfaces are immersed in water the adhesion falls by a factor of 10.

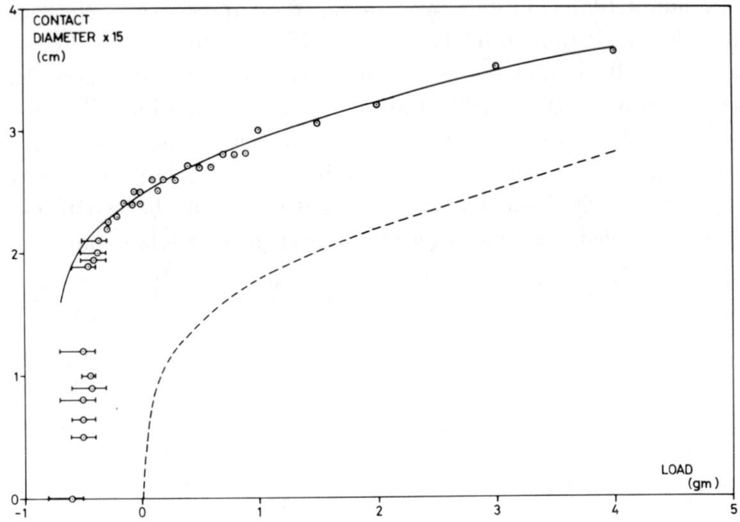

Fig. 6. Contact area results for a smooth rubber sphere (R = 2 cm; E = 0.8 MNm^{-2}) in dry contact with a rubber flat (same modulus) under small positive and negative loads. ⊙ contact results; – – – – Hertz theory; ——— modified theory.

TANGENTIAL SLIDING

For dry smooth surfaces of rubber against glass, the friction is very high due mainly to the large area of contact. A soft rubber sphere lightly loaded, with a somewhat larger than Hertzian contact area due to surface force effects, shows at slow sliding speeds (1 m/sec) a coefficient of friction of about 10. With distilled water, this falls to about 8, with dilute NaCl solution (some double-layer support) to 4, with a viscous silicone (elastohydrodynamic support) to 0.1, and with 0.01 M SDS solution down to values of the order 0.001.

Interferometric observations have been made of sliding of smooth rubber cylinders and spheres over lubricated glass. For example, with a viscous silicone fluid, if a rubber sphere is loaded onto the glass, a bell of trapped liquid is formed as discussed above (see Fig. 2); and upon setting the surfaces in relative tangential motion, the bell lifts up its leading edge and closes on its trailing edge to form a convergent wedge of liquid capable of supporting the normal load. This gives the typical horseshoe shaped contours of Fig. 7. Sliding experiments have also been carried out with 0.01 M SDS solution as the lubricant. Speeds of sliding were adjusted to produce a range of film thicknesses between 20 and 200 nm. For these thin films, the rubber surface acquires almost no distortion, so that friction may be assumed to be due to the viscous resistance of an approximately parallel layer of liquid. Hence, by determining the steady-state frictional force, contact area, sliding velocity, and film thickness, it is possible to calculate the viscosity of the liquid film assuming a Newtonian behavior. The results presented in Fig. 8 show that the friction is extremely low ($\mu \approx 0.001$) and the effective viscosity of the film is the same as the bulk viscosity except for films less than 50 mm (500 Å) thick below which viscosity rises. This rise is probably a result of some penetration of the SDS film by surface asperities.

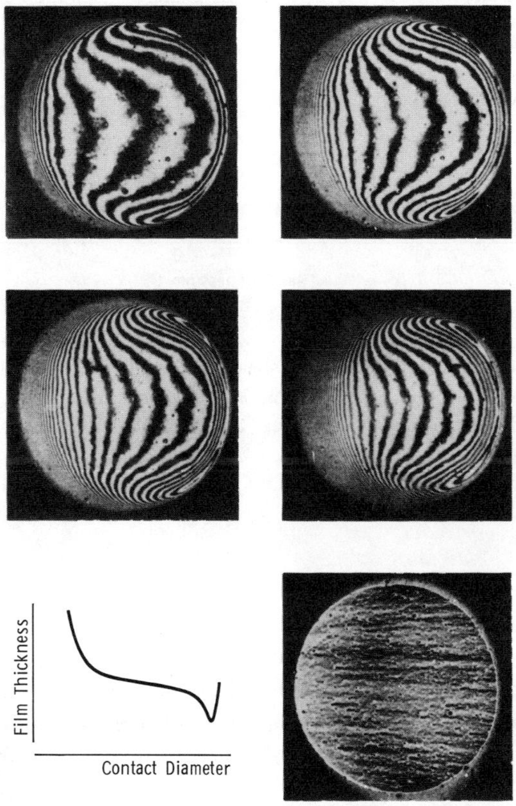

Fig. 7. Interferograms for a smooth rubber sphere sliding over a lubricated glass plate at different speeds. The top four pictures are for sliding in the presence of a 1000 mm^2s^{-1} (1000 cSt.) viscosity silicone fluid at velocities of 0.09, 0.20, 0.32 and 0.47 cm/sec. Each shows a wedge shaped contour at the fluid inlet (LHS) and a "horseshoe" protuberance at the exit. Their centre line profiles are typically elastohydrodynamic in appearance (line diagram) with the characteristic nip at the exit to the contact region. (W = 10 gm; R = 2 cm; E = 0.6 MNm^{-2}; λ = 546.1 nm. The sixth picture (bottom right) shows sliding in the presence of 0.01M SDS solution. The relative velocity between the surfaces was 0.044 cm s^{-1}, and over most of the contact area the film was only about 500 Å thick. In this case the rubber surface shows no characteristic elastohydrodynamic shape and appears more as a bearing pad of very low tilt. Small surface projection are apparent and influence the flow of liquid through the contact (W = 50g; R = 2 cm; E = 0.6 MNm^{-2}; λ = 546.1 nm).

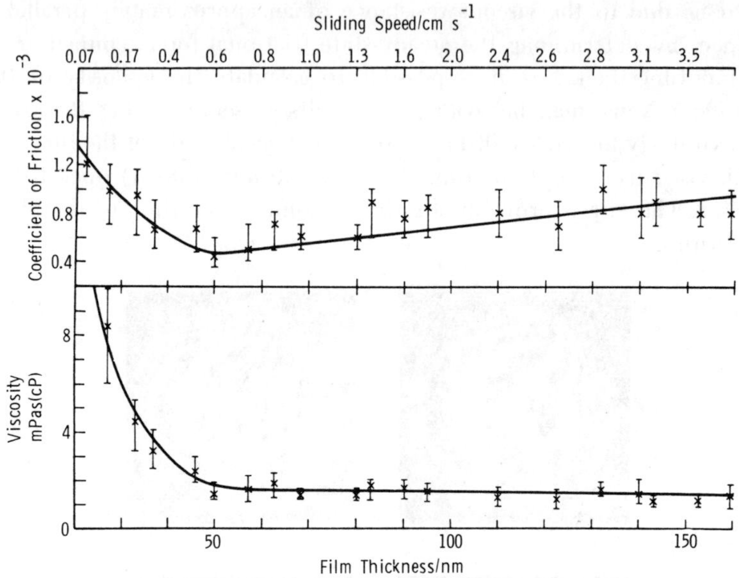

Fig. 8. Friction results for the continuous shear of thin 0.01M SDS films between a rubber sphere (E = 0.6MNm^{-2}; R = 2 cm; W = 50g) and a glass plate. From these the apparent thin film viscosity was calculated and was found to be near the bulk viscosity over the whole range of film thickness down to 50 nm (500Å).

CONTACT SPIN

Relative spin is present between two rolling bodies when they have a relative angular velocity about their common normal at the point of contact. Spin has been observed to lead to a transverse creep motion between the bodies perpendicular to their direction of rolling, and its influence can be interpreted in terms of eccentricity of the axis of spin from the centre of the contact. Consider the contact of a ball rolling on a plate with relative spin present (see Fig. 9). When the ball creeps, it effectively pivots about a point A displaced a distance ϵ from the true centre 0, because tractive forces are not in balance over the front and back halves of the contact zone. If the ball creeps with a radial velocity $\triangle V$, then $\triangle V = \omega\epsilon$ and so $\epsilon/a = \triangle V/a\omega$ which is an expression for the ball eccentricity in terms of its radial creep. One way of realizing spin is to roll three equally spaced balls between two contra-rotating coaxial circular plates as shown in Fig. 9. Assuming symmetrical points of contact, each ball rotates about a horizontal axis whilst the two plates rotate about a vertical axis. This leads to relative spin at each contact point of a magnitude equal to the angular velocity of rotation of the plates.

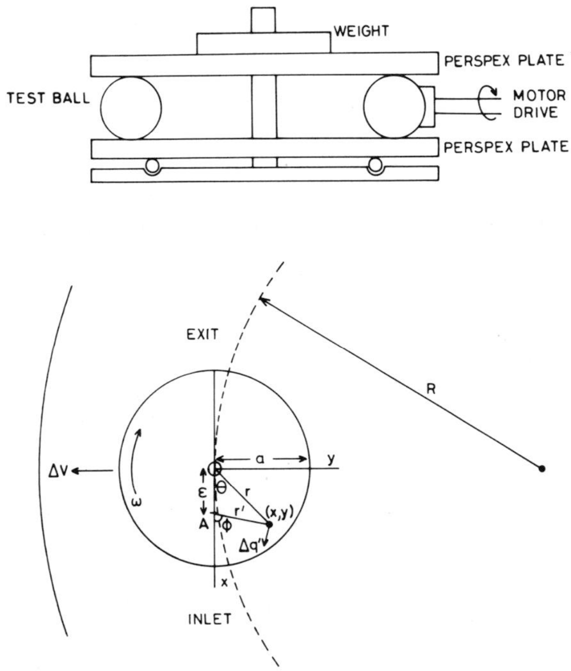

Fig. 9. Apparatus to study the influence of relative contact zone spin upon the motion of a rolling ball (upper diagram) and the kinematics of spin within the contact zone (lower diagram).

Unrestrained smooth rubber or steel balls rolled between perspex circular plates (as in Fig. 9) in dry contact were found to creep radially outwards in a manner consistent with earlier results (26). This result implies that the true centre of pivot A in the contact area is displaced towards the leading edge; that is, ϵ is positive. The forward displacement arises through the elastic distortion of the ball and plate in the contact region together with microslip within the rear part of the contact area. However, in lubricated contact with viscous oil, rubber balls were unexpectedly found to creep *inwards.* This behavior suggests a greater traction over the *back* half of the contact area. Tilt between the contact surfaces generated by the rolling motion may be the reason for this, because such tilt results in a greater viscous resistance to spin over the back half of the contact. If surfaces were contaminated with thin films of grease, the balls always crept out, though less strongly than in dry "clean" contact, and there was no rolling speed effect. These creep results are shown in terms of ball eccentricity at different rolling speeds in Fig. 10. Steel balls adequately lubricated with viscous oils tended to assume a neutral behavior ($\epsilon/a = o$) except at high rolling speed where the plates became starved by centrifugal loss. This led to "dry" conditions with outward creep. Lubricated rubber balls, crept inwards under nearly all circumstances. One exception was when distilled water was used. Presumably, in this case water films were too thin to prevent asperity contact between rubber and perspex so that an outward creep was always observed. The introduction of soap into water, however, minimized the influence of asperity contacts (see above) and increased the bulk viscosity with the result that the balls tended to move inwards. With viscous oils, rubber balls revealed negative speed dependent eccentricities as shown in Fig. 10. Furthermore, if the lubricating oil contained rubber (uncured polyisobutylene dissolved in Shell "Turbo 33") ϵ/a became even more negative.

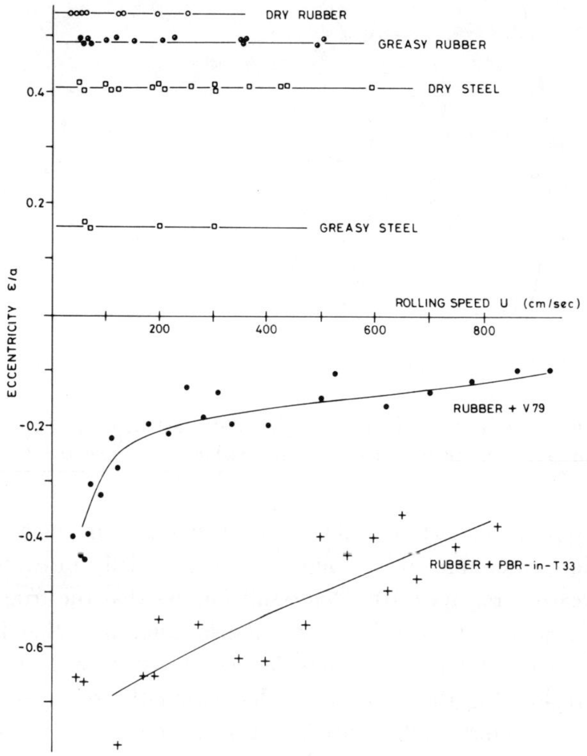

Fig. 10. Change in ball eccentricity with rolling speed for different states of lubrication.

	dry	greasy	Vitrea 79	PBR in T33
6 cm diameter rubber ball	o	◐	●	+
4-1/2 cm diameter steel ball	□	◪		

Oil specifications: Shell "Vitrea 79" has a viscosity of 900mPas (900cP) at room temperature and Shell "Turbo 33" a viscosity of 80mPas. Sufficient polyisobutylene rubber (PBR) was dissolved in the Turbo 33 to bring its viscosity up to 900m Pas. Surface roughness: perspex plates 0.02 μmCLA; metal balls 0.05 μmCLA and rubber balls 1.7 μmCLA. Film thicknesses (with above oils) at U = 100 cm/sec.

Metal balls: : 5.4 μm ($\bar{p}$ = 50 MNm^{-2})
Rubber Balls : 85 μm ($\bar{p}$ = 0.2 MNm^{-2})

$\bar{p}$ is mean contact pressure.

Calculations were made in an attempt to explain the negative eccentricities of the lubricated rubber balls. For a parallel film of lubricant, no uneven contact traction would be expected. The problem was, thus, to find a film shape and thickness at different rolling speeds, and then, by summing viscous tractive forces due to spin, to predict the resultant ball eccentricity. The first approach was essentially a "spinning" tilted pad analysis making use of results contained in an analytic solution (8) for the lubrication of low elastic modulus materials. Such an analysis assumes a square contact area with plane inclined contact surfaces spinning relative to each other. From it values of ϵ/a were predicted at different rolling speeds for typical experimental conditions. The results are shown together in Fig. 11 for comparison with experimental observations. The second way of predicting eccentricity was to make use of film profiles obtained from interferometric data such as that of Fig. 7. The photographs of Fig. 7 show interferograms at different sliding speeds. It is assumed that similar profiles would have been generated in rolling contact. From them viscous forces were summed. Suppose a contact is spinning with an angular velocity ω (see Fig. 9); then $\Delta q' = \omega r'$. The component perpendicular to the radial line r is $\omega r' \cos \phi$ which equals $\omega(x-\epsilon)$. Thus, the traction over the whole contact area is

$$T_y = \eta \int_A \frac{\omega(x-\epsilon)}{h} \cdot dA$$

where η is the film viscosity, and the film thickness, h, is a function of x and y. For the ball to spin free, $T_y = 0$; thus

$$\iint \frac{x}{h} \cdot dx\, dy = \epsilon \iint \frac{1}{h} \cdot dx\, dy.$$

Initially this integration was made only in two dimensions by considering the contact area to be a rectangle with a profile that followed the centre line. This was done graphically by plotting out the functions 1/h and x/h along the centre line, finding the area under each curve, and then dividing to obtain ϵ. The results of this line integration are shown in Fig. 11. There is still a difference between these line interferogram results and the observations, though the trend at higher rolling speeds appears more realistic than the tilted pad analysis. The line results, however, take no account of the side lobes in a point contact. These arguably might reinforce the inward creep because they are regions in the back-half of the contact area where the lubricant film is very thin and therefore the viscous tractive forces should be rather high. To find ϵ/a as a function of area, the integration was performed on a computer. Some results are shown in Fig. 11, but the difference is not great.

References pp. 193–194

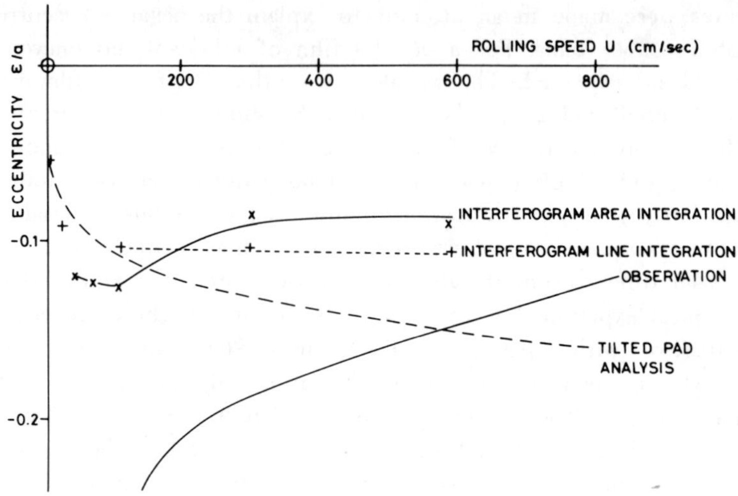

Fig. 11. Comparison of theory and experiment for eccentricity for the case of a 6 cm diameter rubber ball lubricated with the oil shell "Vitrea 79". Points marked + were derived from integration based on film thickness observations along a line through the centre of an interferogram; those marked X were obtained by integration over the entire contact area.

The comparison of predicted and directly measured ball eccentricity (see Fig. 11) shows that contact area tilt clearly contributes to inward creep but can not fully explain the observed results. The three dimensional analysis appears to more nearly follow the measured trend of less negative eccentricity with increasing rolling speed but still only accounts for about half the total observed eccentricity. The analysis, however, takes no account of inertial or rheological effects. Inertial effects caused by spin may distort the rubber surface perhaps making the side lobes asymmetrical. Rheological effects may arise in the lubricant or the contact surfaces. One other factor to be considered which may offer some explanation of these results is fluid starvation. Although theoretical film thickness values were calculated to be greater than the surface roughness of ball and plate combinations, actual values may be much less if the supply of lubricant is inadequate. Under conditions of starvation, solid contact would first occur in the nip at the back of the contact area, and high friction in this region would lead to a strong inward creep.

CONCLUSION

The work described in this paper concerns a series of lubrication studies of smooth rubber contacts. The main experimental technique employed was that of optical interferometry. By this technique it is possible to measure the thickness and study the flow properties of films sandwiched between smooth rubber and glass surfaces down to thicknesses of 10 nm (100 Å) or less. Squeeze film measurements show that films thinner than 40 nm tend to collapse unless they contain free ions in which case a stable film is formed which possesses a modest load-bearing capacity on account of the formation of electrically charged double layers. Under light loads and in the absence of stabilized films, the magnitude of surface attractive forces is comparable to the force pressing surfaces together with the result that the area of contact is greater than the Hertz equation would predict, and there is a measurable force of adhesion. When a smooth rubber sphere is slid tangentially in the presence of a viscous fluid, typical elastohydrodynamic lubrication conditons ensue. Such elastohydrodynamic results have been used to account, in part, for the creep motion of a lubricated rubber ball rolling with spin.

REFERENCES

1. *F. L. Roth, R. L. Driscoll and W. L. Holt, Journal of Research, National Bureau of Standards. 28, p. 439, 1942.*
2. *D. F. Denny, Wear 2, p. 264, 1959*
3. *S. C. Cohen, Ph.D. Thesis, Cambridge University, England, 1965.*
4. *D. Dowson and G. R. Higginson, 'Elastohydrodynamic Lubrication' Pergamon, Oxford, 1966.*
5. *D. Dowson and P. D. Swales, 3rd. International Conference on Fluid Sealing, Cambridge (Cranfield: BHRA), 1967.*
6. *K. Herrebrugh, ASME Journal of Lubricating Technology, 90, p. 262, 1968.*
7. *C. J. Hooke and J. P. O'Donoghue, Journal of Mechanical Engineering Science 14, p. 34, 1972.*
8. *K. P. Baglin and J. F. Archard, Institution of Mechanical Engineers, London, EHL Symposium, Leeds, Paper C3, 1972.*
9. *D. F. Moore, 'Friction and Lubrication of Elastomers' Pergamon, Oxford, 1972.*
10. *A. D. Roberts and D. Tabor, Proceedings of Royal Society, London, A325, P. 323, 1971.*
11. *A. I. Bailey and J. S. Courtney-Pratt, Proceedings of Royal Society, London, A227, p. 500, 1955.*
12. *M. T. Kirk, Nat., London, 194. p. 965, 1962.*
13. *A. Cameron and R. Gohar, Nat, London, 200, p. 458, 1963, Proceedings of Royal Society, London, A291, p. 250, 1966.*
14. *D. Dowson and D. A. Jones, Nat, London, 214, p. 947, 1967.*
15. *C. A. Foord, L. D. Wedeven, F. J. Westlake and A. Cameron, Proceedings of Institution of Mechanical Engineers, London 184 (Pt. 1, No. 28) p. 487, 1969-70.*
16. *R. W. Snidle and J. F. Archard, Institution of Mechanical Engineers, London, EHL Symposium, Leeds, Paper C2, 1972.*
17. *G. R. Paul and A. Cameron, Proceedings of Royal Society, London, A331, p. 171, 1972.*
18. *D. Lee, D. M. Sanborn and W. O. Winer, Journal of Lubricating Technology, Paper 73 – Lub – D, 1973.*
19. *H. Blok, and H. J. Koens, Proceedings of Institution of Mechanical Engineers, 180 (Pt. 3B), p. 221, 1965-66.*
20. *A. D. Roberts, Engineering Materials Design 12, p. 55, 1968.*

21. *A. D. Roberts, Journal of Physics D: Applied Physics 4, p. 423, 1971.*
22. *A. D. Roberts, J. Coll. Interface Sci. 41, p. 23, 1972.*
23. *K. J. Mysels and M. N. Jones, Discussion Faraday Society, 42, p. 42, 1966.*
24. *L. M. Barclay and R. H. Ottewill, Special Discussion Faraday Society, No. 1, Paper 12, 1970.*
25. *K. L. Johnson, K. Kendall and A. D. Roberts, Proceedings of Royal Society, London, A324, p. 301, 1971.*
26. *K. L. Johnson, Proceedings of Institution of Mechanical Engineers, London, 173, p. 795, 1959.*

DISCUSSION

Moore

Thank you very much Dr. Roberts. We have time for one or two short questions.

B. D. Gujrati

I just wanted to ask you what are the real conditions that exist in what you have called dry contacts?

Roberts

You are asking me if in the dry contact case it really was *dry* contact, or whether there was in fact a residual layer of water present?

Gujrati

Yes.

Roberts

We think that there may be even in dry contact, a few residual monolayers of water present, perhaps 10 or 15 angstroms total thickness, but it is very difficult to detect and measure such thin films.

Gujrati

This means that by changing the degree of cleanliness, you could increase the contact area further.

Roberts

It is possible that with cleaner surfaces the dry contact area might be larger.

Gujrati

You have not checked that, I believe.

Roberts

No.

Moore

Any other questions?

Thieme

Maybe it is of interest to tell that Dr. Savkoor of our laboratory made some tests and showed the influence of electric voltage of, say, about 2000 volts, and at very slow sliding speeds, say you have a nominal contact area of 6 $cm.^2$ and a normal load of 3kg. At very low speeds, if you just use normal SBR rubber, then normally you would expect to have a coefficient of friction at low sliding speeds of about 1, but if you put an electric D.C. or A.C. voltage across the contact, you will double the coefficient of friction force.

Roberts

If this is applying a voltage across the contact, it is not the same as the electrical double layer.

Thieme

No, it is not.

Roberts

It may be that you have opposite signs and not like signs on the surfaces, in which case they would attract.

Thieme

Your statement is correct.

Roberts

Now if you, in your case, have perhaps positive rubber and negative glass, they will attract and the coefficient of friction will go up.

Thieme

Yes, due to electrostatic forces, it seems you get a double coefficient of friction, which might be unexpected to occur, but this can be explained due to an increaed normal load.

Roberts

In fact, I carried out some experiments in which a positive soap was absorbed on the rubber surface, the glass remaining negative, with the result that surfaces attracted and friction was higher.

Moore

If I may make a comment, I agree with Prof. Thieme and I support your theory.

Schallamach

Was a gum or a black-loaded SBR compound used?

Thieme

Black loaded.

Schallamach

Moderate voltages (about 200 V) between sliders of black-loaded natural or polybutadien rubber and a smooth, dry track increase the frictional force materially. This effect has been attributed to break down of the carbon black structure in the rubber contacting the track (stress softening), producing a thin top layer of poorly conducting rubber with a large potential gradient across it. The ensuing electrostatic attraction increases the normal pressure. There appears to be no connection between this effect and the double layer effects described by Dr. Roberts.

Roberts

Yes, a different story all together.

Moore

Thank you very much Dr. Roberts.

AN ANALYSIS OF SOME FACTORS THAT INFLUENCE WET SKID RESISTANCE

B. D. GUJRATI and K. C. LUDEMA

University of Michigan, Ann Arbor, Michigan

ABSTRACT

An analysis is given which is expected to serve as the basis for the calculation of the skid resistance of tires on wet roads. The analysis is for a single sphere, sliding and rolling on water covered rubber. The sphere is taken to be a single scaled up protruberance in the road surface texture. Ignoring interaction effects each protruberance is seen to contribute to a total drag force on a sliding tire.

The analysis uses equations of elastohydrodynamics modified to accommodate the viscoelastic properties of rubber. The properties of the rubber are obtained from an identation test. The analysis correlates well with experiments using water on rubber except at low sliding speed. Experiments using fluids that wet rubber better than water does show good correlation over the entire sliding speed of the experiments. The conclusion is then reached that the skid resistance of tires at low speeds is strongly influenced by the poor wettability of water to rubber.

INTRODUCTION

For many years the tire industry has produced tires that will operate well on roads in existence. At the same time, the highway industry constructs roads and provides a certain level of skid resistance on these roads in order to minimize the number of high accident sites. Each activity has operated virtually independently and has developed its own technology.

In spite of all of the past efforts to achieve acceptable skid resistance, neither the tire industry nor the road building industry is able to improve their product in a rational manner from fundamental principles. There has been an accumulation of

References p. 209

considerable data, and there has been the development of many empirical functional models of tire behavior on wet roads, but almost no effort has gone into fundamental considerations in wet tire traction.

Much of the empirical work has been thoroughly done and the results are useful. Tires of many constructions, types, compositions, etc., have been made to operate on a wide variety of roads at a wide range of speeds, loads, angles of headings, inflation pressures, etc. (1). Laboratory studies have also been conducted with simulated tire-road systems. An analysis usually follows such tests with the intent of discovering cross correlations. One such cross correlation is the apparent finding that the damping loss property of the tire rubber interacts with the large texture features of the road surface, whereas the hardness property of the tire rubber interacts with the fine scale texture (2). On the other hand, from the empirical findings, it has not yet been possible to write an equation from which an *optimum* combination of tire rubber properties and road surface texture may be found. Further, it is not obvious that more empirical work will yield more fundamental information.

Whereas the empirical approach to the problem of wet tire traction has limited value, a purely analytical solution is not yet possible. This paper reports an intermediate approach, using analytical expressions, made tractable by results from simple experiments. The analysis begins from the principle that a tire sliding on a wet road is a case of lubricated sliding of rubber over a rough surface. The simplest beginning point in the analysis, and one that takes advantage of previously published analytical methods, is to assume that a road surface texture consists of an array of spheres. To go one step farther, an analysis is presented in the next section of this paper for a single sphere sliding on a flat surface of rubber, immersed in a fluid. This is called the Visco-Elastohydrodynamics (VEHD) analysis since rubber is a viscoelastic material, (i.e., Young's modulus and the internal damping loss are strain rate dependent).

Following the VEHD analysis is a practical application of the resulting equations and a comparison with laboratory tests and with real tire behavior. A discussion follows on the limitation of the VEHD analysis and on the work needed to improve the analysis.

VEHD ANALYSIS FOR A RIGID SPHERE SLIDING ON A WETTED RUBBER FLAT

The resistance to sliding of a rigid sphere on a flat rubber surface, immersed in fluid, has been traditionally thought to arise from two separate causes. They are, 1) the viscous drag in the fluid between the sphere and the rubber; and 2) an energy loss due to "cyclic" straining of the rubber as it is indented by the moving sphere. The latter cause can be the larger of the two when using a fluid which is a good lubricant and when using a rubber of high damping, or hysteresis loss properties.

It will be shown that the above two causes for sliding resistance are not independent. In particular, it will be shown that viscous drag in the fluid film is strongly influenced by the viscoelastic properties of the rubber substrate. The most direct manifestation of viscoelastic properties may be seen in the following way. When a sphere rolls (or slides with perfect lubrication) on a rubber surface at very low speeds, the contact area is large and circular, Fig. 1. There is little resistance to motion. At high speeds, the contact area is small due to an effective increase in rubber stiffness (elastic modulus) at high strain rate. Again, the contact area is circular, and there is little resistance to rolling. At intermediate rolling speeds where the elastic modulus is at an intermediate value and where the damping loss is maximum, there will be a high resistance to rolling. The contact area will be intermediate in size but not circular. It is found that ahead of the rolling axis of the sphere the contact is semicircular, but behind the rolling axis the contact is shortened. The reason is that the rubber does not recover as quickly as is necessary to remain in contact with the rising rear surface of the sphere.

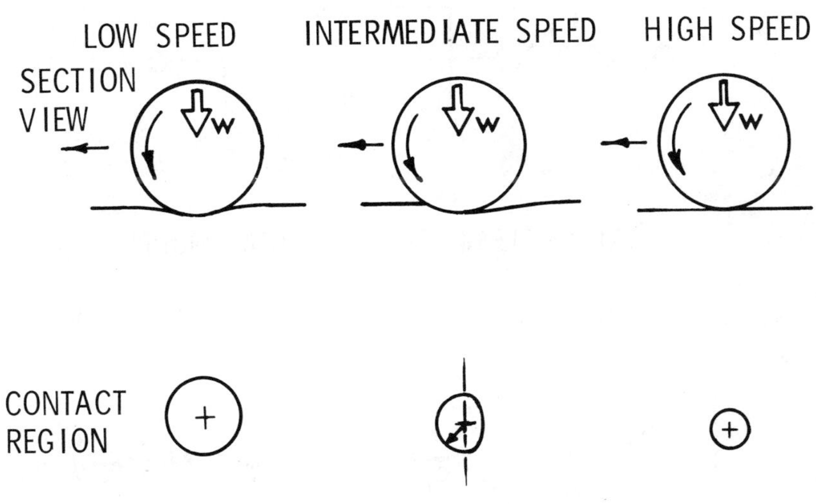

Fig. 1. Indentation and contact area between a sphere and viscoelastic subrate at three speeds of rolling; constant load.

Besides the visible asymmetry of contact, there is an asymmetry of contact pressure. One consequence of such asymmetry of contact is a moment of forces around the rotating axis of the sphere which acts to resist rolling of the sphere. Another consequence of asymmetry of contact area and pressure is that Hertzian conditions cannot be assumed in the equations to calculate viscous drag forces for *sliding* spheres.

References p. 209

It can be seen from the above discussion of contact mechanics that a vital beginning point for the determination of the two causes (viscous drag and damping loss) for sliding resistance is an expression for asymmetry of contact area and for contact pressure distribution. A straight-forward analytical method could take any of several different directions. For example, an expression could be derived for the extent of strain and the rate of strain at every point under a spherical slider. Then it would be necessary to obtain the viscoelastic properties of the rubber over the entire applicable range of strain rate and strain amplitude in the rubber. By the proper mathematical procedures, it would be possible to calculate quantities of interest such as contact area and pressure distribution. However, it would be an exceedingly complex and lengthy analysis.

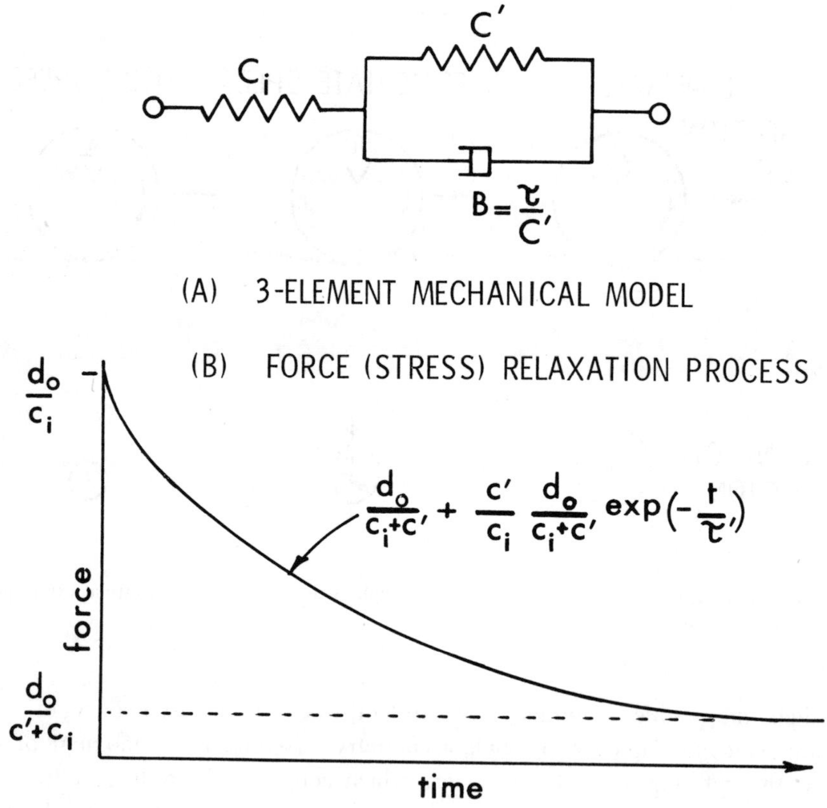

Fig. 2. (a) 3-element Mechanical Model Consisting of Two Springs and One Dashpot; (b) Force (Stress) Relaxation Behavior.

A new and less complex method is proposed below (3). In essence, it involves a transformation of equivalent strain conditions. In the new model, the strain at any point under an indenter or slider it taken to be composed of two parts, a reversibly elastic component and a viscous (internal friction) component. The elastic component acts instantaneously and is instantaneously reversible. On the other hand, the viscous component of strain increases with time. A relationship between the elastic term and the viscous term may be obtained from the 3-element mechanical model as shown in Fig. 2(a).

If the model is instantaneously deformed by d_o and held there for some time, the force that it transmits to its surroundings decreases with time as shown in Fig. 2(b). The characteristic equation of this behavior in terms of elastic compliance, C, is:

$$\frac{1}{C} = \frac{\text{Force}}{d_o} = \frac{1}{C_i + C'} + \frac{C' \exp(-t/\tau')}{C_i \; C_i + C'} \tag{1}$$

where C_i and C' are the compliances of the springs in the assumed model, and τ' is the relaxation time related to the damping rate, B, of the dash-pot such that

$$\tau' = \frac{BC'C_i}{C_i + C'} = \tau \frac{C_i}{C_i + C'} \, . \tag{2}$$

Now, if a sphere is made to indent a rubber half space, the load required to maintain that indentation diminishes with time in a manner similar to the behavior of the model in Fig. 2. The essential features of rubber behavior are similar to the behavior of the three element model; thus, the units for the compliances C_i and C' and the property τ may be used to designate rubber properties. (This does not imply that rubber is made of springs and dash-pots). However, these parameters must be regarded as time-dependent in the model because rubber behaves in this manner.

Given the response of rubber to a controlled indentation, the next step is to calculate the shape of contact during rolling or sliding of a sphere. It is assumed here that it is sufficient to define the asymmetry of contact in terms of a semi-circle of radius, a_f, under the front of the sphere, and a semi-ellipse with the minor half axis of a_r under the rear of the sphere. It is also to be noted that, in general, somewhat more than one half of the total vertical displacement of rubber occurs before the sphere contacts the rubber. Furthermore, the vertical displacement begins ahead of the sphere at a distance estimated (for mathematical convenience as well as to approximate reality) as $100a_f$. The importance of this event consists in the fact that this displacement ahead of the sphere proceeds at a rate of about 0.01 times that of the displacement rate directly under the front half of the sphere. Thus, the advance displacement contains a large creep component of strain and is a significant factor in determining the recovery displacement under the rear half of the sphere. The equations relating these factors are (3):

$$(\text{define}),\ m = 1 - \frac{a_r}{a_f}$$

$$= \frac{C'\,[1-\exp(-\frac{100a_f}{U\tau})]\,\exp(-\frac{100a_f}{U\tau})}{C_i + C'\,[1-\exp(-\frac{100a_f}{U\tau})]} \qquad (3)$$

where U is the relative traveling speed of the slider on the rubber.

It can now be assumed that the contact pressure under the front half of the sphere is Hertzian and that under the rear half is elliptic as shown in Fig. 3. Thus, both the contact area and contact stress distribution are known. By the area moment calculation about the rolling axis of the sphere, a rolling resistance, F_R, due to damping loss alone can be calculated:

$$f_R = \frac{F_R}{W} = \frac{3a_f}{8\,R}\,m \qquad (4)$$

where W is the total load acting on the sphere.

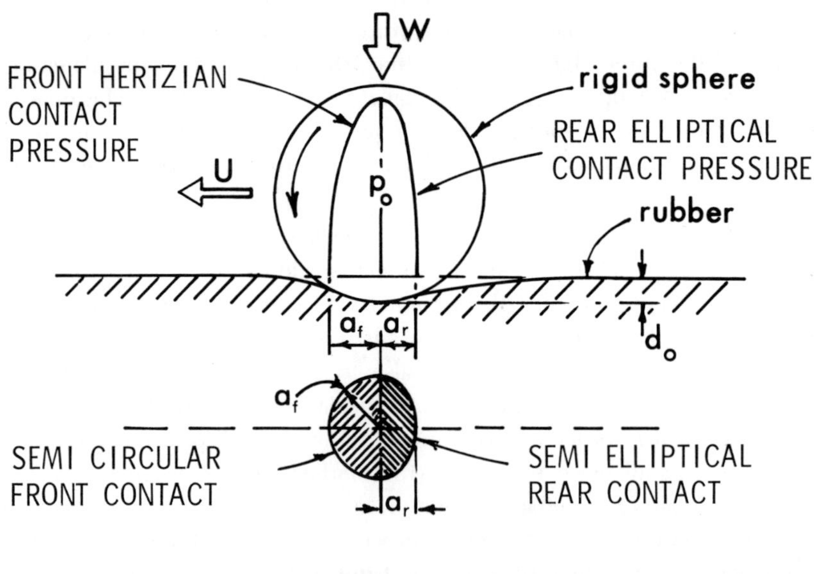

Fig. 3. Contact shape and pressure under a sphere rolling/sliding on rubber.

The equation for calculating the viscous drag in the fluid film may now be developed. This is done by taking the equations for fluid film dimensions from elastohydrodynamics. An equation for a cylinder sliding on a reversibly elastic and soft medium is taken from Hooke and O'Donoghue (4) and altered to correspond to the case of a sphere. The contact flow process under a sphere is regarded as similar to that under an equivalent size cylinder of length $1.7a_f$. The two dimensional flow condition under a sphere is accounted for by considering a leakage factor derived from a classical hydrodynamic analysis (5). The film thickness expression under the sphere is

$$\frac{h_o}{R} = 1.865 \left(\frac{\eta U}{E'R}\right)^{0.6} \left(\frac{W}{E'R^2}\right)^{-0.133} \tag{5}$$

where h_o is the average contact film thickness, η is the absolute viscosity of the lubricant, E′ is the equivalent Young's modulus, and R is the radius of the slider.

In Eqn. 5, Hertzian contact was assumed, and a particular contact pressure distribution then follows from a particular load W. The effect of the revised contact pressure distribution due to viscoelasticity is to produce a different pressure maximum for a particular load. Since the fluid film under a sphere is fixed by the inlet conditions and by the conditions under the front half of the sphere, it is convenient to reflect the asymmetrical conditions simply by replacing W by $\frac{2}{2-m}$ W in Eqn. 5. The new equation for fluid film thickness is

$$\frac{h_o}{R} = 1.70(2-m)^{0.133} \left(\frac{\eta U}{E'R}\right)^{0.6} \left(\frac{W}{E'R^2}\right)^{-0.133} \tag{6}$$

A connection must now be made between, m, E′, W, a_f and C to reflect viscoelastic behavior (in Hertzian form):

$$W = \frac{1}{k}\frac{(a_f)^2}{RC} = \frac{d_o}{C}$$

$$E' = \frac{3}{C\, a_f\, (2-m)} \tag{7}$$

The factor k accounts for the fact that more than half of the total surface deformation d_o occurs in the contact region in the case of rubber. Experimental observations give a value of k as 1.8. For elastic contacts, k is one.

References p. 209

Using Eqn. 6 and the contact area geometry, the viscous drag force F_v may be calculated:

$$f_s = \frac{F_v}{W} = 1.548(2-m)^{0.4}\left(\frac{af}{R}\right)^2\left(\frac{\eta UCaf}{R}\right)^{0.4}\left(\frac{WCaf}{R^2}\right)^{-0.867} \tag{8}$$

Eqns. 4 and 8 express the magnitude of two causes for sliding resistance. It cannot be assured that these two causes are the only causes of sliding resistance where a fluid film exists (persists). Nor has there been proof that Eqns. 4 and 8 are independent, i.e., free of cross coupling. But, to a first approximation, it may be stated that total sliding resistance is the simple summation of Eqns. 4 and 8:

$$f = \frac{3}{8}\frac{af}{R}m + 1.548(2-m)^{0.4}\left(\frac{af}{R}\right)^2\left(\frac{\eta UCaf}{R}\right)^{0.4}\left(\frac{WCaf}{R^2}\right)^{-0.867} \tag{9}$$

CORRELATION OF ANALYSIS WITH EXPERIMENT

The Sliding Experiment – The analysis in the previous section is compared with experiments of the following description. A machine was constructed to hold a flat sheet of rubber in a tank mounted on two transducers. The rubber sheet was 3/4 inch thick and 14 inches square. Above the rubber is a bar 11-1/2 inches long on the ends of which are sliders or rollers of various shapes. The bar is rotated about the midpoint in such a way as to constrain the sliders or rollers to move along the rubber surface in a circular path. One of the transducers under the tank measures the normal force of the sliders/rollers against the rubber, and the other transducer measures the torque due to sliding or rolling resistance. Fluid temperatures from 1°C to 70°C may be held in the tank to any desired depth up to 1 inch. The rotational speed of the bar may be varied over a range to provide a sliding/rolling speed range from 1 fpm to 4000 fpm (0.01 to 45 mph).

The speed range for the experiment appears unrealistically low as compared with tire sliding speeds, toward which the above analysis purports to be directed. However, at moderate vehicle speeds, the actual sliding speed of the rubber against the road at low braking slip is very low. For the purpose of this correlation, the maximum speed used was 1000 fpm. A qualitative extension to higher speeds can be more readily made by lowering the test temperature than by increasing the machine speed. For example, lowering the test temperature from 70°C to 1°C is the equivalent of raising the test speed by a factor of 300 or more, using the well known principles of viscoelastic transforms.

Measurement of Rubber Properties – The two terms in Eqn. 9 may be plotted after appropriate values are substituted. The values of C_i, C' and τ must first be found. These were found by a simple indentation test. A ball of 3/4 inch diameter is made to indent, in a very short time, to a particular depth into a slab of rubber. The variation of indentation *force* with time is measured and plotted as shown in Fig. 4 for polybutadiene rubber. Note that the sphere cannot be indented in infinitesimal time. The decay curve is extrapolated to zero equivalent relaxation time which differs from the zero position of the loading time as shown in Fig. 4. The parameters C_i, C' and τ are different for each sliding speed, U. Each sliding speed is related to a particular loading time in the indentation test by, $t_d = \frac{100a_f}{U}$. Data for each sliding speed may then be taken from an identation test of the corresponding indentation rate, or from a "master" data curve by way of a mathematical conversion. For the example given here, for U = 1000 fpm; $C_i = 0.0021$ in./lb., $C' = 0.0026$ in./lb., and $\tau = 0.22$ sec.

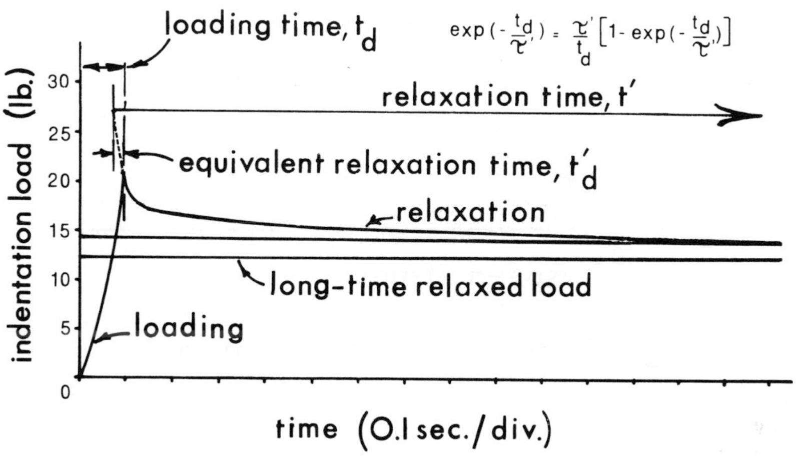

Fig. 4. Indentation force for a ball indented .07 inch onto a rubber sheet (poly-Butadiene) at 70°F.

The Correlation – With the above data, the solution to the second part of Eqn. 9 is shown plotted as curve 1 in Fig. 5. The damping loss part of Eqn. 9 is shown plotted as curve 2 in Fig. 5. The sum of the two curves is shown as curve 3.

The experimental results using sliding spheres of 3/4" diameter are shown as curve 1 in Fig. 6. It may be seen that there is a considerable discrepancy between theory and experiment. Curve 1 in Fig. 6 is attractive from at least one practical point of view, however. If one begins with tire performance data, it is possible to hypothesize the *friction* behavior required to obtain such performance. Such calculations have been done, and the result is plotted as curve 2 in Fig. 6.

Incidentally, the tire performance referred to above is that in straight line braking. The first and simplest observation is that in locked wheel braking, the braking force increases as the sliding speed decreases. A second observation is that, for a constant

References p. 209

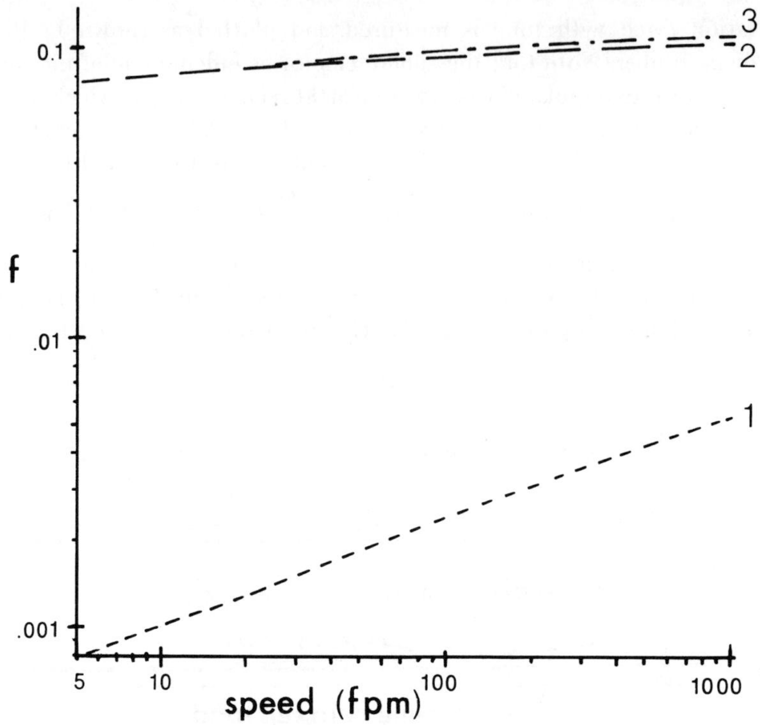

Fig. 5. Plot of separate and combined terms in Eqn. 11.

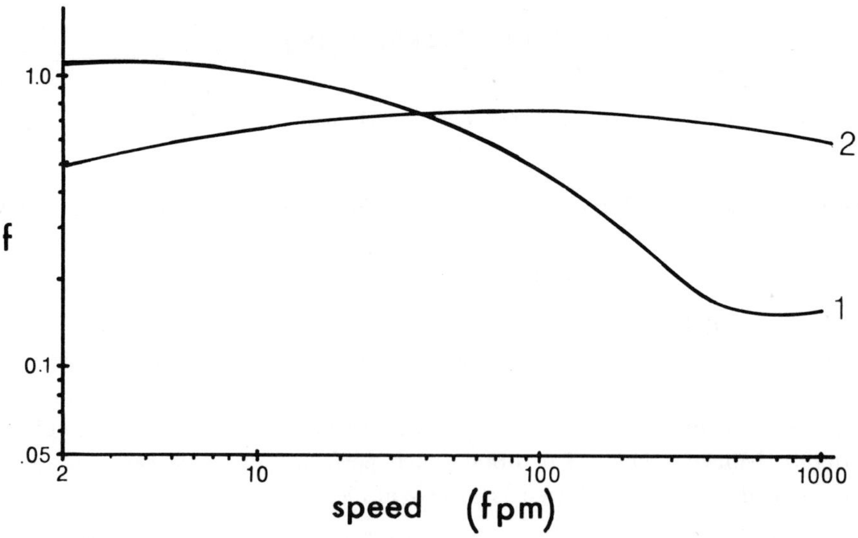

Fig. 6. Experimental results for wet rubber friction. Curve 1, a sphere sliding on poly-butadiene; curve 2, calculated from tire tests.

vehicle speed, the braking force diminishes as a rolling tire decelerates toward the locked up position. The latter behavior encourages the locking up of wheels upon braking with the result that directional control may be lost.

The discrepancy between curve 3 in Fig. 5 and the curves in Fig. 6 requires some explanation. It is usually safe to conclude that if a mathematical model does not predict real behavior, then some assumption in the mathematical model has been invalid. In this case there are two possibilities, namely, either the damping loss properties are not valid, or, the assumption of perfect wettability in the VEHD calculation is not valid. The latter is not possible to evaluate analytically, so an experiment was conducted which will be described below.

The former problem, i.e., inappropriate damping loss properties, will be examined first. The simplest way to verify the damping loss factor as expressed in Eqn. 4, is by way of an experimental measurement of rolling loss. In the machine described above, the rotating bar was fitted with two spherical rollers. The data for a test covering the speed range of 1 fpm to 1000 fpm, for a depth of penetration of 0.07 inch is shown as curve 1 in Fig. 7. It is seen that this curve approximates curve 2 in Fig. 5 which is the plot of Eqn. 4. There is a small difference in magnitude of sliding resistance but the slopes are comparable. To a first approximation, therefore, the difference in slopes between Fig. 5 and Fig. 6 is not due to the rolling loss term in Eqn. 9.

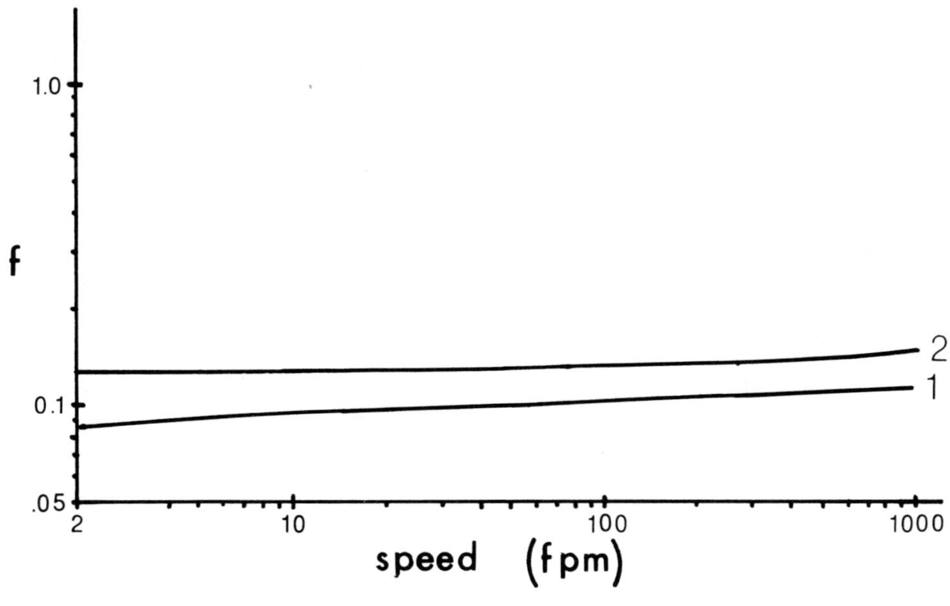

Fig. 7. Experimental data using poly-butadiene; curve 2, sliding test using isopropyl alcohol.

References p. 209

The assumption of perfect wettability in the viscous drag term in Eqn. 9 is less valid. Water does not wet rubber. A manner of expressing this state of affairs is by way of contact angle. The contact angle between water and rubber exceeds 90°. Unfortunately, it is not yet possible to use such information in VEHD equations. On the other hand, to test the importance of wettability, experiments were done with isopropylalcohol in the tank of the test machine rather than water. The data for one such test is shown as curve 2 in Fig. 7. Here the data agrees with theory.

DISCUSSION

Quantitatively, the analysis presented in this paper is found by experiment to be realistic, particularly where perfect wettability exists in the experiment. Where a discrepancy in magnitude exists between analysis and experiment it could be due to other causes not included in the analysis. For example, the surface finish of the materials used in the experiment was not mathematically smooth, as is assumed in viscous drag calculations. Further, the data used in the damping loss calculation is from a test which begins with fully relaxed rubber. On the other hand, in the test where the slider-roller follows a circular path, the rubber is repeatedly strained. The latter more nearly approximates the experience of tire rubber sliding on a road surface.

The discrepancy between the analysis and data from tire performance tests is caused by several additional factors. Road surfaces are, in essence, a large array of protruberances (micro-bumps) of complex contour. The analysis was done for a single smooth sphere. Furthermore, the tire operates on a thin film of water, whereas the analysis assumes a great quantity of fluid.

The subject of wettability brings to mind an early model of tire behavior (6). It is often stated that there are three zones of friction between a tire and the road. The front portion of the contact patch is thought to be a region of a thick water film. In the second portion, it is usually assumed that a water film exists but is too thin to prevent some indentation of the tire rubber by micro-bumps on the road surface. In the third region, there is considered to be dry contact. This point has not been proven experimentally, and it is difficult in principle to see how water can be completely squeezed out from between contacting bodies in a finite time. It now appears that something approaching dryness may exist on the high points of the road surface texture, but this dryness is due to surface forces that are connected with the phenomenon of wetting. Several authors have alluded to the role of wettability before, but usually intuitively (7) or by attempts to correlate surface tension of the liquid with friction results (8). Evidently variables or factors that control surface forces must be considered in future work.

In addition to the new analysis presented, some practical conclusions may be reached from this work. It appears that wettability of water to rubber is the most important factor in controlling the traction of tires on wet roads at lower speeds. From this, one may conclude that if tires were made more water wettable than they are, due to wear or other means of rubber degradation, the braking behavior will be adversely affected. Conversely, if the micro-bumps on road surfaces were non-water-wettable, wet traction might be still more enhanced, up to some limit. Finally, the role of gasoline and other organic solvents spilled onto road surfaces may be predicted without resorting to tests.

The above analysis left out several factors that are important in real tire performance. On the other hand, it includes more factors than are usually considered.

ACKNOWLEDGEMENT

This work was made possible through the financial support of the National Science Foundation and Uniroyal Incorporated, and by a contribution of well characterized rubber from Uniroyal Incorporated, Firestone Tire and Rubber Company, and Dow Corning Company.

REFERENCES

1. *K. C. Ludema and B. D. Gujrati, "An Analysis of the Literature on Tire-Road Skid Resistance," American Society for Testing and Materials, STP No. 541, 1973.*
2. *(a) E. N. Bevilacqua and E. P. Percarpio, "Lubricated Friction of Rubber," Rubber Chemistry and Technology, Vol. 41, 1968, p. 832; (b) B. E. Sabey and G. N. Lupton, "Friction on Wet Surfaces of Tire-Tread Vulcanizates," Rubber Chemistry and Technology, Vol. 32, p. 878, 1964.*
3. *B. D. Gujrati, Ph.D. Thesis at The University of Michigan, 1974. (Unpublished)*
4. *C. J. Hooke and O'Donoghue, "Elastohydrodynamic Lubrication of Soft, Highly Deformed Contacts," The Journal of Mechanical Engineering Science, Vol. 14, p. 34, 1972.*
5. *J. F. Archard and E. W. Cowking, "Elastohydrodynamic Lubrication at Point Contacts," Proceedings, Institution of Mechanical Engineers, Vol. 180, Pt. 3B, p. 47, 1965-66.*
6. *V. E. Gough and D. W. Badger, "Tyres and Road Safety," World Highways, Vol. 17, p. 47, 1966.*
7. *(a) W. F. Kern, "Coefficient of Wet Friction of Tire Treads," Rubber Chemistry and Technology, Vol. 40, p. 982, 1967; (b) H. W. Kummer, "Lubricated Friction of Rubber: Discussion," Rubber Chemistry and Technology, Vol. 41, p. 895, 1968.*
8. *E. Clamroth and W. Heidemann, "Measurement of Rolling and Sliding Friction Coefficients of Tread Compounds," Rubber Chemistry and Technology, Vol. 41, 1968, p. 908.*

DISCUSSION

Moore

Thank you very much Professor Ludema. Any questions?

A. D. Roberts

Does your theory predict a rise in friction at very slow sliding speeds?

Ludema

Yes.

Roberts

This is something like the Streibeck effect?

Ludema

Perhaps you can go into your point just a bit more.

Roberts

I am surprised that with isopropyl alcohol you found a positive gradient the same as you predicted. I found a negative gradient, but we have used different sliding speeds. Therefore, my question, does your theory show an increase in friction at very low sliding speeds? To compare data, it might be better to plot the parameter velocity-viscosity divided by mean contact pressure. This is the Sommerfeld number. I see on your slide friction is only plotted against velocity.

Ludema

In any analysis, there is no change in sense of slope.

Moore

Any other questions?

D. B. A. Phillips

You started off at the beginning of your talk with pictures of rolling spheres and the pressure distribution underneath them. You, without my noticing it, went on to sliding spheres. I cannot work out what happened to the rolling spheres.

Ludema

When you proceed to do some experiments to verify our theory it is useful to separate the two major factors. In order to minimize the viscous drag component and thereby to enhance the damping loss component the rolling sphere is a good approximation. There is little relative motion between the sphere and the rubber and therefore little viscous drag although there will be a small amount. This method of separating the damping loss factor from the viscous drag factor is not totally accurate;

in fact, even in the analysis there are effects which we could not account for. When you slide a sphere instead of rolling it, you have a traction stress on the contact region as well as a normal stress. That traction stress alters the strain field in the rubber which changes the stress distribution. Because of time constraints I did not say all of that in the lecture, however it is all in the analysis in the paper. In summary then, the rolling experiment turned out to be a convenient way of being able to measure the damping loss component in the total friction force.

B. D. Gujrati

I would like to add to Dr. Ludema's statement. In answer to your question, essentially the basic assumption used, in relation to rolling contact and sliding contact is the presence of a fluid film. In sliding, film viscous – drag as well as rubber deformation loss are involved; but with the fluid film present, the shear stresses are so small that the deformation loss part, in the case of sliding, is essentially the same as that in rolling. That is why with reference to the deformation loss, rolling or sliding ball has been used in this paper. Does that answer your question?

Phillips

Yes.

Moore

I would also like to discuss this further, but our time is up. We now formally close the first part of this session which has been devoted to invited lectures from selected experts. The second part which now begins is a panel discussion on the general theme of fundamentals.

PANEL DISCUSSION
"FUNDAMENTAL ASPECTS OF RUBBER FRICTION"

K. A. Grosch
K. C. Ludema
D. F. Moore (Moderator)
A. D. Roberts
A. Schallamach

Moore

Well, so far we have talked about smooth spheres and smooth surfaces and the talks covered different areas. What we try to do now is unify them in some way and provide a discussion which may provoke some arguments, but try to get some consensus. I think we will start in the same order as the speakers. Starting with Prof. Grosch, on your experiments on ice; as we go to the conditions for reproducibility with ice friction, your surfaces are smooth, extremely smooth, etc., and you depart from the realistic situation which we must eventually face, the tire-road situation. I would like Dr. Grosch to comment first on the conditions for this reproducibility. There is some, I would not say controversy, but there is not quite an agreement on what kind of conditions experimentally one should be sure of in order to get this reproducibility.

Grosch

Yes, I think there was some astonishment in the audience when we reported our large friction coefficients which are unusual in tire skid experiments, in particular, that one can reach friction coefficients on ice exceeding three. These coefficients can be achieved only, first of all, if the viscoelastic conditions of the rubber are right and second if the ice is absolutely free of contamination. We achieved this in fact, really, by polishing the ice surface in front of the sliding rubber test piece at all times, since even the smallest contaminations will drastically reduce the friction coefficient.

So in this sense, the experiments are academic because in real life, as in the tire skid experiment, there will always be contamination between the rubber and the ice surface.

The real point we wanted to make is that, if you have contact between ice and rubber, then the friction is adhesive in nature. It is no different to the friction that is obtained on clean glass and no different to the friction between rubber and highly polished steel, for instance. Very near the freezing point, the ice can no longer be considered to be solid in the real sense of the word. The cohesive forces are very small, and the adhesive forces between ice and rubber suffice for the ice to melt. This is not the pressure melting which occurs, for instance in skating. The very fact of bringing rubber close to the ice suffices for the ice to disintegrate, providing perfect lubrication. Around 0°C large friction coefficients can, therefore, never be achieved. Below about −5°C proper friction sets in. There will be rubbers that cannot achieve high friction even at −5°C because at that temperature their viscoelastic condition is not conducive to produce large adhesional forces.

I would like to pick up the point of adhesion and dry contact on lubricated surfaces. We obviously are going to have some slight differences of opinion. First of all, I was very pleased to notice from Dr. Roberts' experiments that in the presence of soap the friction decreased and no real contact could exist between rubber and track. When clean water was the lubricant regions of perfect contact and regions of pockets of water were obtained. This seems to me a direct confirmation of the experiment I tried to show on a rough track saturated with water, where we demonstrated the existance of adhesion exactly in the same way as on a dry track and completely obliterated it when we added a small proportion of wetting agent to it. In the experiments with a sphere sliding on a lubricated rubber track, too, the rising part of the master curve (Fig. 9 upper right in the printed version) coincides exactly with the rising part of the master curve obtained on dry glass. Now one can argue about the word "dry" but such a close correspondence of the absolute friction values on dry and lubricated surfaces suggests to me that we have rubber adhesion forces. Whether there is still a thin layer of water, is then really not very important any more.

With increasing lubrication film thickness at higher sliding speeds, shear forces decrease, and the friction decreases. But at low sliding speeds, there is something which in its physical behaviour is identical to dry adhesional friction.

Roberts

May I comment on Dr. Grosch's remarks. We tried to measure the thickness of the water film in "dry" contact. You saw in my pictures a very black contact area with pockets of trapped water. Experiments showed that the pockets of trapped water in time escaped, suggesting that, in fact, the rubber was not in absolute contact with the glass. It is very difficult to measure films thinner than 50 angstroms, but we would make a guess at 30 angstroms for the separation between rubber and glass under these conditons. Of course, when the water film is so thin, it may, by surface force ordering acquire rigidity so that when sheared it doesn't look so unlike dry contact.

Experiments carried out on smooth rubber spheres showed that the coefficient of friction for a dry sphere which had never been in contact with water was about 10. In the presence of water, the friction was still about 8. So the water lowered the friction a little, but not very much.

Moore

I think the distinction between dry and wet contact is artificial to some extent. There is no cut off point. There is a transition, especially when we are dealing with rubbers. Now in metal-on-metal contact it may be a more clean cut distinction, but I think for example with a lubricant, its chain length is comparable to that of rubber. Thus, for water-lubricated rubber contacts, there must exist a gradual transistion from dry to wet contact.

Grosch

Just one more remark. Whether one can speak of rubber adhesion on lubricated tracks is to me defined experimentally, when viscoelastic effects of lubricated friction can be measured in the same way as they can on dry smooth clean glass surfaces. Perhaps this might be a suitable definition for adhesion.

Ludema

There are several problems involved in this topic that Karl just mentioned. There is a viscoelastic response *in* the solid which polymer chemists and others define in a certain way. There is a viscoelastic response in a liquid which is traditionally a viscous substance but in thin films may act as an elastic membrane. Thus, there can be a viscoelastic behavior in a lubricant film as well as in the rubber substrate. I have not seen a total analysis of this problem. However, I do support the Grosch approach using $a_T V$, in spite of the fact that some people do not quite believe in it. It is a transform, after all! I support it mostly because I have measured the friction of polymers over a fairly wide range, and I always see those viscoelastic transforms. They are always there; it is real. That is, one can take data for the lubricated case and the unlubricated case, and transform the data by the so-called Williams-Landel-Ferry transformation.

Moore

Could we move on now to the next subject? We want to cover quite a number of topics, and we now refer to Dr. Schallamach's talk on wave detachment.

Schallamach

In following up my talk, I should like to mention in passing another manifestation of waves of detachment which first brought them to my notice. During experiments with a transparent rubber slider on a transparent hard track, tiny flashes of light were seen to travel rapidly across the contact area.

I should, furthermore, add to my description of the phenomena that waves of detachment are best observed with very clean surfaces. This finding may be connected with the results of early experiments which had shown that one pass of a rubber slider over a glass track greatly reduced the friction force in subsequent passes. The indications were that sliding rubber leaves behind a film with lubricating properties which modifies the results. The question may be legitimately asked as to what is actually measured in ordinary friction experiments, particularly when a circular track is used continously.

Moore

I think the reproducibility that you get with smooth surfaces is generally accepted to be much lower than in the case of rough surfaces, since there is some film in there that is difficult to remove for one reason or another. If I can go back to the comments that were made earlier to the effect that one could model all road protruberances by spheres. I maintain they are not spherical, in fact some of them are pointed and there are differences in behavior. If you get very sharp asperities, there is a tendency to penetrate the films which you may have been talking about, Dr. Roberts. And you almost get down to Dr. Schallamach's needle experiments to point out an extreme situation. This is equally important as the experiments on spheres by Prof. Ludema.

Schallamach

Even soft rubber is scored by a wet Carborundum track. Also, needle-scratching experiments with lubricated rubber produced cuts instead of leaving the interrupted traces found on dry rubber. It certainly looks as if sharp asperities can penetrate a water film.

Ludema

That something was high viscous drag in a very thin film of water on top of those very small spheres.

Moore

Are there any questions from the audience.

C. E. Bleil

It seems to me that the thin films you are considering are about 30 angstroms thick; for ordinary water this is only about 10 times the space charge region that one would find in the adjacent solid, and it is many times smaller than the ordinary Debye double layer that you would have if the water extended to a considerable thickness. If that is the case, then the charge that would be collected in this very thin layer should, indeed, give you a very rigid and very thin film of water on that surface. So, it sort of supports Dr. Roberts' argument. I would like to ask a question of Dr. Schallamach. It seems that you talked about the light that was generated in your first experiment. Was that light because of light scattering or was that because of tribo-electric phenomenon due to adhesion?

Schallamach

I think the light flashes mentioned by me were just reflections from the flanks of the waves.

Hays

The question of the flashes of light has been raised, and I'm wondering what surface temperatures (thinking in terms of Dr. Blok's surface temperature phenomenon) one might have when you rub some material against rubber, and if perhaps those very thin layers might be at appreciable temperatures. I should address this question to somebody doing thin film work, perhaps Ken Ludema.

Ludema

I would defer to Alan. Have you ever calculated the temperature rise in those thin films?

Roberts

No.

Ludema

I have not either.

Moore

Any other questions?

E. Kontos

I have a question for Dr. Roberts. Did you attempt to change in any way the surface polarity of the two phases that you brought in contact, the glass or the rubber, by having different species being absorbed? You mentioned something like an ionic soap, and I will appreciate it if you can explain further. What happened?

Roberts

Yes, I did this very experiment using a soap which gave a positively adsorbed charge on the rubber. The soap was tetradecyltrimethyl ammonium bromide (TTAB). Its positive ions adsorb quite well on rubber, but only slightly on glass. One is left with the situation of positive rubber, negative glass and they snap together.

Gujrati

I wish to direct my question to Dr. Schallamach with reference to his experiments between a slider and a rubber surface where he observes these moving folds. As Dr. Grosch and others have said, that is surely due to some adhesion phenomenon. Now if there is adhesion at these folds between the slider and the rubber surface, have you observed whether these cycles of adhesion, and then band breaking occur in the rubber substrate or at the interface between the rubber surface and the countersurface. Is there any evidence about this?

Schallamach

Rubber and the other frictional member appeared to separate in the interface, and not in the substratum.

Gujrati

Well, that indicates that there is no adhesion process as we normally think.

I want to make a comment about the original discussion on the transition from lubricated to dry contact conditions. From many of the discussions, it seems to me that the transition occurs smoothly from a lubricated condition to dry contact, if you may call it, due to the surface forces which come into play suddenly and become dominant under conditions of thin films. Here we have a balance between the hydrodynamic fluid-flow forces created and the surface forces, and at small speed conditions, these forces try to counter each other. Thus, there is a transition due to completely different phenomena which is introduced by the surface forces. I believe that this may be considered to be a part of the mechanism which leads to the transition from lubricated to dry contact conditions; if we still can call it "dry." Thank you.

Moore

We will now pass on to another subject that we could discuss briefly, and that is friction as an energy dissipation mechanism, rather than adhesion and hysteresis. There seems to be some controversy as to whether adhesion and hysteresis originate from the same source, or whether they are in fact separate. There are two schools of thought. Dr. Schallamach has pioneered molecular stick-slip theory of adhesional friction which is well accepted, but hysteresis does not require a similar theory to explain its nature. There are those who believe that both mechanisms may be due to the same cause since both are proportional to the tangent modulus, and therefore, both viscoelastic. It is not necessary, however, to attribute both adhesion and hysteresis to the same mechanism. I would like to open this topic for discussion. Would you care to follow up on that?

Schallamach

Before an adhesional bond between rubber and a solid can be broken, the rubber in the vicinity of the bond must be strained. There are therefore two sources of energy loss, and hence of friction: the adhesional energy and the energy expended in straining the rubber. The latter component depends on the viscoelastic properties of the rubber and, as such, on the forces between segments of the polymeric molecules. The adhesional energy, however, must be equated with surface energy and has no obvious relation to viscoelastic properties. Its effect on friction is to limit the maximum strain, and so the energy dissipation in the rubber around the bond.

Grosch

Hysteresis is the commonly accepted word for the energy loss in a stress-strain cycle which occurs in a viscoelastic material. Since rubber adhesion has a viscoelastic component, it is bound to have associations with hysteresis. In both cases, energy is lost as Dr. Schallamach said, as rubber molecules are moved past each other, or at least disturbed.

Moore

May I just make a brief statement? Viscoelastic mechanisms are a result of a molecular activation mechanism. I am trying to trace the hysteresis mechanism back farther beyond its physical visualization, to a more fundamental origin.

Grosch

Prof. Ludema threw some doubts on the transformation technique of friction but I still believe it is a very powerful tool, not only to get practical results but also to throw some light on the basic mechanism of friction. Look at the simple master curve.

For a gum rubber on a smooth surface it appears that it has a single distinct maximum. If several rubbers of different glass transition temperatures are examined, it appears that the shape of the master curve is similar for all, only the position on the $a_T v$ axis differs. A comparison with the position of the maximum of the loss modulus on the a_T- frequency axis shows that the $a_T v_{(max)}$ and the $a_T f_{max}$ values are related by a length which has the same magnitude for all rubbers, about 60 Å. It has been termed jump distance in the literature. Its exact physical meaning may be obscure; it, nevertheless, shows that the adhesional friction is molecular in nature; the complexes which are affected are of the size of about 60 Å.

If the master curve is obtained on rough carborundum, it shows in addition a second maximum at $a_T v$ values which relate, again by a length, to the position of the maximum loss factor on the a_T- frequency axis. In this case the length is about 10^{-2} cm and is again the same for all rubbers. It is of the same dimensions as the coarseness of the track. In fact, the maximum position can be shifted by using carborundum of different coarseness. The fact that it is associated with the loss factor provides further strong evidence that this type of friction, termed deformation friction, is a hysteretic loss mechanism, i.e. energy is lost in stress – strain cycles around the asperities of the track.

Moore

If I may just make a final comment. The hysteresis process, I think, must be due to some molecular mechanism in more fundamental terms, and in this sense definitely would have a common origin.

SESSION III

TIRE TRACTION
THE ROLE OF THE TIRE

Session Chairman
S. K. CLARK

University of Michigan
Ann Arbor, Michigan

INTRODUCTORY REMARKS (S. K. Clark)

Mr. Hays and Dr. Browne, ladies and gentlemen, it is a distinct pleasure to be here today. For my part, I would like to express my congratulations to the participants of yesterday's sessions for the most informative and entertaining meetings, and I can only hope that our efforts today will measure up to the standards they set. I stand before you in a little bit of awe. In thinking over my introductions to the speakers for this session, and in thinking over these brief remarks, it struck me just how much we cannot yet do analytically in the field of mechanics as applied to tires. The goal of mechanics in tire technology is, given a handbook of material properties, a mold design drawing, and post-cure inflation and drum set specifications, to be able to calculate all the performance characteristics of the tire, including its wear rate and tractive characteristics on a given surface. Needless to say, we have a little way to go to achieve that goal. Nevertheless, the material properties of the constituents can be measured readily. We have available rather elegant shell theories, constructed, it is true, primarily for shells of fixed geometry, but nevertheless, elegant. We understand the mechanics of orthotropic or anisotropic elastic bodies. Yet truly adequate means of calculation for the simplest tire properties still elude us, and we must fall back on rather crude models involving strings, beams, and springs. But part of the trouble lies in the fact that contained in the problem of analytical representation of a tire are many difficulties which have been dealt with separately in mechanics but almost never altogether. Some of these are: highly anisotropic, cord-reinforced materials; non-linear elasticity in some components; appreciable hysteresis effects; large deformations of shells; frictional forces not describable by simple frictional laws; and permanent set and heat contraction of materials. For the moment we must content ourselves with small pieces of the overall problem. We wait expectantly for attempts to adopt finite element techniques to the calculation of tire deformations, yet even here, we sense limits. These are based on the knowledge that not all of the physics of friction is clearly formulated, so that the interaction of shear forces, normal forces, and tire deformation is still not analytically clear. In spite of all the difficulties and the failures, progress has been and is being made. Today we have many excellent young engineers and scientists engaged in applying mechanics to tire technology. Many of these are in the audience today. We welcome them and we hope that they share with us the excitement of working on a truly difficult problem in mechanics which has great practical application both in terms of safety and economics. Today we are fortunate to have four papers in this session which summarize and in some cases, condense some of the properties of the tire itself as it pertains to traction and point out some explanations for observed phenomena.

TREAD COMPOUND EFFECTS IN TIRE TRACTION

R. F. PETERSON, JR., C. F. ECKERT, C. I. CARR

UNIROYAL Research Center, Middlebury, Connecticut

ABSTRACT

Changes in practical compounds have given appreciable improvements in friction on wet surfaces but only minor improvements on dry or icy surfaces. The wet skid resistance of a tread compound is determined primarily by its hardness and hysteresis. Improvements in skid resistance are usually made with a concurrent, predictable, loss in wear resistance for practical compounds. Tire materials research devotes its efforts to development of factors which will permit gains in traction with minimal losses in wear. A typical tire compound is examined for the effects of the ingredients on improvements in wet traction, i.e., raising the glass transition temperature of the polymer or the extender oil, increasing the fineness of the carbon black, and lowering the level of curatives.

INTRODUCTION

The tread compound must meet a number of different tire requirements. It must have good skid resistance on dry, wet, water-covered, snow-covered and ice-covered roads in addition to providing high wear resistance, a comfortable ride, and good durability. The various kinds of traction have been covered in detail in other portions of this symposium and will be covered again from the point of view of the rubber chemist. Our basic question is: what changes can be made in the tread recipe which will improve the traction of the tire?

Because of the large drop in friction coefficients which occurs on wet roads, and the resultant rise in the probability of a skid, we have directed most of our efforts to a study of the factors affecting wet friction. Accordingly, most of this paper will be

References pp. 237–238

concerned with tread compound effects on wet traction; however, we will also briefly discuss friction on dry and icy roads.

Even the most drastic changes in the tread recipe will not provide large improvements in skid resistance. Table 1 shows the relatively minor role of the tread compound, as estimated by Allbert and Walker, in determining tire friction on wet roads.(1)

TABLE 1

Factors Influencing Wet Friction

Factor	Variability due to that factor
Road:	
Road surface characteristics	Up to 5:1
Water depth (range 0.05 inches – 0.3 inches)	Up to 3:1
Vehicle:	
Speed (range 30-80 mph)	Up to 10:1
Braking system (Locked-wheel vs. antilock braking)	Up to 3:1
Tire:	
Tread pattern design	Up to 4:1
New tire vs. smooth tire	Up to 8:1
Changes in compound	Up to 1.8:1

Table 1 might be taken to indicate that the choice of tread compound is of only minor importance. This, of course, is not correct. The average driver is not aware of road surface characteristics, nor is he likely to slow down by more than a few miles per hour on a wet road or even a water-covered road. He is likely to drive on well-worn tires. Antilock braking devices are not yet in common use. Of all the factors in Table 1, the tire manufacturer can alter only the tread compound and the tread pattern.

PHYSICAL PROPERTIES AFFECTING RUBBER FRICTION

Most of our laboratory work on wet friction has been carried out using the British Pendulum tester, or portable skid tester (2). Results obtained on 8 different stocks using this instrument have been shown to correlate well with stopping-distance measurements (3). Fig. 1 shows the correlation between 20 mph stopping-distance rating and portable skid tester readings on a slippery road surface.

Kummer and Meyer have found that skid resistance can be separated into four major components (4) as shown in Fig. 2. These four components are deformation, adhesion, wear and tearing. We must ask: what physical properties of the rubber can alter the magnitude and relative contributions of these types of friction?

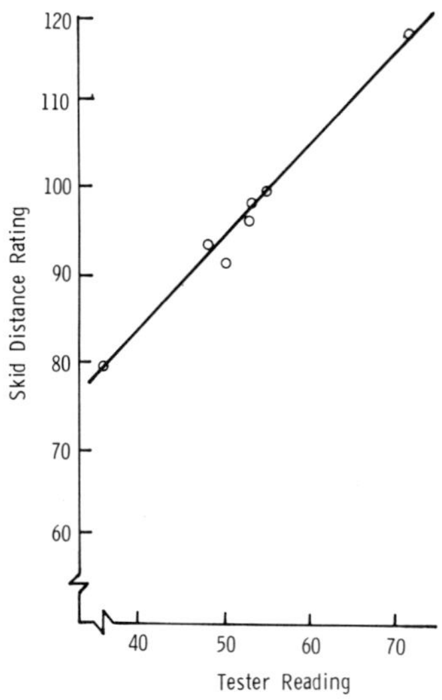

Fig. 1. Agreement between portable skid tester reading and stopping distance test on wet, slippery asphalt road.

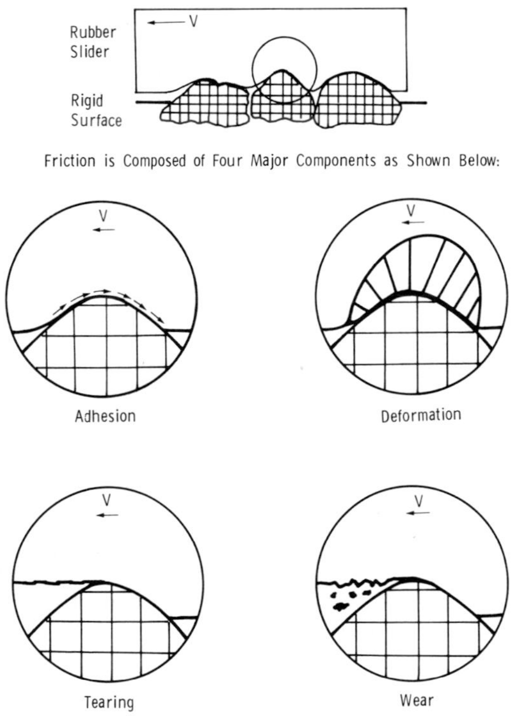

Fig. 2. Skid resistance separated into four major components (adapted from Kummer).

References pp. 237–238

Wet skid resistance on our standard slippery road, a smooth asphalt road, can be accurately predicted by the equation:

$$\text{W.S.R.} = 71.3 - 0.40\ R_{32} - 0.18\ D \quad r^2 = 0.93$$

where R_{32} indicates Bashore resilience (ASTM D-1054) at 32°F, D indicates hardness (Shore A durometer), and r^2 is the correlation coefficient.

Other roads have regression equations of the same general form. For actual stopping-distance measurements at our test facilities in Texas, again on a smooth asphalt road, the following equation was derived:

$$\text{Skid Distance (ft.)} = 91.6 + 0.61\ R_{32} + 0.77\ D \quad r^2 = 0.99$$

In Table 2 we see the success of this equation in predicting stopping distance.

TABLE 2

Observed and Calculated Stopping Distance
Wet Asphalt Road

Rubber	Rebound, 32°F	Hardness	Stopping Distance, (ft.) Calculated	Stopping Distance, (ft.) Observed
BR	62	62	177	180
NR/BR	46	57	162	163
EPDM	32	63	160	160
SBR/BR	33	55	155	158
SBR	34	52	152	151
Butyl	9	40	132	130

Stopping distance seems to be determined by the hysteresis and modulus of the tread stock. Any change in the composition of the tread stock which will raise hysteresis, or lower hardness, will decrease stopping distance.

This regression equation indicates that the major component of wet skid resistance is deformation, i.e., deformation of the rubber by large-scale asperities. A softer rubber will be deformed more by a given asperity, and a high-hystersis rubber will be capable of absorbing a greater percentage of the energy produced in such deformations. To reiterate, softness and high hysteresis are desired qualities.

The association of low hardness with good wet skid resistance may be true only for stopping-distance tests. In cornering tests, low hardness, or low resistance to deformation, will mean that a tread element may easily buckle and lose contact with the road, giving lower cornering skid resistance. Similarly, the tire designer may wish the tread rubber to be as hard as possible so that the tread pattern is not squeezed shut or otherwise deformed by slip, thus hindering water drainage.

Our regression equations usually show the best fit to measured skid resistance when the temperature of the rebound test is slightly below the temperature of the skid test. This is consistent with time-temperature principles (5): the frequency of the rebound test is below the average frequency developed in passing over asperities. A 10° downward shift in temperature is approximately equivalent to a tenfold increase in frequency. Applying this rule, we find that the principal frequencies produced by large-scale deformations are in the range 100-1000 Hz, as measured by the amount of shift necessary for best fit. The same result is also obtained by actual measurements of asperity size.

We have also found, using the portable skid tester, a component present in both wet and dry friction resulting from the deformation of rubber by very small asperities; that is, produced by hysteresis at very high frequencies. Higher high-frequency hysteresis leads to better skid resistance. This component may be the "adhesion" component of friction reported by other workers (4, 6).

High coefficient roads also develop tearing or wear, sometimes termed abrasive friction. This component appears to be associated with the difficulty of rupturing a sample. Higher tensile strength, or tear strength, will generally lead to higher abrasive friction.

THE TREAD RECIPE

A typical tread recipe is given below:

Polymer	100
Extender oil	40
Carbon black	70
Zinc oxide	3
Stearic acid	2
Antioxidants	1.5
Accelerators	1.5
Sulfur	2

We will examine the effect that varying these components has on skid resistance.

If skid resistance were the only quality required of a tire, it would be relatively easy to make tires with substantially better skid resistance. Unfortunately, a tread compound requires a compromise in many properties. The most difficult qualities to obtain simultaneously are high skid resistance and high wear resistance, since high wear resistance is usually associated with low hysteresis (7).

Effect of Polymer on Skid Resistance – The dynamic properties of the tire compound i.e., modulus and hysteresis, at service temperatures are determined, to a large extent, by the difference between the service temperature and the glass transition temperature (T_g) of the polymer, the temperature at which the polymer is changing most rapidly from a glassy to an elastomeric state.

References pp. 237-238

In general, the glass transition temperature increases as the percentage of atoms in side chains increases. Both polyethylene and polybutadiene, which have no atoms in side chains, have a very low T_g, below −100°C. Table 3 shows the effect of side chains on glass transition temperature.

TABLE 3

Effect of Side Chains on Glass Transition Temperature

Polymer	Abbreviation	Side Chain/ Backbone	T_g	Repeating Unit
Polybutadiene	BR	0	−105	
Polyisoprene (Natural Rubber)	NR	0.25	− 70	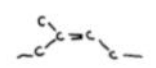
Ethylene-Propylenediene Copolymer	EPDM	0.3	− 56	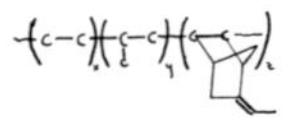
Styrene-Butadiene Copolymer	SBR	0.3	− 55	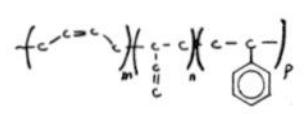
1, 2 Polybutadiene	1, 2 BR	1.0	− 11	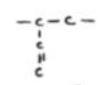
Polyisobutylene	Butyl	1.0	− 70*	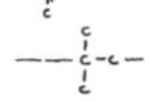
Polystyrene	—	3.0	+100	

At ambient temperatures, hysteresis is higher for elastomers of higher T_g, as shown by Fig. 3. Tread recipes made using SBR as the polymer will have substantially higher skid resistance than recipes using polybutadiene.

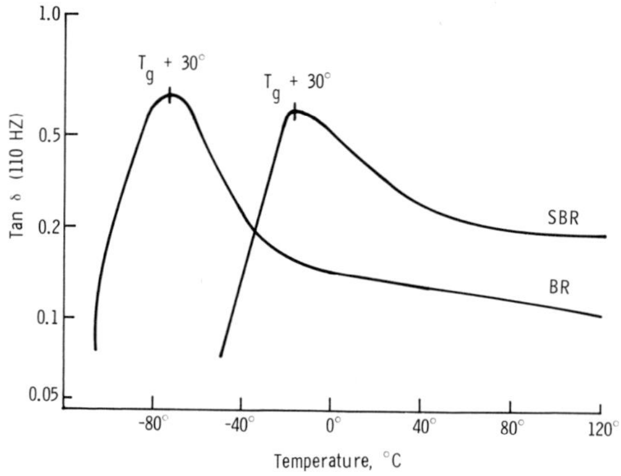

Fig. 3. Hysteresis (Tan δ) vs. Temperature for BR(T_g = −105°) and SBR(T_g = −56°) tread stocks.

Since high hysteresis is associated with a high glass transition temperature, one might try to obtain very high skid resistance through use of a polymer whose T_g is near tire operating temperatures. Such a material would actually have a very low skid resistance. As a polymer nears its glass transition temperature, it becomes harder and more brittle. High hardness, as we have stated, will hurt wet skid resistance. In actual practice, elastomers with a T_g above –40°C are not used in tires because of potential low-temperature stiffness.

One way of adjusting the glass transition temperature is to raise the styrene content of a butadiene-styrene copolymer. Table 4 shows the effect which these changes will produce on wear and wet skid. (8)

TABLE 4

Effect of Styrene Content on T_g, Abrasion Resistance and Skid Resistance

Polymer	% Styrene	T_g	Rebound, 0°C	Wet Skid, Concrete	Abrasion Resistance
BR	0	–105	42	76	240
6:4 SBR:BR	14	– 76	26	94	122
SBR 1500	23.5	– 56	12	100	100
Solution SBR	25	– 50	10	102	97
SBR 1516	40	– 30	6	111	76

As the percentage of styrene is increased and the T_g is increased, wear resistance will go down and skid resistance will go up.

For styrene-butadiene copolymers, the T_g is also dependent on the microstructure; i.e., the relative fraction of cis, trans and vinyl groups, in the butadiene. In Fig. 4, we have shown wear resistance and skid resistance as a function of T_g for a number of diene polymers differing in cis, trans, vinyl and styrene content (7). In this study, we found the following correlation coefficients:

Wear resistance with rebound: 0.92
Skid resistance with rebound: –0.99

Butyl rubber is an exception to the rule relating T_g to the proportion of atoms in side chains. This material has a T_g equal to that of natural rubber, but has much higher hysteresis and hence, good skid resistance. Skid resistance and wear resistance of butyl are about equivalent to those of a high styrene SBR.

Blends of butyl rubber and SBR are not cure compatible, and are deficient in important physical properties. When the polymer is lightly chlorinated, blends of this new material with SBR give properties suitable for testing in tread compounds, and a range of hysteresis can be obtained. Table 5 shows wear and skid resistance for blends of chlorobutyl with SBR (9).

References pp. 237–238

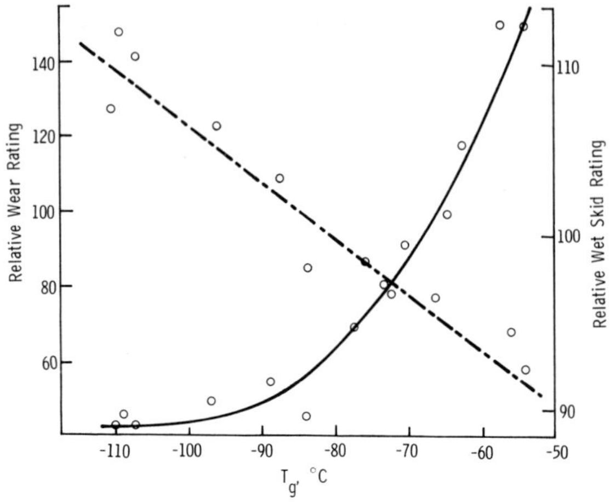

Fig. 4. Dependence of skid resistance (solid line) and wear resistance (broken line) on polymer T_g.

TABLE 5

Hysteresis, Wear Resistance, and Skid Resistance for Chlorobutyl/SBR Blends

% Chlorobutyl	Pendulum Rebound, 73°F	Wet Skid, 20 MPH, Dorset Pebble*	Wear Rating*
20	46	96	105
30	42	104	96
45	37	108	88
60	34	120	85

*Skid and wear ratings are given relative to an SBR/BR control and not to an all-SBR control.

Hysteresis, wear resistance and skid resistance show the same relations found in styrene-butadiene copolymers.

To summarize: the choice of polymer largely determines the hysteresis and glass transition temperature of the tread compound, which in turn determines skid and wear resistance. High hysteresis gives better skid resistance and lower wear resistance.

Extender Oil – Dynamic properties of tire tread stocks are not determined solely by the T_g of the polymer; they can also be affected by the amount and type of extender oil used. The oils usually added as extenders are of low molecular weight compared to that of the polymers and are, therefore, not cured into the 3-dimensional network formed when the compound is cured. This has two consequences important for wet skid resistance. First, the network is not as tight as it would be if oil were not present. Oil-containing compounds are, therefore, softer than non-oil-containing compounds. Second, the extender may act as an energy sink since energy may be transferred out of the polymer network into the oil and, hence, lost from the network. Oil-containing compounds will, therefore, have higher hysteresis than non-oil-containing compounds. Table 6 shows these effects in a standard SBR polymer at constant black level (10).

TABLE 6

Effect of Oil on Some Physical Properties

Amount of Oil	0	50
Shore A Hardness	78	57
Rebound, 32°F	27	12
Tensile Strength	3640	2850

More carbon black is usually added to oil-polymer mixtures in order to keep tensile strength and hardness at acceptable levels. Table 7 shows the effect of oil level on physical properties of a cis-BR polymer as black level is increased along with oil content (10).

TABLE 7

Effect of Simultaneous Changes in Oil and Black Level on Physical Properties

Recipe:			
Polymer	100	100	100
Black	50	70	90
Oil	0	40	80
Properties:			
Rebound	60	36	23
Hardness	71	67	49
Tensile Strength	2160	2630	2040
Wet Skid	100	110	127

The increase in hysteresis and lowering of hardness produces better wet skid resistance. Physical properties are not substantially lowered by adding 40-50 parts of oil, but will decrease rapidly for oil amounts above this level. Optimum wear-skid balance is, thus, obtained at about 40-50 parts of oil and 70-75 parts of black.

The choice of extender oil can also significantly affect wet skid resistance. The glass transition temperature of the final compound is determined not only by the polymer T_g, but also by the T_g of the extender oil. The use of oils of high T_g will

References pp. 237–238

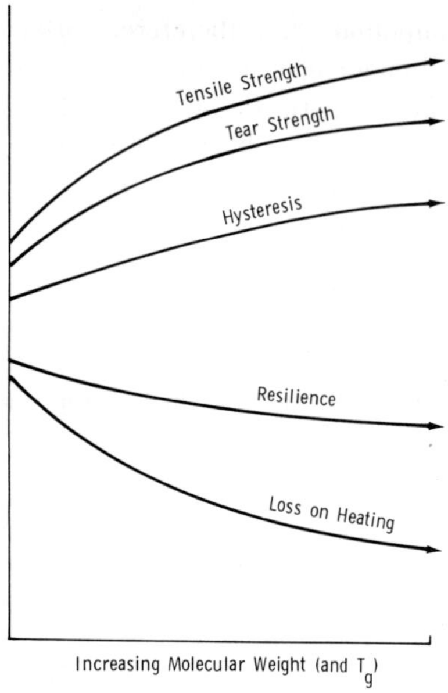

Fig. 5. Effect of extender molecular weight on some physical properties.

raise skid resistance by raising hysteresis. Hardness and tensile strength will also be raised. Fig. 5 shows schematically the effects which can be expected from using higher molecular weight extenders, which generally have higher T_g (11).

Table 8 shows actual tire test results in which the oil type was varied in an SBR/BR blend (12).

TABLE 8

Effect of Oil T_g on Wear and Wet Skid Resistance

Oil	Sundex 790*	Dutrex 985+	Dutrex 998+
Oil T_g	–52	–33	–2
Compound T_g	–73	–66	–56
Rebound, 32°F	27	22	17
Hardness	55	56	57
Tensile Strength	2600	2650	2450
Wet Skid Resistance	100	104	108
Wear Resistance	100	86	77

*Sun Oil Co.
+Shell Oil Co.

The usual inverse relation between wear resistance and wet skid resistance is found as hysteresis is increased.

Carbon Black – Carbon black is a reinforcing filler which, when added to either natural or synthetic rubber, greatly increases abrasion resistance and also raises hysteresis. For this reason, the proper choice of carbon black level and type can simultaneously raise both skid resistance and wear resistance.

Table 9 shows the effect of black loading in SBR on hysteresis (as measured by loss tangent, tan δ) and other properties which might affect wet skid resistance (10, 13, 14). As black loading is increased, hysteresis is also increased, raising wet skid resistance. Abrasion resistance is improved with higher black loadings but reaches a maximum near 50 phr. An increase above this level will continue to raise hysteresis and hardness, but will no longer raise tensile strength. The best balance of wear resistance and wet skid resistance is achieved near this optimum black loading, which is well-known.

TABLE 9

Effect of Black Level on Some Physical Properties

Parts Black	Tan δ	Hardness	Tensile Strength	Abrasion Resistance
0	0.10	40	220	–
10	0.15	43	500	10
20	0.20	48	1200	30
30	0.23	60	2000	50
50	0.30	66	3360	100
75	0.40	78	3200	85

Some improvements can be achieved through changes in black type. Carbon blacks are classified by their average particle size (of the carbon black unit rather than of the aggregrate) and by structure level. For carbon blacks composed of the same-size unit, the degree of aggregation, i.e., the structure, is higher when the units are connected together into long chains (15).

Table 10 summarizes the effect of black fineness at constant structure level in a SBR/BR recipe containing 70 parts of black and 40 parts of oil (10). As the data show, use of a finer black produces a compound with higher tensile strength, hardness and hysteresis.

Table 11 shows the effects of black structure at constant particle size in the same recipe (10). The use of a higher structure black will increase hardness, leaving tensile strength and hysteresis unchanged. The greater hardness (resistance to deformation) might be expected to help wear resistance, and is known to hurt wet skid resistance.

References pp. 237–238

TABLE 10

Effect of Black Fineness

Black	Average Particle Size, Å	Rebound, 32°F	Hardness	Tensile
N 110	100	20	56	3140
N 220*	200	23	54	2860
N 330*	300	23	53	2680
N 440	400	23	52	2450
N 550	500	30	50	2190
N 660	600	36	46	1610

*Average particle size of blacks used in tread recipes.

TABLE 11

Effects of Black Structure

Black	Particle Size	Structure	Rebound, 32°F	Hardness	Tensile
N 219	200 A	Low	25	49	2790
N 220	200 A	Normal	23	54	2860
N 242	200 A	High	24	59	2810

Use of a small particle, high structure black would, therefore, be expected to have the following qualitative effects on wet skid resistance and wear:

Higher hysteresis — higher skid resistance, lower wear resistance
Higher hardness — lower skid resistance, higher wear resistance
Higher tensile strength — higher skid resistance, higher wear resistance

The net effect of these changes was determined by a tire test as shown in Table 12 (16).

TABLE 12

Particle Size	Structure	Rebound	Hardness	Wet Skid	Wear Resistance
200 A	Normal	23	57	100	100
100 A	High	24	59	99	111
<100 A	High	20	62	99	120
<100 A	High	19	57	101	92

A substantial improvement in wear resistance was achieved at equivalent skid resistance through use of a finer, high structure black. This wear resistance could then be traded for improved skid resistance at equivalent wear by raising the polymer T_g. At wear = 100 relative to control, skid resistance of 106-108 could be expected.

The high wear resistance shown in Table 12 was achieved only through the use of special mixing techniques and long mixing times since the finer blacks are much more difficult to disperse into rubber.

The last line of Table 12 shows results obtained using normal mixing techniques and an extremely fine black. Hardness has not increased, showing that the reinforcing capabilities of the black were not being fully utilized, and some black remained undispersed. An actual loss in wear resistance resulted.

In summary, the changes in skid resistance which can be achieved through proper selection of black are small, but with proper processing conditions, these changes can be achieved without loss of wear resistance. For this reason, changes in black structure and fineness are being studied in detail.

Other Ingredients – The other ingredients in a tire tread compound are either antioxidants, which do not affect skid resistance, or are substances connected in some way with the vulcanization reaction. These latter materials will be considered as a group. Hardness and hysteresis can be altered considerably by changing the level of curatives. As the cure network is tightened, hardness will increase, and hysteresis will decrease, since the tighter network leaves fewer opportunities for energy to transfer out of the network. Tensile strength goes through a maximum with increasing curative level.

Table 13 shows these effects quantitatively (10, 17) for a SBR/BR blend. Optimum sulfur level in this recipe is about 2 parts. At this level, tensile strength and wear resistance are at a maximum. Decreasing sulfur level will, indeed, lower hardness and raise hysteresis, resulting in increased skid resistance. However, the loss in wear resistance outweighs the gains in skid resistance. This approach does not appear promising.

TABLE 13

Effect of Sulfur Level on Wear and Wet Skid

Amount of Sulfur	0.5	1	2	4	8
Tensile Strength	1070	2530	2770	2300	2160
Hardness	40	48	53	63	69
Rebound, 32°F	31	32	34	36	38
Wet Skid, P.S.T.	109	104.5	100	96	93
Wear Resistance	$\ll$80	93	100	80	$\ll$80

Material Properties Affecting Dry Skid Resistance – Dry skid resistance has become increasingly important in recent years as the Federal Government has taken more of an interest in minimum stopping distance. We will, therefore, touch briefly on the material properties which may affect dry skid resistance. In our experience, even drastic changes in the tread recipe have produced at most a 15% range (from worst to best) in dry skid stopping distance.

This small range in dry skid resistance may be due to the very high temperature developed at or near the surface of the tire. A temperature rise of 500°C or more has been calculated (18, 19) which is probably sufficient to cause thermal decomposition of rubber. At this temperature, there are few if any differences between tread stocks. Changes in black content or type which lead to higher tensile strength or hysteresis do appear to produce slightly higher dry friction coefficients. Changes in polymer T_g or oil T_g, which affect wet skid resistance, do not have a significant effect on dry skid coefficients.

Material Properties Affecting Friction on Ice – Although friction on wet roads is, we believe, fairly well understood, the mechanisms producing friction on ice-covered roads are less well established. Published work (20, 21, 22) has concentrated on "dry" ice; that is, ice at temperature of –5°C or lower, where there is no water or dirt layer on top of the surface. On such a surface, friction coefficients can be quite high and can approach values found on dry roads.

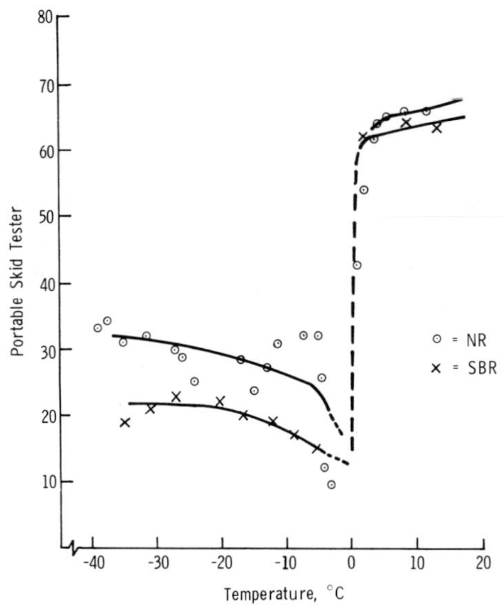

Fig. 6. Skid resistance (portable skid tester) on icy road and wet road with oil-extended natural rubber and oil-extended SBR.

Near the freezing point, however, the surface becomes wet and very slippery. Friction coefficients on wet ice are often 0.2 or lower, but are very difficult to measure reproducibly.

Some advantages can be gained on wet ice by using tread stocks composed of oil-extended natural rubber. Large asperities are masked in ice-covered surfaces, leaving only the very small asperities present in the ice itself. These small asperities produce very high-frequency deformations in the rubber. Oil-extended natural rubber has a T_g around −70°C and a maximum in hysteresis (for the frequencies involved in skidding on wet ice) near 0°C. This material, therefore, shows somewhat higher traction than other rubbers with either higher or lower T_g.

Fig. 6 shows portable skid tester coefficients vs. temperature for two tread stocks: oil-extended natural rubber and oil-extended SBR (22). Friction is lowest just below 0°C, and coefficients around 0.2 are found. Well below 0°C coefficients are higher, and the natural rubber stock is superior.

SUMMARY

In summary, we have identified the major properties affecting wet friction as hysteresis and hardness, and discussed changes in the tread recipe which would raise hysteresis, or lower hardness, thus, raising traction. Wear resistance, however, is usually lowered at the same time that skid resistance is enhanced. Some of these changes are: raising the glass transition temperature of the polymer or the extender oil, increasing the fineness of the carbon black, and lowering the level of curatives.

REFERENCES

1. *S. J. Allbert and J. C. Walker, rubber Chemistry and Technology, 41, p. 753, 1968; and Proceedings Institute of Mechanical Engineers (Automobile Division) 180, 2A, p. 105, 1965-66.*
2. *ASTM Test E303-66 T.*
3. *C. G. Giles, B. E. Sabey and K. H. F. Cardew, Rubber Chemistry and Technology, 38, p. 840, 1965.*
4. *H. W. Kummer, Rubber Chemistry and Technology, 41, p. 895, 1968.*
5. *E. P. Percarpio and E. M. Bevilacqua, Rubber Chemistry and Technology, 41, p. 870, 1968, and succeeding papers.*
6. *E. Clamroth and W. Heidemann, Rubber Chemistry and Technology, 41, p. 908, 1968.*
7. *R. N. Kienle, E. S. Dizon, T. J. Brett, and C. F. Eckert, Rubber Chemistry and Technology, 44. p. 996, 1971.*
8. *K. H. Nordsiek, Kautschuk und Gummi, 25, p. 87, 1972.*
9. *R. C. Keller, Tire Science and Technology, 1, p. 190, 1973.*
10. *R. Peterson, Jr., Uniroyal unpublished work.*
11. *F. M. MacMillan, in M. Morton, ed., "Introduction to Rubber Technology," p. 162, Reinhold, 1959.*
12. *C. F. Eckert, Uniroyal unpublished work.*

13. A. R. Payne, in G. Kraus, "Reinforcement of Elastomers," Interscience, p. 112, 1965.
14. B. B. Boonstra and E. M. Dannenberg, Rubber Chemistry and Technology, 22, p. 774, 1956.
15. S. G. Main and R. J. Kirschbaum, Presentation to the Southern Ohio Rubber Group, Dayton, O., April 6, 1973.
16. H. E. Haxo, Uniroyal unpublished work.
17. C. E. Scott and V. M. Chirico, Paper #18, Division of Rubber Chemistry, 94th Meeting, September 10-13, 1968.
18. R. R. Hegemon and J. J. Henry, Wear, 24, p. 361 1973.
19. W. Viehmann, Rubber Chemistry and Technology, 31, p. 925, 1958.
20. W. Gnorich and K. A. Grosch, Journal I.R.I., p. 192, 1972.
21. E. Southern and R. W. Walker, Nature Physical Science, 237, p. 142, 1972.
22. K. A. Grosch, Rubber Age, 99, p. 63, 1967.

DISCUSSION

Clark

Thank you very much Dr. Peterson. I am sure Dr. Peterson would be happy to answer questions.

W. B. Straub

You alluded to a standard surface. What is that surface?

Peterson

As you know, there is no single "standard" surface. The regression equation in the text is for one of our test surfaces at Laredo, a slippery asphalt road surface containing substantial large-scale roughness and not too much small-scale roughness. This surface has a similar regression equation to many other slippery road surfaces.

F. R. Davis

How were the glass transition temperatures determined in your data? Are they static or dynamic?

Peterson

We usually determine glass transition temperature by DTA (differential thermal analysis). We have also found excellent agreement between T_g by DTA and loss tangent peaks determined using the Vibron DDV-II. Highest loss tangent is usually found about 30° above T_g by DTA.

Grosch

I was interested in your multiple regression equation between the skid resistance, the rebound, the resilience, and the hardness. Have you investigated what the correlation coefficient would have been if you had taken these items singly and how much correlation there existed between rebound and hardness?

Peterson

Our regression equations are determined using a set of stocks which was chosen to be orthogonal in rebound and hardness and hence the correlation between rebound and hardness is zero. A one-term regression equation in rebound gives an r^2 of about 0.7, and most of the remaining variance is accounted for by a term in hardness.

D. C. Mahone

Could you reproject slide one. You indicate that by altering the pavement surface characteristics the skid resistance can be improved fivefold, and that by altering rubber compounds in the tires the skid resistance can be improved twofold. I contend that while pavement surfaces can be built that would differ by a ratio of 5 to 1 in skid resistance, highway departments have already realized most of the possible improvements in surface characteristics and would be hard pressed to improve the skid resistance on even below average friction roads by twofold. Also, I suspect that much of what you see as a potential twofold improvement in rubber compounds has already been realized and is being produced in stock automobile tires.

Peterson

Table I, to which you have referred, is taken from a paper by Allbert and Walker. The 1.8:1 difference in that table refers to the difference between the worst compound known and the best. Current tread compounds differ by only 15-20% in skid resistance from the very impractical butyl tire. To answer your question, we cannot hope to give a twofold improvement in skid resistance with compounds now under development since standard tread recipes are actually of fairly high skid resistance. Our problem is thus similar, as you have pointed out, to that of a highway engineer. Road friction values are already sufficiently high that a fivefold increase in average friction coefficient is not possible.

J. M. Long

What was your laboratory abrasion tester?

Peterson

The abrasion data of Table IX were obtained by Cabot Corporation using a modified Lambourn abrader.

Clark

Thank you Dr. Peterson.

THE ROLE OF THE TREAD PATTERN – A BLEND OF THE SIMPLE AND COMPLEX

R. N. KIENLE

UNIROYAL, Inc., Detroit, Michigan

ABSTRACT

Dry traction and wet traction respond to tread design in opposite fashion. Traction coefficients obtained under dry conditions are lowered by the presence of a tire tread pattern, but are increased by a pattern under wet conditions. The lowering of the friction coefficient on dry surfaces is related to the amount of void area in the total contact patch. Under wet conditions, however, the vital role of the tread pattern is to facilitate the removal of water from between tread and road surface. This is accomplished by providing localized high pressure points and sharp edges for rapid contact and by providing channels for either water drainage or to serve as water reservoirs. The role of row number, row width, sipes, groove distribution and groove angle are reviewed. The complexity of the tread surface picture has so far defied detailed analytical dissection. However, a rudimentary view from a hydrodynamic approach is outlined.

TIRE TREAD DESIGN

The role of tread design in tire traction provides an interesting study of dichotomies. It combines a unique blend of simple with complex.

Contrast the different effect of tread design under dry and wet conditions. A tread design of any form is detrimental to dry skid and traction. This response is singularly simple. However, tread design plays a vital and complex role in wet skid and traction.

Best dry traction is obtained when a tire has no tread design at all. This is axiomatic to race drivers, and evidence is readily found in the literature. The data presented comes from the work of J. H. Cox (1) (Table 1).

References p. 254

TABLE 1

Dry Coefficient of Friction Effect of Tread Design*

Tire	Surface Coefficient	Tire Coefficient
Bald	0.87	0.91
	0.77	0.88
5 Straight Grooves	0.87	0.84
	0.77	0.79

*From J. H. Cox

Only one factor of the tread design is known to affect the coefficient of friction (Table 2). It appears that the greater the percentage of the total tread contact area that is groove void area, the lower will be the dry coefficient of friction. The comparison in Table 2 is between two tires of similar construction. One has a standard rib and sipe design, and the other has the lug design of a mud and snow tread. The open design lug tire has a lower dry coefficient.

TABLE 2

Dry Coefficient of Friction Effect of Groove Area
(0.69 Coefficient Surface)

Tire	Mph	% Road Area In Contact Patch	Coefficient of Friction	
			Peak	Slide
Standard 7 Rib	20	0.83	0.76	0.66
	60	0.83	0.87	0.73
M&S Lug	20	0.65	0.73	0.66
	60	0.65	0.77	0.62

One might expect the tread element stiffness to affect dry traction. However, as can be seen in Table 3, the coefficient of friction is not changed as antiskid depth decreases. In this experiment, the tire tread was ground off while maintaining the initial crown radius and sharp tread element edges. In wet skid and traction, the effect of tread design is extremely complex. Tread design is vital to obtaining good wet skid and traction at any speed above a modest one. The design effect is so complicated that it has defied mathematical analysis.

The lubricating action of the water reduces adhesion of tire to road. The aim of the tread design is to remove the water and establish areas of "dry" contact. Of course, in the limiting sense, this is not done. Some finite layer of water remains.

TABLE 3
Dry Coefficient of Friction
Effect of Antiskid Depth

Skid Depth Inches	Speed Mph	Coefficient of Friction	
		Peak	Slide
0.360	20	0.87	0.86
	60	0.92	0.96
0.180	20	0.90	0.90
	60	0.93	0.97
0.063	20	0.90	0.92
	60	0.92	0.95

The tread contact patch can be divided into three separate areas: A, B, and C (See Fig. 1). In the front area (A), no contact exists between tire and road. In area (B) partial contact is obtained. In area (C) general contact between tire and road has been made. The tread design affects the percentage of the footprint area lying within (A) and (B) by providing channels for water drainage, by providing water reservoirs, and by providing a complex pattern of high and low pressure points to assist in cutting through the water. Naturally, the surface roughness as well as the tread design contributes to the pressure distribution and water dispersal.

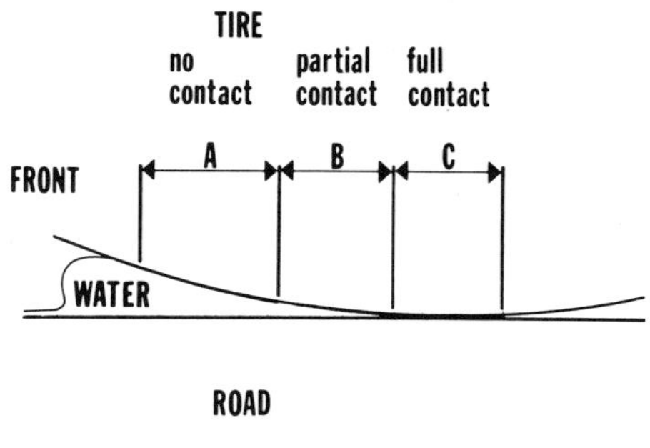

Fig. 1. Tire contact of three zones on wet surfaces.

It is the downward or normal pressure of the tire that is the main driving force for water removal from under the rib. Without even considering road surface as a variable, one finds a complex pressure distribution from rib to rib and around each sipe of a tire tread. Fig. 2 indicates the complexity of the static pressure pattern that exists on the tread face of a loaded bias belted tire. The downward pressure varies from rib to rib across the tread face and from point to point within a single rib. In addition, tire movement causes additional intricate pressure variations. Superimposed upon all this are the pressure distributional changes caused by wear patterns. Obviously, the deformation of both the entire tire structure and the single tread elements influences

References p. 254

the effectiveness of the latter. It is not surprising that tread design has remained an art and that mathematicians and physicists, to date, have not completed a detailed analysis (2).

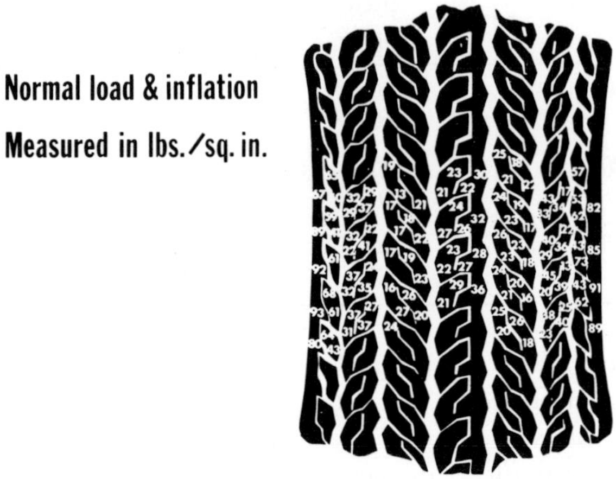

Fig. 2. Typical pressure print of bias belted tire.

While detailed analysis is lacking, there are a large number of empirical studies of tread designs upon wet skid and traction (2, 3, 4, 5). They all support the premise that the role of the design is to remove the water from between tire and road.

The time required to accomplish this is extremely important. While dry traction coefficients are relatively insensitive to speed, wet coefficients usually drop drastically as speed increases (4, 6, 7). The drop is more precipitous as water depth is increased (6).

Several experimenters have demonstrated that increasing the number of rows improves the wet skid resistance (3, 4, 5). Clear evidence of this is contained in results obtained by Allbert and Walker (3) who used radial tires having respectively 4, 5 and 6 grooves. (See Fig. 3.) Note in Fig. 3 the decrease in coefficient with speed and the fact that the effect of groove number is more pronounced as the speed is increased (contact time decreased). Maycock indicated that a further increase in rib number above 7 showed no significant effect (4). Allbert and Walker as well as Kelley felt further improvements in coefficients were obtained with more grooves (3, 5). The point of discussion has little practical significance however since when many ribs are present, the ribs are so narrow that the tire becomes deficient in wear resistance and lateral stiffness (5).

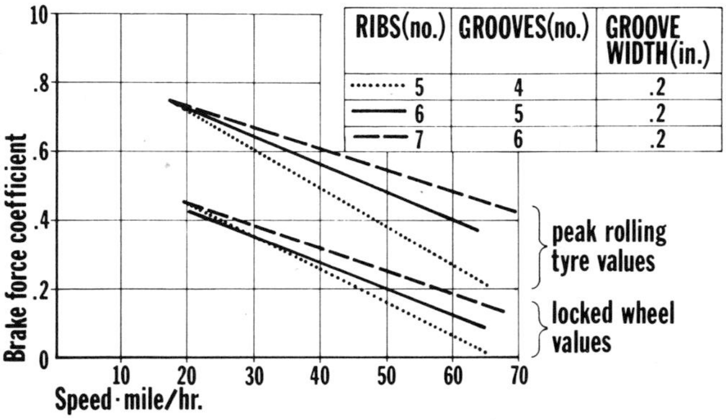

From Allbert & Walker

Fig. 3. Effect of number of grooves on traction.

Maycock has also investigated the effect of groove width on skid resistance. (See Fig. 4.) He tested tires with five equally spaced 0.5 inch wide ribs. Increasing the groove width increased the coefficients. The effect diminished with successive increases and approached an asymptote above a groove width of 0.2 inch. The groove width at which the asymptote is closely approached should vary with water depth and road surface.

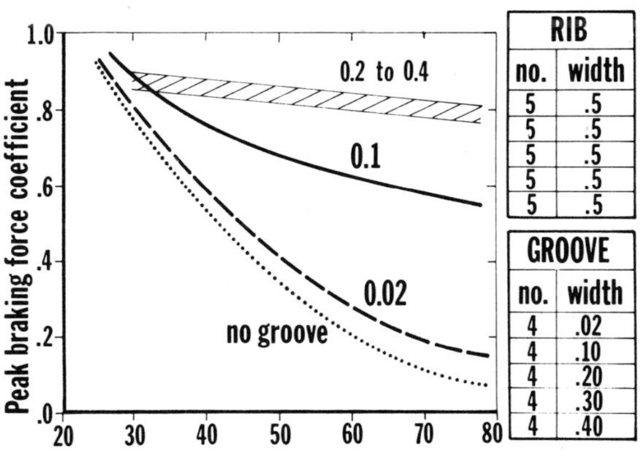

From Maycock

Fig. 4. Effect of groove width.

Skid studies have been made in which the tire tread width was varied while keeping the number of tread rows and the ratio of groove width to tread rib width constant (5). The change in coefficient with tread width was not a simple one. (See Fig. 5.) In these experiments, Kelley used 5 ribs equally spaced on the tread (5). The tires were

References p. 254

compared to an ASTM control tire and there was formed an optimum tread width for traction. I believe that this optimum indicates that two competing factors are affecting the tire traction. As the tread width increases, both the groove volume and the tread rib widths increase. Increasing the groove volume assists the water removal. This explains the initial *increase* in traction. But further increase in tread width causes the ribs to become so wide that the water cannot be removed completely from under them. At this stage, a decrease in coefficient results. The large groove volume can no longer be effectively used.

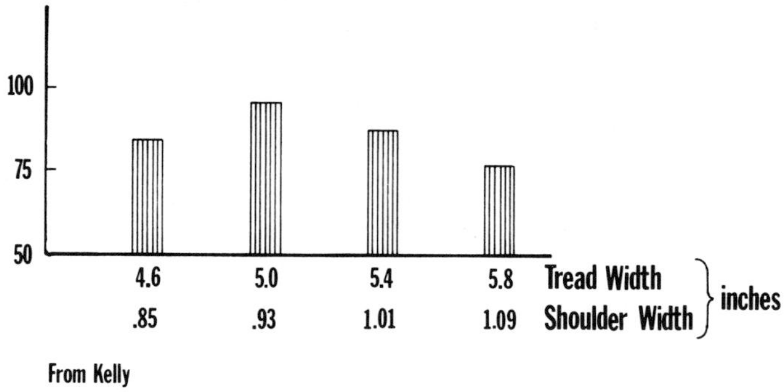

Fig. 5. Effect of tread width on stopping traction (vs. ASM contact tire).

There is a general belief that the groove should be as straight as possible so as not to impede water flow (2). However, Kelley did not find this to be so in his experiments (5). He has postulated that the wiping action of rib edges at an angle to the direction of tire motion must compensate for decreased water flow in tires with zig-zag grooves.

It has been proposed that sipes can act to assist traction in three ways: as wipers, water reservoirs, and pressure points. Obviously, sipes subdivide the tread surface providing multiple cutting edges.

There is conflicting data in the literature as to the effect of siping.

Sometimes sipes appear to increase traction coefficients, and sometimes they do not. Fig. 6 from Maycock (2) indicates that the variation in effect may be due, in part, to road surface. In his study, Maycock compared the peak coefficients of friction for (1) bald, (2) plain ribbed and siped, and (3) ribbed tires on two surfaces, gravel and asphalt. On the gravel surface, the sipes have a positive effect on traction while on the asphalt surface they do not. Apparently, under appropriate conditions, the effect of sipes can be huge. Kelley reports that on a special three rib tire that he prepared, the lateral sipes could improve the coefficient 220% (5).

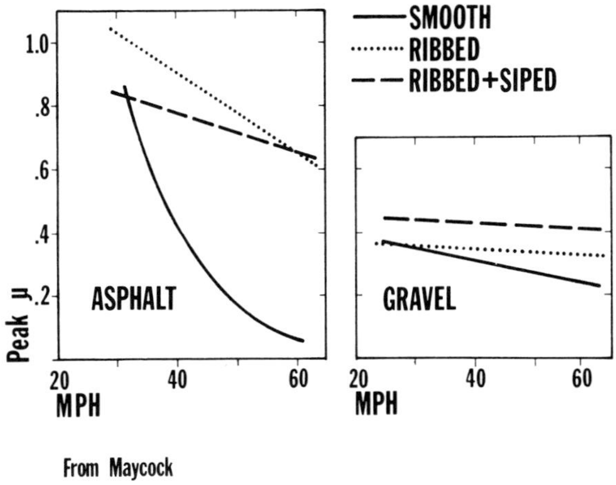

Fig. 6. Effect of siping upon various surfaces.

Studies of the effectiveness of siping may be further confused by comparing experiments performed using different types of siping. In commercial tires, most sipes do not open to the groove and are not of full antiskid depth. Such sipes do not possess sharp cutting edges, channel water into the groove, or give the tread elements much flexibility. Tests of such sipes should obviously not be compared with tests made using full depth sipes cut into the groove.

The advent of radial ply tires has led to the use of open tread designs with cross grooves. Cross grooves are neither groove nor sipe and must be considered separately. The effectiveness of cross grooves has been recognized for a long time, and they have been used advantageously in mud and snow tread designs.

Cross grooves help by equalizing the distribution of water between the circumferential grooves of the tread. The effectiveness of cross grooves is particularly noticeable in reducing loss of traction at high speeds.

Figs. 7 and 8 may be used to compare the loss of peak and slide coefficients with speed. In each case a full rib tire is compared to a tire with cross grooves. In both figures the peak coefficients are plotted against the slide coefficients over the range of speeds from 20 to 60 mph. For each tire, this ratio is assumed to be a straight line function of speed. The length of the vector indicates the amount of coefficient change. The angle of the slanted line indicates the relative change in peak coefficient compared to slide. In Fig. 7 belted bias mud and snow tires and highway tires made of the same basic construction are compared (6). The two tires have about equal friction coefficients at 20 mph. However, both coefficients for the highway rib design tire drop more with speed than those for the mud and snow tire. A similar effect can be seen in Fig. 8 which compares two radial tires. This comparison is not as exact as the

References p. 254

first one since construction differences (aspect ratio and percent groove void) do exist between tires. Again, the coefficients of the tire with cross grooves change less with speed.

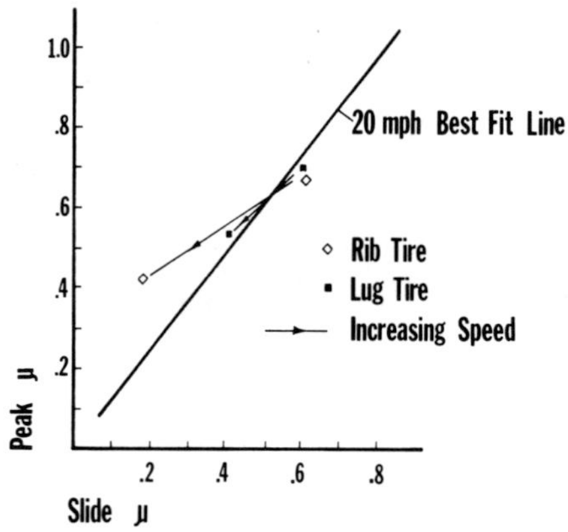

Fig. 7. Comparison of speed dependence – traction coefficients of lug and rib tread.

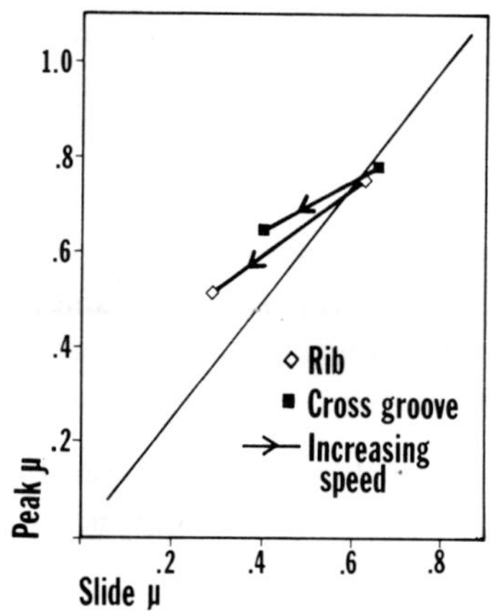

Fig. 8. Comparison of speed dependence – radial tires with and without cross grooving.

Most studies assume that the coefficients change linearly with speed. This assumption is not true, of course. Straight line plots should not be obtained unless log coordinates are used. W. Klamp of UniRoyal used such plots in his studies (unpublished) and found two intersting design effects. He studied the effect of groove angle upon wet coefficients and obtained the results shown in Fig. 9. He used a very smooth concrete surface (0.25 coefficient) to accentruate design effects. Grooves without sipes were cut across the tire at 0, 30°, 50°, 70° and 90° to the circumferential direction. He studied the change of coefficients over the speed range from 10 to 50 mph. The highest coefficients were obtained on tires with groove angles carved at 50° to the circumference. Although the tires with grooves at the highest angles had slightly lower coefficients, the 0° and 30° grooved tires were drastically lower in coefficient. There is a sharp break in all curves somewhere between 20 and 35 mph. With the higher groove angles, the speed at which this break in the curves occurs is higher.

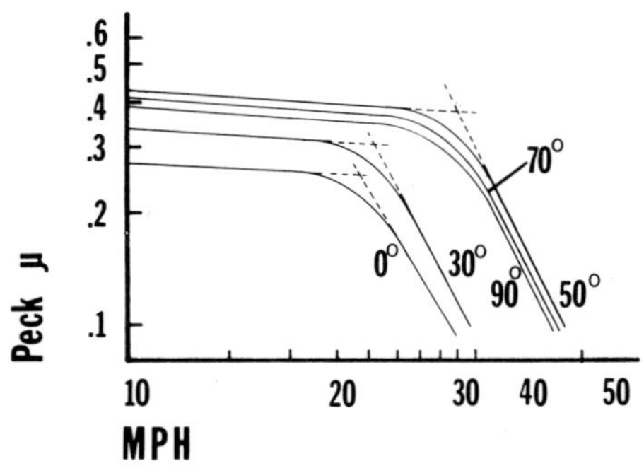

Fig. 9. Effect of groove angle upon coefficient (μ) on wet concrete.

As mentioned before, an exact mathematical definition of the principles governing the role of the tread design has eluded investigators, most of whom have approached the problem by studying tread design principles. As an alternate approach, the design effects may be examined in terms of water flow principles. An intial rough analysis indicates that such an approach may explain many of the findings and, therefore, be quite rewarding.

The discussion given here is limited to the study of the traction conditons under which peak coefficients are obtained. The water can be assumed to be standing stationary on the road. Although the tire rotates, actually the tread surface does not move greatly relative to the water or the road surface underneath. Therefore, it is the downward or normal pressure of the tire that causes water movement. Any portion of the tire is only in contact with the road for a short time. In this interval, the water must be accelerated from standing. Thus, the water can only be moved so far, and the distance it can be moved will vary with tire rotation speed. These times and distances can be calculated. For this calculation, assume a contact length of a passenger tire to be 8 inches. The time that any individual element of a tire tread would be in contact with the road is given in Table 4. The time interval decreases as speed increases.

References p. 254

TABLE 4

Time of Contact for a Tread Element Traveling Through 8 Inch Contact Patch

Mph	Contact Time (sec.)
20	2.3×10^{-2}
30	1.5×10^{-2}
40	1.1×10^{-2}
50	0.90×10^{-2}
60	0.76×10^{-2}

There are two classical formulas which describe the flow of water.

Stefan's formula describes those factors which control the movement of water squeezed between two plates for a finite length of time. (See Fig. 10.) One can apply this formula to the design of tread patterns. The distance that water can be moved during the tire contact time interval dictates how wide a tire rib should be. By using the contact time intervals calculated for various tire speeds (Table 4), a water viscosity of 2.04×10^{-4}, an initial water depth on a skid pad of 0.06 inch, the assumption that road contact occurs when the water thickness has been reduced to 0.006 inch, and average experimental pressures for tire ribs, one can use Stefan's formula to estimate the effective width for a rib (Table 5). One actually calculates a radius (R) representing the area of water that can be squeezed under the above conditions. This radius is, of course, 1/2 of the rib width effectively cleared of water in the contact time.

The static contact patch of a passenger tire contains local pressures of from 100 through 15 psi. However, average rib pressures range from 70 to 30 psi. The calculations indicate that under shoulder ribs which have downward pressure of about

$$t : 3/4\, N\frac{R^2}{P}\left(\frac{1}{(h_2)^2}-\frac{1}{(h_1)^2}\right)$$

t : Time

P : Pressure

R : Radius

N : Viscosity of liquid

h_1 : Initial water thickness

h_2 : Final water thickness

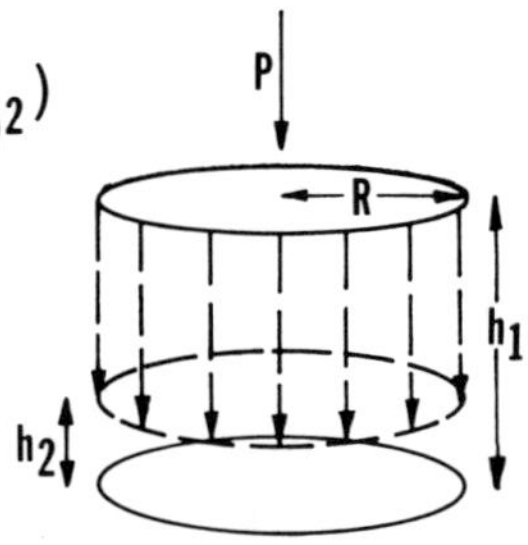

Fig. 10. Stefan's formula.

TABLE 5

Distance Drained in Time Available (8 Inch Contact Patch)
Calculated from Stefan's Formula

Normal Psi	Inches Drained at Speed (Mph)				
	20 Mph	30 Mph	40 Mph	50 Mph	60 Mph
70	0.62	0.50	0.43	0.39	0.37
50	0.52	0.42	0.36	0.33	0.30
30	0.41	0.33	0.28	0.25	0.23
15	0.29	0.23	0.20	0.18	0.17

70 psi water can be moved approximately 0.37 inch when the tire is free rolling at 60 mph. This means effective edge rows can be 0.74 inch wide. On the other hand, according to this calculation, in the center rows which only average 20 or 30 psi of downward pressure, the water can only be moved 0.2 to 0.25 inch when the tire is moving at 60 mph. If such a rib is wider than 0.4 to 0.5 inch water will be trapped under it and will not be completely removed.

Such calculations are in keeping with experimental findings. They help explain why Kelley found that the wider tread tires (with wider ribs) showed a decrease in coefficient of friction. One cannot say why Maycock chose rib widths of 0.5 inch in his groove volume study. However, it was a wise choice according to these calculations, which also explain why tires with higher numbers of tread ribs have higher skid resistance. The rib width decreases as the number increases. One can also see why, after seven or eight rows, when the rib width is less than 0.5 inch, the coefficient improvement levels off. The final film thickness, h_2 of Stefan's formula is drastically reduced when calculating examples of contact with smooth surfaces. In such cases, one would expect to find R reduced as a result. This is in keeping with the fact that Kelley found frictional improvement for rib numbers as high as twelve. He was using a very smooth surface in his tests.

$Q : .61\, a\sqrt{2gh}\, t$

t : time

a : area of orifice

h : pressure head

Q : volume of water drained

g : gravitational acceleration

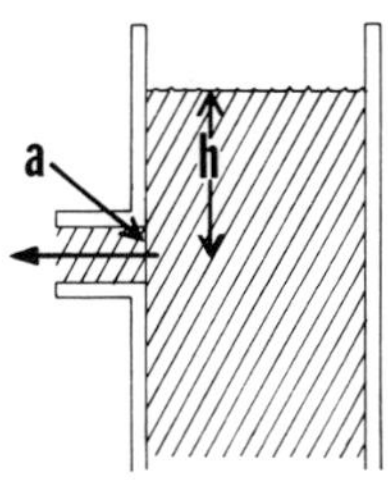

Fig. 11. Weir orifice formula.

The second classic formula used to explain water flow is the Weir orifice formula as shown in Fig. 11. This formula may be used to calculate the orifice size needed to drain a given volume of water or the volume of water that a given orifice can drain in any period of time with a given pressure head. Using this formula, one can predict the effectiveness of sipes and grooves in removing water.

TABLE 6

Distance That Can Be Drained By a Sipe of a New Tire
(Water Depth = 0.06 Inches)

Normal Psi	Distances (Inches)		
	20 Mph	40 Mph	60 Mph
30	0.34	0.24	0.20
50	0.38	0.27	0.22
70	0.42	0.29	0.24

Most modern sipes do not open into the groove and are, therefore, of little use as a water drain. The normal sipe is about 0.04 inch wide. If they were of full tread depth (0.360 inch) and open into the groove, they would effectively drain some of the water from the rib centers. The Weir orifice formula may be used to calculate how much water could be removed. Normal rib pressures for the head pressure and appropriate time intervals can be substituted into this formula to calculate the volume of water that can be forced through such a sipe. Then, by using again a water depth of 0.06 inch, a circular area with radius (R) can be obtained that can be drained through such a sipe (Table 6). At 60 mph only a 0.20 to 0.24 inch radius area of water can be drained through the sipe opening.

The Weir formula can also be used effectively to calculate how much the groove can drain. An average groove is 0.17 inch wide and 0.360 inch deep when new. For a vehicle traveling at 60 mph with normal 30 psi pressure on a tire groove, a groove can only rid water from 1.02 inch of its length. Obviously, at this speed, a tire must contain any water in the groove area rather than "squirt" it out. If water fills the groove, back pressure would force the rest to remain under the tread rib inhibiting road contact.

However, if more time is available, as when tires are traveling at slower speeds, enough water could be expelled from under the tread through the groove openings to help traction significantly. If water can be moved over 8 inches in the time available, theoretically all water can be expelled through the groove opening. To the extent that the distance water can be moved is less than 8 inches, then the groove must act as a reservoir.

The extent to which the groove must act as a reservoir can be estimated. To do this, one must know the average velocity of the water (Table 7). If one considers the normal pressure to be the only factor accelerating the water during the contact time interval, one can calculate average velocities obtainable by the water beneath the tread of tires traveling at different rates. Rather than integrate the acceleration curve to obtain the average velocity, an average velocity was assumed that corresponded to the velocity attained at 5/8 of the time available for acceleration by the various pressure heads. The pressures seen on a static pressure print as shown by Fig. 2, range from 13 to 93 psi. The calculations were made using extreme pressures close to these.

TABLE 7

Estimated Average Velocity of Water Moved Under the Tread for Various Row Pressures and Speeds

Normal Psi	Average Velocity (ft./sec.)		
	20 Mph	40 Mph	60 Mph
100	103.5	51.8	34.2
70	72.5	36.3	23.9
50	51.6	25.9	17.4
30	31.1	15.6	10.3
15	15.6	7.8	5.2

Using these average velocities and the time available, the distance that water can be moved was estimated (Table 8). At low speeds of 20 mph or less, most of the water in the grooves can be effectively forced far enough to be rejected from underneath the contact patch. However, at speeds over 40 mph, there is no longer enough time to accelerate the water the eight inches required to remove it from underneath the tread ribs.

TABLE 8

Estimated Distances Water Can Be Moved at Varying Speeds and Pressures

Normal Psi	Distance Moved (Inches)			
	20 Mph	30 Mph	40 Mph	60 Mph
100	28.6	13.1	6.84	3.12
70	20.0	9.15	4.79	2.18
50	14.2	6.60	3.42	1.59
30	8.58	3.94	2.06	0.94
15	4.31	1.97	1.03	0.47

References p. 254

The critical speed range for effective water removal is between 20 and 40 mph. Interestingly, this is the range at which the slope changes on the plots shown in Fig. 9.

Certainly this cursory examination indicates that a more detailed study of the hydrodynamics of water movement should be undertaken.

CONCLUSIONS

In summary, the effect of tread design upon wet and dry traction possesses several dichotomies. Dry skid and traction of tires is hurt by any tread design. Designless tires give the highest dry coefficients of friction. Dry coefficients are controlled by the percentage of the contact patch actually in contact with the road. Tread designs are particularly effective in improving wet traction. The role of the design is to help eliminate water from between the tread and road surface. While the pressure distribution is too complex to allow exact analysis of the interaction of tread design with the road surface, many simple generalizations can be found. The more tread ribs that the tire possesses, the higher is the traction coefficient. There is an optimum tread width for a given number of ribs. The effectiveness of widening grooves is asymptotic. The effect of sipes is not clear from test data. Sipes appear particularly effective when the tire is tested on smooth surfaces, when the sipes open in the grooves, and when the sipes are of full tread depth. The angle of the tread groove has an effect upon the coefficient of friction. Tires with grooves at angles over 50° to the circumference have higher traction coefficients than tires with circumferential grooves. The increased coefficients are particularly noticeable at higher speeds. Lug tires have the most constant values for coefficient of friction over the normal range of driving speeds. Experimental observations can be explained through the application of basic hydrodynamic principles.

REFERENCES

1. *J. H. Cox, Automotive Engineers Conference, SAE Paper #970D, Detroit, Mich., Jan. 1965.*
2. *G. Maycock, Proceedings of Institute of Mechanical Engineers 180, p. 122, 1965.*
3. *B. J. Allbert and J. C. Walker, Proceedings of Institute of Mechanical Engineers 180, p. 105, 1965.*
4. *G. Maycock, Road Research, Laboratory Report, LR 122, 1967.*
5. *J. D. Kelley, Jr., Automotive Engineers Conference, Detroit, Mich. SAE Paper #680138, 1968.*
6. *R. N. Kienle, Automotive Engineers Conference, SAE Paper #730281, Detroit, Mich., 1973.*
7. *J. K. Meades, Road Research Laboratory Report, LR73, 1967.*
8. *W. Klamp, Uniroyal Work, Unpublished.*

DISCUSSION

Clark

Thank you very much Dr. Kienle. We have time for a few questions.

I. R. Ehrlich

When you say, a 50° angle of the rib, exactly what type of a pattern does that look like? Would you draw us a diagram please.

Kienle

Let's say the tire patch is 6" wide and 9" from front to back. The 50° angle is with respect to the centerline or circumferential line.

D. F. Moore

I was wondering about the validity of using equations without viscosity for flow through sipes. I do not think that is correct. I think it is generally agreed that under a tire viscous effects predominate. You must therefore include that in finding the tread effects on water films.

Kienle

Yes, this is probably true. There is a "fudge" factor in this equation, the 0.61, which in this classical work is for orifice side effects. For small openings as when you get to the size of sipes, however, there may be a term other than 0.61 as was used in weir opening.

Moore

Perhaps viscous and surface tension effects predominate in that range.

W. Bergman

In all your investigations of tread effects, on what surfaces do you conduct your tests? I ask this because the effect of surface micro-texture is a very essential contributor to the overall effect. It is an effect of the same type as that which the tread has, essentially.

Kienle

Well, many of the slides that I showed were from other people's work. References are in the paper, you can go back to them – Kelly, Allbert and Walker, etc. The experiments that Walter Klamp ran, were done on a very fine ground terrazzo tile-type surface with a 0.2 coefficient. Our other experiments were carried out on the asphalt surface, exactly the same one which we have in Laredo, which does have quite a bit of surface roughness.

Clark

Thank you very much Dr. Kienle.

SOME RECENT INVESTIGATIONS INTO DYNAMICS AND FRICTIONAL BEHAVIOR OF PNEUMATIC TIRES

H. B. PACEJKA

Delft University of Technology, The Netherlands

ABSTRACT

A concise review of dynamic tire response to both in-plane and out-of-plane wheel motions is followed by discussions on special subjects. First, the possible self-excited in-plane motion of a wheel of which the axle is suspended with respect to the steady-moving car body is discussed. The influence of several parameters such as the rate of change of effective rolling radius with tire deflection, suspension angle, and tire torsional and slip stiffness is indicated. Second, the influence of tire inertia upon out-of-plane tire performance is elucidated on the basis of theoretical results. The experimentally observed considerable reduction of the first natural frequency of the out-of-plane motion of the tire about a diameter, due to wheel rotational speed, is analyzed. The last portion deals with a theoretical explanation of the creation of a loop in the quasi steady-state cornering force characteristic which appeared to occur on wet slippery roads with tires exhibiting certain wear patterns.

INTRODUCTION

Tire behavior may be subdivided into in-plane (symmetrical) and out-of-plane (anti-symmetrical) performance. Due to the symmetrical structure of the tire-wheel system, interaction between in-plane and out-of-plane motions may be neglected when these motions remain small. Separate treatment of both modes of motion becomes possible in that case.

At large deflections, interaction will occur. For instance, large amplitudes of the vertical motion of the wheel axle will considerably influence the average cornering stiffness of the tire (1). At large slip angles, the tire radial spring rate and deflection change. In this article these interaction effects will not be considered.

References p. 277

In-plane dynamic behavior of the tire-wheel system has recently become more important due to increased interest into dynamic properties of suspension-steer systems. It plays an important role in the forced excitation of steering vibrations caused by tire non-uniformity, wheel imbalance, and road unevenness.

Tire out-of-plane dynamics is of crucial importance in the same problem of steering vibrations or wheel-shimmy. Anti-symmetric tire properties determine the degree of self-excitation or negative damping about the kingpin.

In the discussion of problems which are related to the above mentioned areas, dynamic tire properties will be assumed to be linear. Material damping will be neglected. The last problem to be discussed is not of a dynamic nature, but is associated with the slow cyclic slip angle variation. Especially large slip angle amplitudes are of interest, as the frictional behavior is found to be responsible for the unusual variation of shear force vs. slip angle. This problem is essentially non-linear.

IN-PLANE DYNAMIC TIRE RESPONSE TO VERTICAL AND LONGITUDINAL AXLE MOTIONS

The discussion will be restricted to a uniform tire, rolling over a perfectly flat road. Small motions are considered in order to limit the analysis to linear response. We refer to (1) for a more complete treatment of the subject matter.

The tire model to be considered is depicted in Fig. 1. The model embodies the capability of the tire to distort in the radial and the tangential directions. The elastic tread elements enable the tire model to exhibit a deformation slip in the longitudinal direction. When the tire is distorted tangentially with respect to the wheel-rim due to a longitudinal force F_x, a large part of the tire is not deformed or is at least much less deformed than the lower portion of the tire near the contact zone. This is true when the frequency of excitation is much less than the lowest frequency of the tread band in-plane vibration.

The angular vibration of the wheel plus the effective (upper) portion of the tire mass, with respect to the footprint portion of the tread band, shows a relatively low natural frequency and plays an important role in our analysis.

Another tire property which is essential in the interaction between longitudinal and vertical motions is the variation of the effective rolling radius as a function of vertical tire deflection. A well known theory states that the effective rolling radius differs from the free outer radius because of the tangential compression of the tread band due to radial tire deflection.

When the rolling resistance is neglected, the following equations govern the in-plane response of longitudinal (F_x) and vertical (F_z) force to longitudinal (x) and vertical (z) axle displacements.

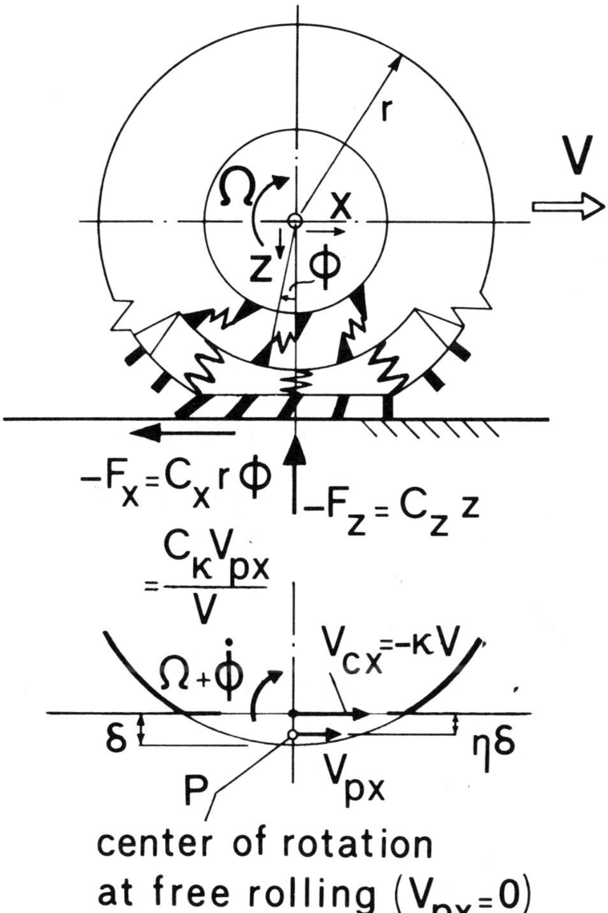

Fig. 1. In-plane tire model showing capability of carcass distortion and longitudinal slip.

The forces may be divided into a constant (average) and a variable part:

$$F_x = F_{xo} + \overline{F}_x, \quad F_z = F_{zo} + \overline{F}_z$$

where (1)

$$F_{xo} = 0, \quad F_{zo} = -W_o$$

where W is the load and the index o denotes the original situation or reference value.

References p. 277

The longitudinal force is a result of the longitudinal slip of a point P, located a distance $\eta\delta$ below road level (Fig. 1) which is the center of rotation at free rolling.

$$F_x = C_\kappa (-V_{px}/V) = C_\kappa \left\{ \kappa + (\Omega + \dot{\phi})\, \eta\, \delta / V \right\} \tag{2}$$

where C_κ denotes the longitudinal slip stiffness, V_{px} the speed of point P, V the speed of travel, Ω the rotational wheel speed, ϕ the torsion angle of the lower tread band portion with respect to the wheel-body, and δ the vertical tire deflection. The longitudinal creep or slip at road level is defined as:

$$\kappa = -\, V_{cx}/V \tag{3}$$

where V_{cx} denotes the longitudinal slip speed at road level (Fig. 1) at point C, the contact center. We have:

$$V_{cx} = V + \dot{x} - (r - \delta)\,(\Omega + \dot{\phi}). \tag{4}$$

At a particular instant, points P and C may be considered to be attached to an imaginary body which is deflected over an angle ϕ with respect to the wheel-body. Under steady state conditions, these points may be thought to be fixed to the wheel body. It may be noted that at free rolling ($F_x = 0$) and

$$\kappa = \kappa_o = -\, \eta\, \delta_o / r_{eo} \tag{5}$$

with an effective rolling radius

$$r_{eo} = V/\Omega_o = r - (1-\eta)\,\delta_o\,. \tag{6}$$

Analogous to (1) we define:

$$\kappa = \kappa_o + \bar{\kappa}, \quad \Omega = \Omega_o + \bar{\Omega}, \quad \delta = \delta_o + \bar{\delta}\,.$$

From (6) it follows that the effective rolling radius changes with tire deflection at a rate determined by the factor $(1 - \eta)$. If η is close to unity, the effective radius will hardly change with deflection. This is the case with radial steel belted tires. Bias-ply tires may have η values of the order of 0.5. When δ is neglected with respect to r and the relations are kept linear in the variable quantities, we obtain the following expression for F_x:

$$\begin{aligned} F_x &= C_\kappa \left\{ \bar{\kappa} + \eta\bar{\delta}/r + \delta\,(\bar{\Omega} + \dot{\phi})\delta_o/V \right\} \\ &= C_\kappa \left\{ (-\dot{x} + r\bar{\Omega} + r\dot{\phi}) \,/\, V - (1-\eta)\,\bar{\delta}/r \right\}. \end{aligned} \tag{7}$$

The force may also be expressed in terms of the torsion angle ϕ. With C_x denoting the longitudinal stiffness of the standing tire (not including the stiffness of tread elements), we obtain:

$$F_x = -\, C_x r \phi\,. \tag{8}$$

The relation of F_x to the rotational acceleration of the wheel and effective tire mass (total polar moment of inertia: I_y) reads:

$$rF_x = -I_y\dot{\Omega}\,. \tag{9}$$

With the aid of the foregoing equations, we obtain the following differential equation for F_x as a response to the axle motions, x and z $(=\bar{\delta})$:

$$\frac{1}{C_x}\ddot{F}_x + \frac{V}{C_\kappa}\dot{F}_x + \frac{r^2}{I_y}F_x = -\ddot{x} - (1-\eta)\frac{V}{r}\dot{z}\,. \tag{10}$$

The amplitude ratios of output F_x and inputs x and z as a function of excitation frequency ω when the natural frequency $\omega_{\phi o}$ is taken to be $\omega_{\phi o} = \sqrt{(C_x r^2/I_y)}$ become:

$$\left|F_{x,x}(\omega)\right| = \frac{C_x\omega^2}{\sqrt{(\omega^2_{\phi o} - \omega^2)^2 + (VC_x/C_\kappa)^2\,\omega^2}} \tag{11}$$

$$\left|F_{x,z}(\omega)\right| = \frac{C_{cx}\,(1-\eta)\,(V/r)\omega}{\sqrt{(\omega^2_{\phi o} - \omega^2)^2 + (VC_x/C_\kappa)^2\,\omega^2}}\,. \tag{12}$$

Fig. 2 shows the above relations in graphical form for the following parameter values:

$$I_y = 0.6 \text{ kgm}^2,\ C_x = 5 \times 10^5 \text{ N/m},\ r = 0.32 \text{ m},$$
$$\eta = 2/3,\ C_\kappa = 4000 \text{ N},\ \omega_{\phi o} = 92\pi \text{ rad/s}. \tag{13}$$

At low values of the frequency of excitation, ω, the response of the amplitude of F_x to x varies approximately quadratically with ω and, therefore, will become important in the higher frequency range. This is in contrast to the response to z which varies approximately linearly with ω near the origin as follows:

$$\left|F_{x,z}(\omega)\right|_{\omega\to o} = (1-\eta)\,I_y\,V\omega/r^3 \tag{14}$$

indicating the importance of speed, rotating mass and the factor $(1-\eta)$. The maximum of $\left|F_{x,z}\right|$ is achieved at the natural frequency $\omega_{\phi o}$. The maximum value turns out to be independent of speed V. Here, the influence of the longitudinal slip stiffness C_κ is greatest (Fig. 2).

Experiments with full scale tires confirm the trend predicted in Fig. 2 of the tire response to vertical axle motions.

The response of the vertical force, F_z, to the axle motions x and z is treated in a relatively simple manner by neglecting possible dynamic influences. The transfer functions then will read:

$$F_{z,x} = 0,\ F_{z,z} = -C_z\,. \tag{15}$$

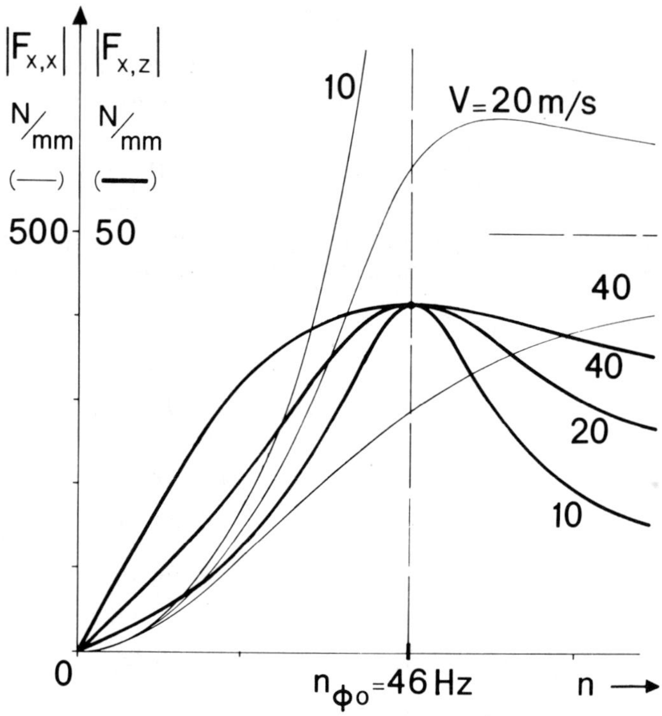

Fig. 2. Amplitude response of longitudinal force, F_x, to longitudinal axle motions, x, (thin curves) and to vertical axle motions, z, (thick curves). The parameter is the speed of travel, V.

SELF-EXCITED WHEEL BOUNCE

The wheel-suspension system to be considered is depicted in Fig. 3. The suspension angle α gives rise to a constraint relation between vertical and longitudinal axle motions:

$$x = z \tan \alpha . \tag{16}$$

This constraint is necessary for the unstable motion to arise. The subsequent analysis makes use of the tire model discussed in the previous section.

The self-excited oscillation has been demonstrated in Professor S.K. Clark's laboratory at the University of Michigan. The phenomenon occurred with a small model tire of bias-ply construction rolling over a rotating drum. The pivot of the trailing arm was located at road level and the self-excitation showed up only at one range of speed.

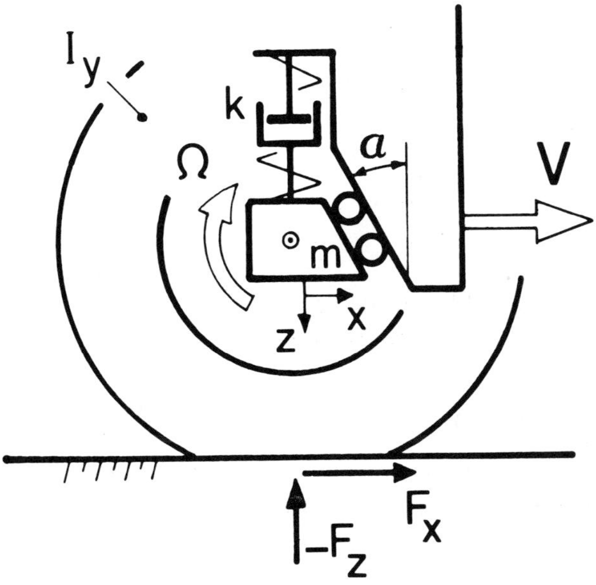

Fig. 3. Wheel-suspension system showing inclined wheel-axle guidance (angle a).

The vertical motion of the translating mass, m, is governed by the following equation

$$m^* \ddot{z} + k\dot{z} = F_x \tan a + F_z \tag{17}$$

with effective mass

$$m^* = m/\cos^2 a. \tag{18}$$

For relatively small a, this effective mass may be considered as constant.

In combination with Eqs. (10), (15) and (16), the motion is fully described. We will introduce a number of dimensionless quantities:

$$\sigma = \frac{C_x}{C_z}, \quad \tau = \frac{\omega^2_{\phi o}}{\omega^2_{zo}}, \quad \gamma = \frac{C_\kappa}{rC_x},$$

$$\beta = \frac{k}{2m^* \omega_{zo}}, \quad \bar{V} = \frac{V}{r\omega_{zo}}, \quad \bar{\omega}_o = \frac{\omega_o}{\omega_{zo}} \tag{19}$$

References p. 277

in which appear the natural frequencies of the undamped translation and rotation:

$$\omega^2_{zo} = \frac{C_z}{m^*}, \quad \omega^2_{\phi o} = \frac{C_x r^2}{I_y} \tag{20}$$

The characteristic equation reads ($\bar{p} = \bar{\sigma} + i\bar{\omega}_o$):

$$\begin{gathered} \bar{p}^4 + (\bar{V}/\gamma + 2\beta)\bar{p}^3 + (1 + \tau + 2\beta\bar{V}/\gamma + \bar{\sigma}\tan^2 a)\bar{p}^2 + \\ \left\{2\beta\tau + \bar{V}/\gamma + (1-\eta)\sigma\,\bar{V}\tan a\right\}\bar{p} + \tau = 0 \end{gathered} \tag{21}$$

The conditions for stability, according to Routh-Hurwitz, reduce for the undamped system ($\beta = o$) to the following inequality:

$$\frac{1}{(1-\eta)\gamma} - (1-\eta)\gamma\sigma + \frac{\tau - 1}{\tan a} + \sigma \tan a > 0 \tag{22}$$

In case Eqn. 22 is not satisfied, oscillatory instability will arise. Usually, $0 < \eta < 1$ and $\tau > 1$. For negative a, the system becomes unstable at sufficiently large values of $(1-\eta)\gamma$. Positive a, however, does not ensure stability. Large values of γ may

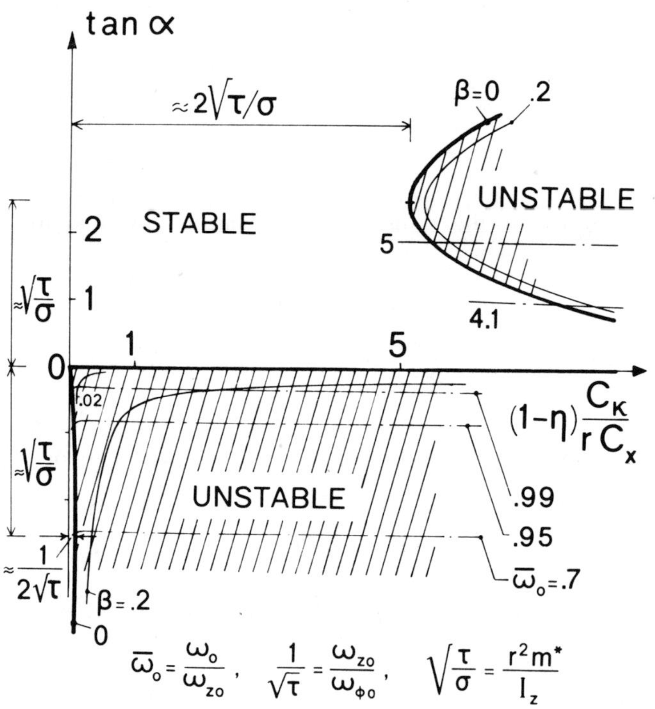

Fig. 4. Boundaries of unstable vertical wheel motions of system shown in Fig. 3.

destablize the system. A tire with η sufficiently close to unity will not be capable of showing self-excited wheel bounce.

Fig. 4 presents the stability boundaries in the parameter plane with ordinate $\tan a$ and abscissa $(1 - \eta)\gamma$. The calculations have been carried out for the following parameter values:

$$\sigma = 2.5 \quad \text{and} \quad \tau = 16 \tag{23}$$

The boundaries for the case with damping (damping ratio $\beta = 0.02$ and 0.2) hold for $\eta = 2/3$ and $V = 1$ (with $r \simeq 0.3$ m and $\omega_{zo} \simeq 20\pi$ rad/s we get $V = 20$ m/s $= 72$ km/h).

The two separated unstable areas appear to represent self-excited vibrations of different modes. For positive a, the frequency of the vibration on the verge of instability (ω_o indicated in the figure) turns out to be of the order of the natural frequency of the torsional deflection, $\omega_{\phi o}$. For negative a, the frequency is closer to the much lower natural frequency associated with the radial deflection, ω_{zo}.

It may be noted, that a tire with the parameters specified in (13) which has a rather low value of C_κ and for which $(1 - \eta)\gamma = 1/6$, needs only little damping to avoid instability. The model aircraft tire which experienced instability had a bias-ply carcass (small η) and showed relatively stiff continuous ribs (large C_κ).

OUT-OF-PLANE DYNAMIC TIRE RESPONSE TO LATERAL AND STEER MOTIONS OF THE WHEEL AXLE

The antisymmetrical response of pneumatic tires has been a subject of study during the last three decades. Relatively recently, the mass of the tire has been included in the analysis. This extension of the well-known kinematic theories (2) is of importance, not so much to explain the phenomenon of self-excited shimmy vibrations, but to achieve a far better quantitative description of real tire behavior.

The theory of the dynamic antisymmetric response of a stretched string type tire model with mass is of considerable complexity (3). An approximate dynamic theory (4) developed recently as an extension to an earlier and simpler approximation of the dynamic influence (1), is of greater practical use. The agreement with experimental results turns out to be good (4).

The essential approximation introduced in this theory is to neglect those harmonic components of the inertial force distribution along the tire circumference that have wavelengths shorter than the circumference. This means, that the elastic deformation associated with this lateral inertial force distribution varies according to a plane. The position of this plane with respect to the wheel center plane is defined by the lateral and (two) angular displacements ($\tilde{y}_t$, $\tilde{\gamma}_t$, $\tilde{\psi}_t$).

References p. 277

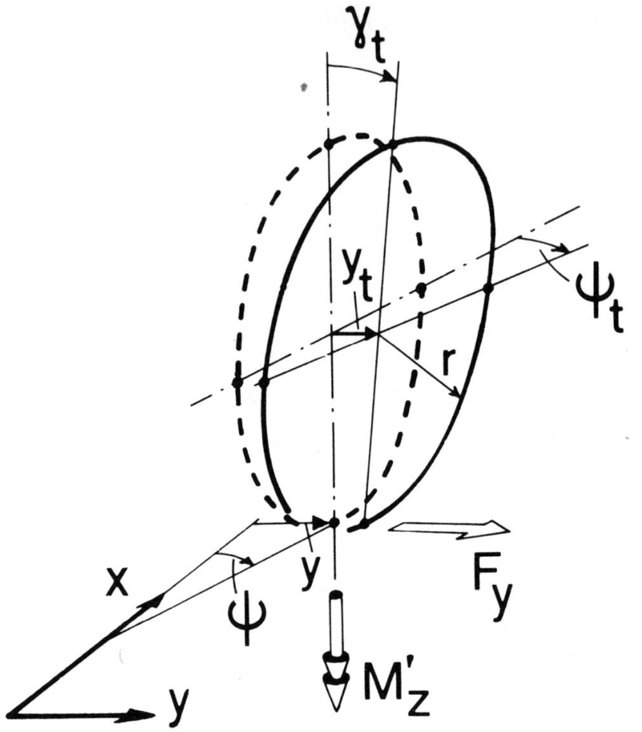

Fig. 5. Zero and first order harmonics of tire lateral distortion (= average plane distortion).

Besides these inertial forces, we have the external ground force F_y and moment M_z' (see Fig. 5) which cause the tire to distort. The zero order plus first harmonic of this "static" distortion corresponding to an inclined plane is defined by the displacements $(\bar{y}_t, \bar{\gamma}_t, \bar{\psi}_t)$. Higher harmonics do not produce a resulting force or moment acting on the wheel rim.

With C_y, C_γ, C_ψ denoting the stiffness of tire tread band with respect to wheel-rim and r the tire radius, the following relations apply:

$$\bar{y}_t = F_y/C_y, \quad \bar{\gamma}_t = -F_y r/C_\gamma, \quad \bar{\psi}_t = M_z'/C_\psi \tag{24}$$

The sum of dynamic and static deformations is denoted by (y_t, γ_t, ψ_t).

$$y_t = \bar{y}_t + \tilde{y}_t, \quad \gamma_t = \bar{\gamma}_t + \tilde{\gamma}_t, \quad \psi_t = \bar{\psi} + \tilde{\psi}_t \tag{25}$$

The input motion of the wheel-plane is defined by (y, γ, ψ). Camber angle γ will be kept zero in our analysis. The input motion added to the total deformation of the tire (y_t, γ_t, ψ_t) yields the total displacement of the tire-plane with respect to a spatial axes system. These displacements produce the inertial forces mentioned above.

With m_t, I_t and I_{tp} denoting effective tire inertia parameters and Ω the speed of revolution of the wheel, we obtain for the dynamic deformations:

$$\begin{aligned} \tilde{y}_t &= - m_t (\ddot{y} + \ddot{y}_t) / C_y, \\ \tilde{\gamma}_t &= - \left\{ I_t \ddot{\gamma}_t + I_{tp}\, \Omega\, (\dot{\psi} + \dot{\psi}_t) \right\} / C_\gamma, \\ \tilde{\psi}_t &= - \left\{ I_t (\ddot{\psi} + \ddot{\psi}_t) - I_{tp}\, \Omega\, \dot{\gamma}_t \right\} / C_\psi \, . \end{aligned} \tag{26}$$

The ground force and moment produce static deformations on top of these dynamic deformations. The generation of F_y and M_z' is considered to be the result of an effective input motion which is formed by the actual wheel-plane motion plus the dynamic deformation. By taking this effective input motion as an input to the massless kinematic theory, the approximate response of force and moment is obtained.

Instead of taking the motion of the effective wheel-plane, we may suffice with the motion of the effective line of intersection with the road plane. (cf. Fig. 6). Its location is defined by:

$$y_e = y + \tilde{y}_t - r\tilde{\gamma}_t, \; \psi_e = \psi + \tilde{\psi}_t \, . \tag{27}$$

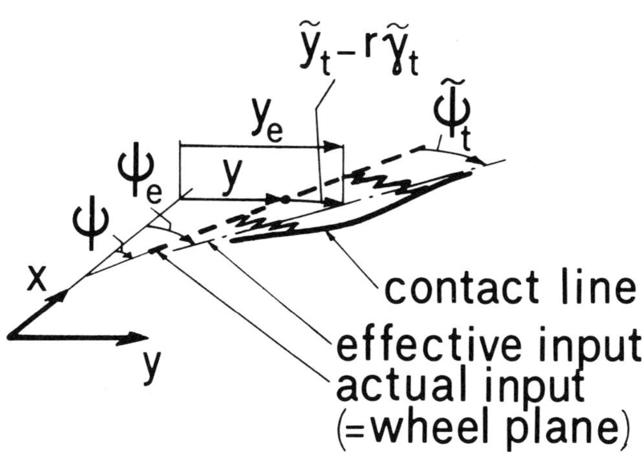

Fig. 6. Actual and effective input motion (= position of line of intersection of (effective) wheel plane and road plane).

References p. 277

We shall employ the excellent approximation of the theory of the massless string-type tire model due to von Schlippe and described in (1, 2, 3). We introduce the following quantities: 2a, the contact length between tire and road; σ, the so-called relaxation length; and C_F and C_M, the cornering and aligning stiffnesses, respectively. The frequency response functions can be written in vectorial form as a function of frequency ω:

$$(F_{y,y_e}, F_{y,\psi_e}) = \frac{C_F}{\sigma + a}\left\{(-1,0) + 1/2\,\frac{1+e^{-2i\omega\, a/V}}{1 + i\omega\sigma/V}(1,\sigma + a)\right\}$$

$$(M'_{z,y_e}, M'_{z,\psi_e}) = \frac{C_M}{a}\left\{(0,-a) + 1/2\,\frac{1-e^{-2i\omega\, a/V}}{1 + i\omega\sigma/V}(1,\sigma + a)\right\} \quad (28)$$

For practical purposes, simpler expressions may be used (3,4).

A tire which shows a finite length and width of the contact area will also generate longitudinal forces. These forces form a couple M_z^*. The sum of M_z' and M_z^* give the total moment M_z. According to a theory given in (1, 3), the frequency response function for M_z^* reads:

$$(M^*_{x,y_e}, M^*_{z,\psi_e}) = -\frac{\kappa^*}{a}\left\{1 - \frac{1 - e^{-2i\omega\, a/V}}{2i\omega a/V}\right\}(0,1) \quad (29)$$

with κ^* representing a tire constant.

For applications in vibratory problems associated with wheel-suspension-steer systems, it is of greater interest to know the response of force and moment acting on the assumedly rigid wheel with tire mass included. The equivalent force and moment under this condition are obtained as follows:

$$F_{y\ eq} = F_y - m_t\ddot{y}_t$$

$$M_{z\ eq} = M_z' + M_z^* - I_t\ddot{\psi}_t + I_{tp}\,\Omega\,\dot{\gamma}_t \quad (30)$$

This result should correspond to the force and moment measured in the wheel hub, $F_{y\ hub}$ and $M_{z\ hub}$, after these have been corrected for the inertial force and moment acting on the assumedly rigid wheel plus tire. With inertia parameters of wheel plus tire we find:

$$F_{y\ eq} = F_{y\ hub} + m_{wt}\ddot{y}$$

$$M_{z\ eq} = M_{z\ hub} + I_{wt}\ddot{\psi} \quad (31)$$

For the following set of parameter values measured on a radial ply steel-belted tire, $a = 0.063$ m, $\sigma = 0.377$ m, $r = 0.322$ m, $C_F = 49{,}000$ N, $C_M = 1400$ Nm, $\kappa^* = 104$ Nm2, $C_\psi = C_\gamma = 18{,}100$ Nm, $C_y = 360{,}000$ N/m, $m_t = 5.7$ kg (total mass: 8.5 kg), $I_t \simeq 1/2\ I_{tp} = 0.295$ kgm2. The amplitude and phase of the equivalent moment as a response to steer angle have been computed and are shown in Fig. 7a as a function of excitation frequency $n = \omega/2\pi$. Experimentally obtained data up to a maximum of 8 Hz have been added. The phase lag occurring in the lower frequency range is responsible for the creation of self-excited wheel shimmy. The influence of tire mass is evident. Without tire mass, the same response would occur at equal wavelengths, $\lambda = V/n$, irrespective of the value of the speed of travel V.

The wavelength remains constant when the motion is excited by say a wheel imbalance ($\Omega = \omega$, $\lambda = 2\pi r$). The corresponding response is shown in Fig. 7b.

The resonance-frequencies, which at speeds close to zero approach a value somewhat larger than the natural frequency of the free non-rotating tire, n_{oo}, appear to separate at increasing speed. Vibration experiments performed on a free tire (not contacting the road) for the purpose of obtaining the values of certain tire parameters, gave a clear indication that a considerable reduction of the resonance-frequency may arise when the wheel spins. The next section gives a brief analysis of this phenomenon.

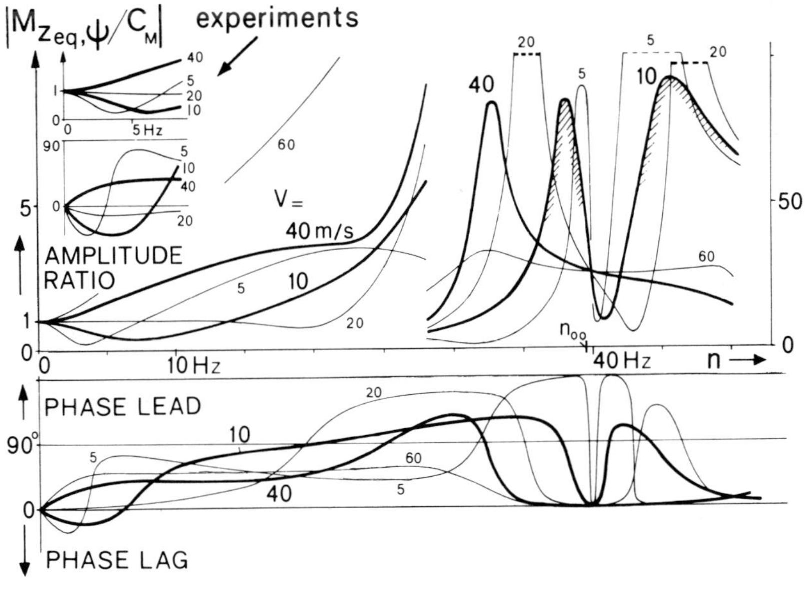

Fig. 7a. Amplitude and phase response of aligning torque to steer angle as a function of frequency of excitation. Parameter is speed of travel, V. Experimental results are given up to a frequency of 8 Hz.

References p. 277

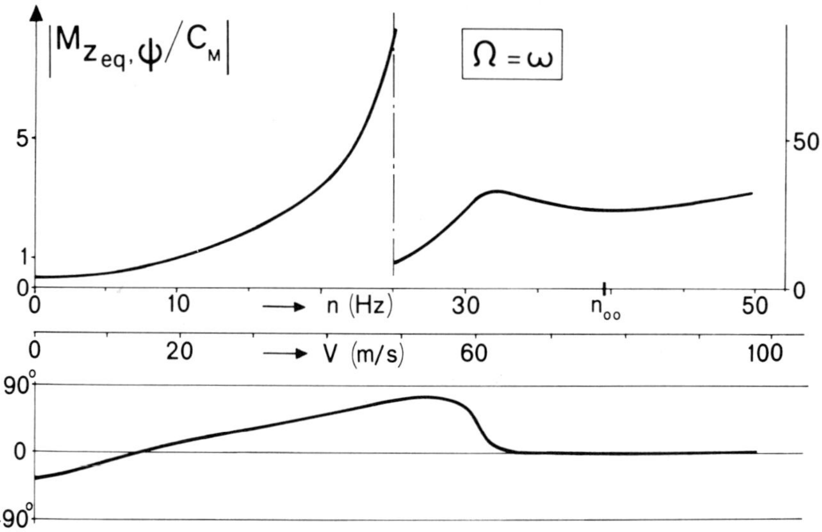

Fig. 7b. Amplitude and phase response as a function of frequency (the frequency being that of wheel revolution).

REDUCTION OF NATURAL FREQUENCY DUE TO WHEEL ROTATION

The response of a free rotating tire that is subjected to vibrations of the wheel-plane is, of course, governed by the general equations presented in the preceding section. The difference and considerable simplification in the analysis for the free rotating tire is the absence of ground force and moment. As a result of this, the static deformations vanish, and the last two equations of (26) become:

$$\begin{aligned} I_t\ddot{\gamma}_t + I_{tp}\,\Omega\,(\dot{\psi} + \dot{\psi}_t) + C_\gamma\gamma_t = 0, \\ I_t\,(\ddot{\psi} + \ddot{\psi}_t) - I_{tp}\,\Omega\dot{\gamma}_t + C_\psi\psi_t = 0\ . \end{aligned} \tag{35}$$

The free vibration of the tread band with respect to the rotating wheel-rim ($\psi = 0$) shows two natural frequencies, ω_o, which are the roots of the frequency equation:

$$\left(\frac{\omega_o}{\omega_{oo}}\right)^4 - \left\{\left(\frac{I_{tp}}{I_t}\right)^2\left(\frac{\Omega}{\omega_{oo}}\right)^2 + 2\right\}\left(\frac{\omega_o}{\omega_{oo}}\right)^2 + 1 = 0 \tag{36}$$

where

$$\omega_{oo} = \omega_{\psi oo} = \sqrt{C_\gamma/I_t} = \sqrt{C_\psi/I_t} \tag{37}$$

represents the natural frequency of the non-rotating tire ($\Omega = o$). It can be shown that with increasing Ω the natural frequencies separate more and more from the value

ω_{oo}. The lower natural frequency, which is the more important one, appears to be reduced for increasing values of Ω (Fig. 8).

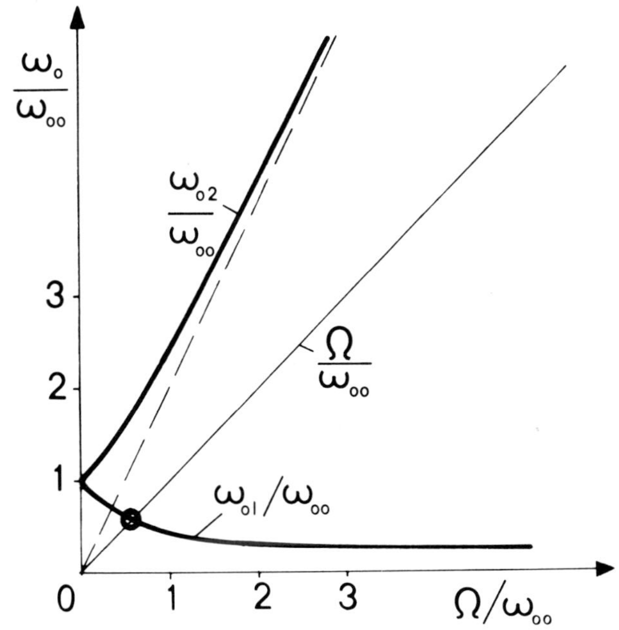

Fig. 8. Natural frequencies, ω_{o1} and ω_{o2}, as a function of rotational speed, Ω, for a free tire (not contacting the road). Camber and steer motions of the tire with respect to the rim are considered. The point where Ω reaches the value of ω_o, indicates the resonance frequency found through excitation by wheel imbalance.

When an imbalance weight excites the vibration, resonance will occur at a frequency equal to the frequency of wheel revolution, $\omega_o = \Omega$. In this particular case with the aid of this equality, the only real root of (36) becomes:

$$\omega_o = \omega_{oo} \sqrt{\frac{1 - I_{tp}/I_t}{1 - (I_{tp}/I_t)^2}} \,. \tag{38}$$

For the case where the effective polar moment of inertia $I_{tp} = 2\, I_t$, expression (38) reduces to

$$\omega_o \simeq 0.6\, \omega_{oo} \,. \tag{39}$$

References p. 277

This considerable reduction of around 40% is in excellent agreement with experimental findings. For a radial steel-belted tire, a reduction from about 40 Hz to about 25 Hz has been found.

From the equations of motion (35), it can be deduced that at $\Omega = \omega_o$, the deformation angle, γ_t, lags 90 degrees behind the angle ψ_t and 180 degrees behind the excitation angle, ψ. This agrees with experimental evidence. At $I_{tp} = 2\ I_t$, the amplitudes of γ_t and ψ_t are equal to each other. The 90 degree phase lag of γ_t with respect to ψ_t implies that an alternating deformation of the tire takes place.

SIMILARITY RULES FOR TESTING SMALL SCALE MODEL TIRES

The equations governing the dynamic performance of tires that have been derived above, may be written in dimensionless form. As a result of this, certain dimensionless numbers are recognized. These should be kept constant in order to ensure similarity between real and experimental situations.

For a small scale tire model for which geometrical similarity is to be preserved (including that of the contact zone), the ratios between various lengths must remain unchanged. If we are dealing with stiffness parameters of the mathematical tire model in the case where they almost exclusively depend on the inflation pressure, p_i, the ratio between these parameters will hardly be effected by the geometrical scaling. If p_i is not mainly responsible for the stiffness, the ratio of material elastic moduli and inflation pressure should be kept constant.

In addition, we have the following numbers, which ensure kinematic and dynamic similarity, respectively:

Kinematic Conditon:

$$\frac{\omega r}{V} = \text{constant} \tag{33}$$

Dynamic Condition:

$$\frac{\omega^2 r^2 \rho}{p_i} = \text{constant} \tag{34}$$

where ρ denotes the mass density of the tire. The kinematic condition simply states that the wavelength of the path on the road must be scaled down at the same rate as each other length parameter. The dynamic condition implies that the excitation frequency must be regarded in terms of a natural frequency of tire and model. At fixed ρ/p_i, a reduction of radius r would require an increase in excitation frequency ω while the speed V should remain unchanged. If p_i can be reduced without seriously violating the stiffness similarity condition, V and ω may be kept in an acceptable operating range.

QUASI-STATIC CYCLIC BEHAVIOR ON SLIPPERY ROADS

As has been reported earlier (5), a tire may show a non-unique cornering force vs. slip angle characteristic. An original plot is shown in Fig. 9. This phenomenon has been shown to occur on very smooth wet surfaces. The tires tested exhibited tread elements (ribs) with worn edges. This type of wear promotes the formation of a water film underneath the elements.

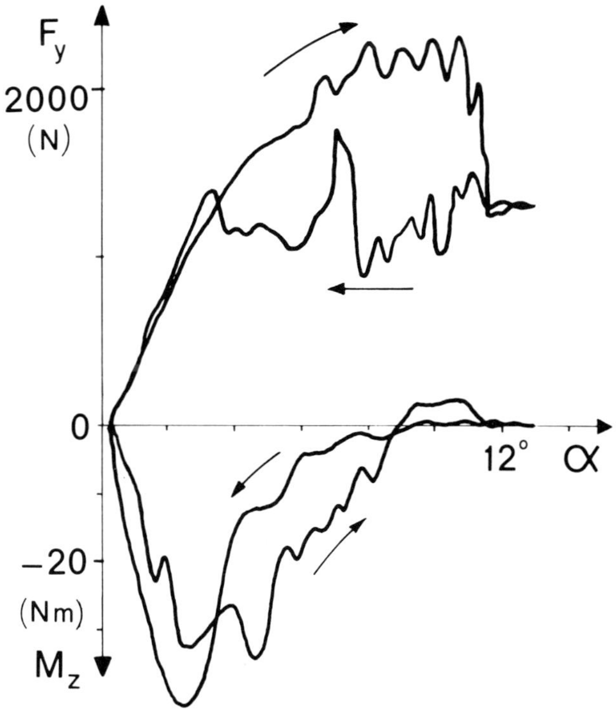

Fig. 9. Loop in quasi steady-state cornering force-slip angle relationship. Test data: Road: wet, smooth asphalt. Tire: bias-ply 5.60-13, load 400 kgf, inflation pressure 1.7 kgf/cm^2, worn edges of tread pattern ribs. Speed: 40 km/h.

The explanation of the phenomenon is not self-evident. It is rather obvious that, as a start, our model of the friction curve should drop with increasing sliding speed beyond a low speed maximum. Extensive computation reveals that this is not a sufficient conditon to get at the lower branch of the cornering force characteristic.

A second condition has been found to be necessary. The fact that the tires, which were capable of showing the loop, exhibited worn edges brought up the idea that

References p. 277

possibly a coefficient of friction, μ, arises which depends upon the orientation of the tread element relative to the road surface.

The orientation is directly related to the lateral deformation of the tire and, consequently, to the cornering force F_y. For the sake of simplicity, we have chosen a function (of speed) for the coefficient of friction which abruptly changes to an alternative representation when a critical value of the cornering force $F_{y\ crit}$ is passed (see Fig. 10).

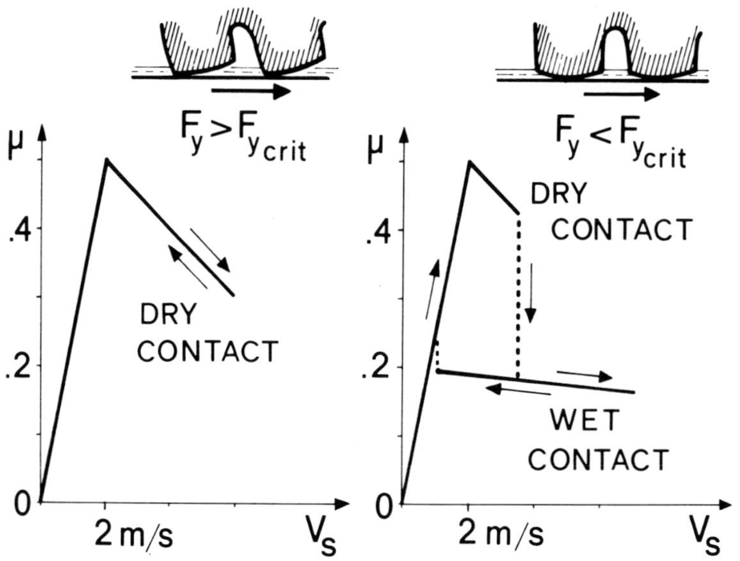

Fig. 10. Two different hypothetical wet friction functions (friction coefficient vs sliding speed). Validity depends on orientation of tread element with respect to road surface.

Mathematically, we take the following simplified friction functions (of sliding speed V_s) for the two ranges of the side force.

$$\begin{aligned} \mu &= A_s V_s \quad \text{if } V_s < V_{so} \\ \mu &= \mu_o - B_s V_s \quad \text{if } V_s > V_{so} \end{aligned} \tag{40}$$

with parameters different for each of the distinct force ranges.

The friction properties as postulated will be applied in connection with a one-dimensional model of tire lateral elasticity represented by a Green's or influence function G (ξ,x). A lateral concentrated load acts at coordinate x and results in a lateral deformation measured at coordinate ξ (see Fig. 11).

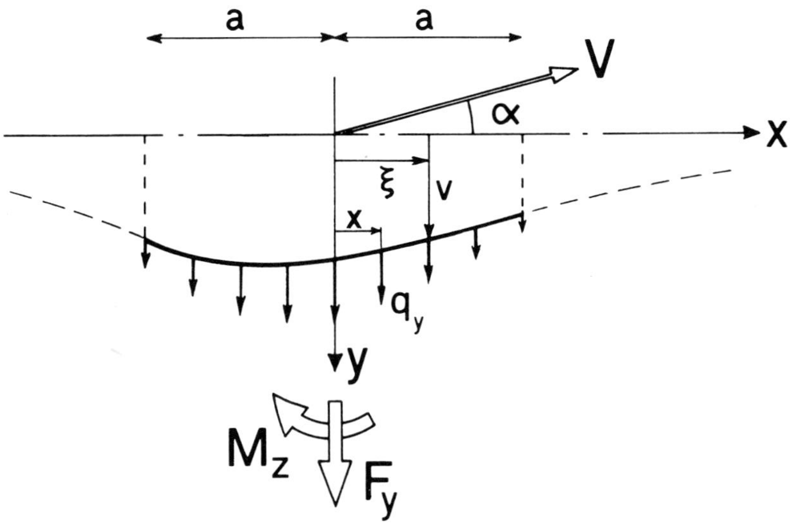

Fig. 11. Tire shear force model.

The lateral deformation, v, which is the result of a lateral force per unit length, q_y, continuously distributed over the contact length (2a), is found by means of the integral:

$$v(\xi) = \int_{-a}^{a} G(\xi,x)\, q_y(x)\, dx. \tag{41}$$

The Green's function may be obtained using the method of Savkoor (6). We shall assume a Green's function which is a symmetrical function of $x - \xi$ and which is approximated by the first two terms of the polynomial expansion of this symmetrical function:

$$G(\xi,x) = h + g(x - \xi)^2. \tag{42}$$

This approximation entails that the shape of the lateral distortion of the contact line, v (ξ), becomes parabolic as well, irrespective of friction functions and vertical pressure distribution:

$$v = a_o + a_1\xi + a_2\xi^2. \tag{43}$$

To simplify the calculations, a uniform pressure distribution, $q_z = q_o$, is chosen. The lateral shear force per unit length reads:

$$q_y = \mu q_o. \tag{44}$$

References p. 277

In the contact zone, two ranges may be distinguished. In the front portion, the speed of sliding is small and does not exceed the value V_{so}. In the trailing range, the sliding speeds exceed this critical value. The point of contact where transition from one range into the other takes place, is located at coordinate x_o.

The sliding speed V_s depends on both the slip angle α, and the gradient of the lateral distortion, v:

$$V_s = V\,(\sin\alpha + \partial v/\partial\xi) \tag{45}$$

With the use of the foregoing expressions, the integral equation (41) assumes the form (for small α):

$$a_o + a_1\xi + a_2\xi^2 =$$

$$q_o\int_{-a}^{x_o} \{h + g\,(x-\xi)^2\}\{\mu_o - B_s V\,(\alpha + a_1 + 2a_2 x)\}\,dx \tag{46}$$

$$+\,q_o\int_{x_o}^{a} \{h + g\,(x-\xi)^2\}\,A_s V\,(\alpha + a_1 + 2a_2\,x)\,dx$$

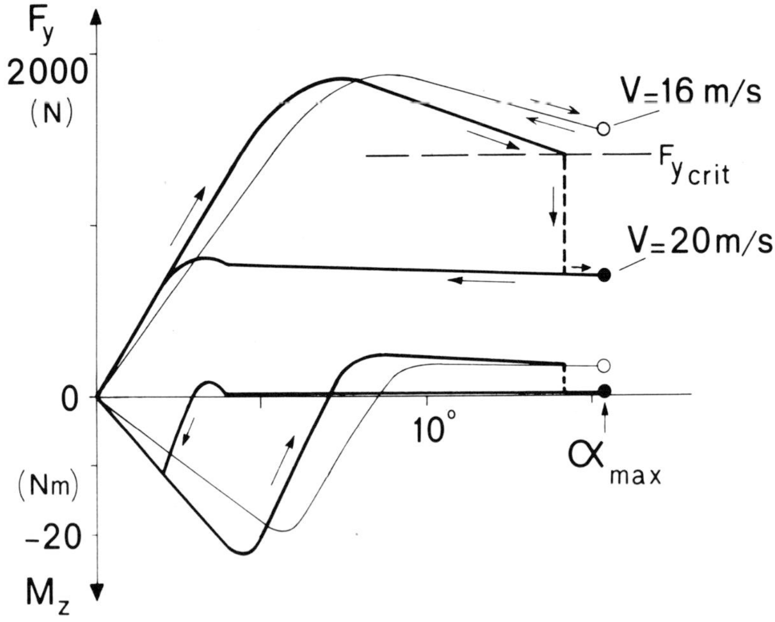

Fig. 12. Calculated side force characteristic showing jump to lower branch for a particular combination of high speed and slip angle.

Equating the coefficients of equal powers of ξ in the left and right member of this equation, we obtain expressions for the unknown quantities a_0, a_1 and a_2.

Once the distortion is known, side force and aligning torque may be computed:

$$F_y = \int_{-a}^{a} q_y \, dx, \quad M_z = \int_{-a}^{a} q_y \, x \, dx \tag{47}$$

The critical value of the side force $F_{y\ crit}$ determines which portions of the friction-speed characteristics (which were calculated with both possible friction functions) are valid. A jump from the upper branch to the lower one will occur when F_y drops below $F_{y\ crit}$.

Fig. 12 shows the loop which occurs at a speed of 20 m/s with a hypothetical tire and friction model configuration with parameters:
$a = 0.1$ m, $g = -0.000125$ 1/Nm, $q_o = 20{,}000$ N/m, $F_y > F_{y\ crit}$: $\mu_o = 0.6$, $V_{so} = 2$ m/s, $A_s = 0.25$ s/m, $B_s = 0.05$ s/m, $F_y < F_{y crit}$: $\mu_o = 0.2$, $V_{so} = 1$ m/s, $A_s = 0.25$ s/m, $B_s = 0.005$ s/m.

At the lower speed of 16 m/s, a loop is not established for the α amplitude considered. This agrees qualitatively with experimental results (5).

REFERENCES

1. *H. B. Pacejka, "The Tire as a Vehicle Component, Yaw and Camber Analysis," Chapter 7.5 of Mechanics of Pneumatic Tires, ed. S.K. Clark, pp. 757-839, NBS Monograph 122, Washington, D.C. 1971.*
2. *B. von Schlippe and R. Dietrich, "Das Flattern eines bepneuten Rades," Bericht Lilienthal Gesellschaft 140, 1941. English translation: NACA TM 1365, p. 125, 1954.*
3. *H. B. Pacejka, "Analysis of the dynamic response of a rolling string-type tire model to lateral wheel-plane vibrations," Vehicle System Dynamics 1, pp. 37-66, 1972.*
4. *H. B. Packejka, "Approximate Dynamic Shimmy Response of Pneumatic Tires," Vehicle System Dynamics 2, pp. 49-60, 1973.*
5. *H. C. A. van Eldik Thieme, "The Tire as a Vehicle Component, Cornering and Camber Experiments," Chapter 7.3 of Mechanics of Pneumatic Tires, ed. S. K. Clark, pp. 631-693, NBS Monograph 122, Washington, D.C., 1971.*
6. *A. R. Savkoor, "The Lateral Flexibility of Pneumatic Tires and its Application to the Lateral Rolling Contact Problem," SAE Paper No. 700378, FISITA/SAE Congress, 1970.*

DISCUSSION

Clark

We have a few minutes set aside for questions.

G. Potts

I am wondering about how you obtained the experimental natural frequencies of the tire.

Pacejka

For the experimental assessment of the camber or steer mode natural frequencies of the tire with respect to the wheel axle, we have used a balancing machine. The wheel can rotate freely around the shaft of the machine (by means of ball-bearings). The wheel is restrained from rotation with respect to space by means of a fixation with a piece of string. The shaft is suspended very rigidly in the vertical direction and very softly in the horizontal (lateral) direction. On both sides of the wheel, a disc is attached to the shaft. To both discs an imbalance weight is mounted. These weights are positioned 180 degrees out of phase. By rotating the shaft, the (now non-rotating) wheel is vibrated about a vertical axis. We have measured the vibrations of tire tread – band, rim and shaft with respect to space. The amplitude responses of the tire with respect to space, tire with respect to shaft, tire with respect to rim and rim with respect to shaft have been plotted. For a radial steel corded tire, we found for the resonance frequencies of these respective motions: 38.5 Hz, 37 Hz, 40.5 Hz, 37 Hz. This indicates that the peak frequency of the vibration of the relatively heavy shaft plus discs and wheel gives an indication of the natural frequency of tire plus rim being approximately 1.5 Hz too high.

By fixing the wheel to the shaft, we obtain the situation of the wheel rotating with the same frequency as the excitation frequency. In this case the shaft showed a resonance frequency of 25 Hz. The natural frequency of the tire plus rim will be a little lower.

Potts

This would seem to involve an interaction of the suspension system and the tire. Can you estimate what this interaction would be when you go from this suspension to a real suspension of a car. How would it change?

Pacejka

The mass of the shaft plus discs and wheel is large and the natural frequency in its suspension is very low. Of course, the situation will be entirely different with a real wheel suspension system. Such a system is very complex and shows a number of natural frequencies. The fact that the tire natural frequency comes closer to, for instance, the steer natural frequency will certainly increase its importance in the total performance. What the precise effect would be, requires deeper analysis.

M. R. Barone

Have you considered the larger amplitude response of this vertical in-plane effect, in other words, the possible coupling of the in- and out-of-plane vibrations at larger amplitudes? Do you think this is a possibility of significance or not?

Pacejka

You are referring to the coupling between vertical axle motion and lateral force?

Barone

And yaw moment.

Pacejka

In chapter 7.5 of my section in Clark's book on mechanics of pneumatic tires this problem has been analysed. I may refer to Fig. 7.5.48 of that section. Only the change of side force at constant slip angle with time-varying load has been considered.

A different problem is the response of the tire force and moment to camber changes which motion may be coupled to (large) vertical axle motions. This problem has not yet been analysed although the equations (24) – (30) presented in the present paper would allow for this without much alternation ($\gamma \neq 0$).

Barone

Well, I have that book but I do not believe that the coupling stability analysis is discussed there.

Clark

We will have to hold our further questions until later during the panel session. I am sure that what we have just heard represented rather a considerable amount of work, and I think we owe a vote of thanks to Dr. Pacejka for the paper.

The printed program lists a contribution from Eric Gough who many of you know, either personally or by reputation. For many years he was with the Dunlop Company and I think was quite well known for his ability to present very interesting papers, to generate comments, and if I might say, occasional disagreements at technical meetings. The organizers very kindly requested that he send over a paper, and as many of you know, he is not able to attend due to the fact that he is confined essentially permanently to his home with multiple sclerosis. He sends his best regards to many of his friends here, and regrets that he cannot be present, but did ask me if I would read at least part of the written contribution which will be under his name in the printed volume.

A TYRE ENGINEER LOOKS CRITICALLY AT CURRENT TRACTION PHYSICS

V. E. GOUGH

North Devon, England

ABSTRACT

The author discusses several points in the present theory of tyre traction which he feels need critical analyzation and notes several important areas of traction physics which should present challenges to today's research engineer:

1. Steady state sliding friction of tyre tread materials and energy loss concepts.
2. Tyres on vehicles in reference to topic 1.
3. Peak friction on braked wheels.
4. Micro and macro topographical road features and their consequences in tyre behaviour.
5. Tread design features.
    a. Slots, cuts, and sipes.
    b. Bulk modulus of tread rubber-trapped water globules.
    c. Three zone contact concept.
6. Tyre losses under torque conditions and the effective moment arm.
7. Tread reversion phenomena.

STEADY STATE SLIDING FRICTION OF TYRE TREAD MATERIALS AND ENERGY LOSS CONCEPT

Most laboratory measurements of friction of various rubbers or polymer blends have been made under steady state sliding at low speeds. It is commonly noted that the initial values are lower than the final steady state; or put another way, movement between rubber and ground is necessary to achieve the maximum friction pertinent to the materials and operating conditions.

References p. 297

The above mentioned steady state values have been shown to correlate with the viscoelastic properties of the polymer. Because viscoelastic mechanical or physical properties have been shown to follow the W-L-F transformation when speed of deformation and temperature of operational conditions are altered, the W-L-F transformation is widely assumed to be a valuable means of estimating the frictional behaviour at speeds up to one or two decades above the speeds of the laboratory determinations. Under locked wheel or high slip-conditions, particularly the former, it appears that this might be a legitimate inference.

Now the results of low speed laboratory tests conducted over a range of temperatures and at the limited range of speeds possible on the laboratory apparatus, when integrated into a continuous curve using the W-L-F transformation, produce a master curve. This "master curve" shows that the friction is a maximum at a particular position on the speed scale, the location depending on the temperature. This maximum value of the friction coefficient is of the order of 2 to 2-1/2 on clear dry smooth surfaces. So far, I am not suggesting any criticism of the thinking except perhaps to ask whether the W-L-F transformation, in which the speed changes are on a logarithmic scale, is to be taken as an exact law or as just a very good and useful approximate law indicating that the same factor or similar factors which influence the mechanical cum physical behaviour of the polymer have a strong influence on the molecular properties of the surface, which determine the friction of a clean polymer surface?

It is often thought, and in fact, originally claimed that these findings proved that the frictional force developed was due to internal hysteresis losses in the polymer; more recently it is often thought that the hysteretic characteristics of the surface and energy losses at a molecular or atomic level are the cause of polymer friction.

TYRES ON VEHICLES AND RELEVANCE OF LABORATORY FINDINGS DISCUSSED IN THE PRECEDING SECTION

While the polymer molecular or like characteristics are strongly related to – and indeed may be accepted as the factor which determines – the steady state sliding frictional forces obtained under a given set of velocity, temperature and type of mating surface, the friction is not directly due to energy losses either at macroscopic, microscopic, molecular or atomic or electrical levels.

Consider the following logical argument:

With locked (non-rotating) braked wheels, no energy is dissipated in the brakes. The brake transmits torque.

With braked but rotating wheels, some energy is dissipated in the brake, some within the tyre structure and some at the tyre-ground interface.

The total energy to be dissipated – primarily if not wholly, transformed into heat – to bring a given vehicle to rest from a given speed is determined by vehicle mass and initial speed. (For this discussion ignore small factors such as air resistance).

The proportion of the energy dissipated and turned into heat in the brake depends on the wheel "slip ratio". The remainder is dissipated primarily at the tyre-ground interface – the increase in tyre internal losses due to the transmission of torque are, at most, two or three times the free rolling resistance of the tyre, which is low being equivalent to a drag force between 1 and 2 percent of the load on the tyre.

Under conditions of braking in which the tyre rotates, including that of peak retardation which occurs at wheel slips of between 2 to 12% approximately, depending on the tyre-ground characteristics and conditions (the higher figure arising under good surface retardation conditions, the lower figures under very slippery conditions), 85 to 95% of the energy is transmitted (by the force transmission characteristics of the tyre-ground interface) to the brakes where it is dissipated as heat.

To take a specific case of approximately 5% wheel slip, about 6 or 7% of the energy is lost in the tyre system, mostly at the surface of contact of tyre to ground – say 5% there, and 1 or 2% within the tyre structure, leaving 93-94% in the brake. The tyre transmits force through its contact with the ground to apply the torque to the brake.

It follows at once that any theory which ascribes frictional grip or tangential force transmission of polymer to road surface as being *entirely* due to energy *loss* within, or on the surface of the polymer, whether it be couched in terms of heat generation by mechanical vibrations or other distortions or in terms of molecular, atomic or electrical phenomena, will not stand a moments' critical inspection. If the energy to be dissipated to retard the vehicle is lost in this region as is demanded by such theory, how does it get to the brake to appear as heat? This all engineers know but some physicists tend to ignore.

One can only conclude that while hysteresis loss effects as such do play some (small) part in frictional retardation, it is not hysteresis but force transmission by traction of the surface-to-surface adhesion type which is the fundamental factor.

In the case of locked wheels, no energy is lost in the brake, but rather it is lost at the tyre to ground interface. In the case of *maximum* braking with rolling wheels, approximately 95% of the energy is lost in the brake so only 5% of the energy is lost in the tyre-ground system which is performing the function of a transmitter of force.

Polymer molecular characteristics determine bulk characteristics such as modulus and hysteresis and also surface characteristics, so there may be (or most likely is) a correlation between hysteresis and friction: it is not a "Causation" – it could well be the same or very similar features of "entanglement" of "free ends" or "cross-links"

References p. 297

between the "long chain molecules" which determine the two features. Clearly, friction is not caused by losses at the tyre to ground interface, and so friction is due primarily to a form of surface-to-surface adhesion capable of transmitting traction forces without necessitating an energy loss mechanism.

A further point of interest in this matter is that in traction, changes due to polymer changes are greater in magnitude at the peak than at full sliding although the number of cycles of vibratory movement in the distance of one contact length is twenty times as great because in the former case the "slip" is about five percent.

In support of this last point, I offer the following data from my discussion of Allbert & Walkers' and Maycocks' papers (1). It shows one effect of a change of tread material from natural rubber to that of high grip rubber (so-called effect of high hysteresis) as revealed by the variation of the ratio of change in peak coefficient $\Delta\mu_\rho$ (at low, around 5% slip) to change in locked wheel sliding coefficient $\Delta\mu_s$ (100% slip).

$$\frac{\Delta\mu_\rho}{\Delta\mu_s}$$

	Mean Value	Lowest Value	Highest Value
At 30 mph	2.0	1.5	Around ∞
At 60	4.0	2.0	∞

These values cover 21 comparisons made over bald, ribbed and multislotted tread patterns and a wide range of road surfaces of various mixtures of macro characteristics and micro characteristics. As can be seen from the above data, in no case was the effect of change of tread material as great at 100% slip as at low (tyre rotating) slip (peak coefficient). Consequently, there is not even one case which supported the idea that the hysteresis losses as a means of dissipating energy is the cause of the better retardation.

To sum up – much that the rubber physicists tell the engineer about the phenomena of friction will not stand critical inspection.

PEAK FRICTION ON BRAKED WHEELS

It has now been well established both on roads and in the laboratory that during the rising part AB of the braking force slip curve from zero slip through low "slips" (see Fig. 1) the curve is dominated by the tyre properties and is substantially "slip distance dependent" and substantially independent of speed. It is dependent more closely on "slip ratio" than vehicle speed.

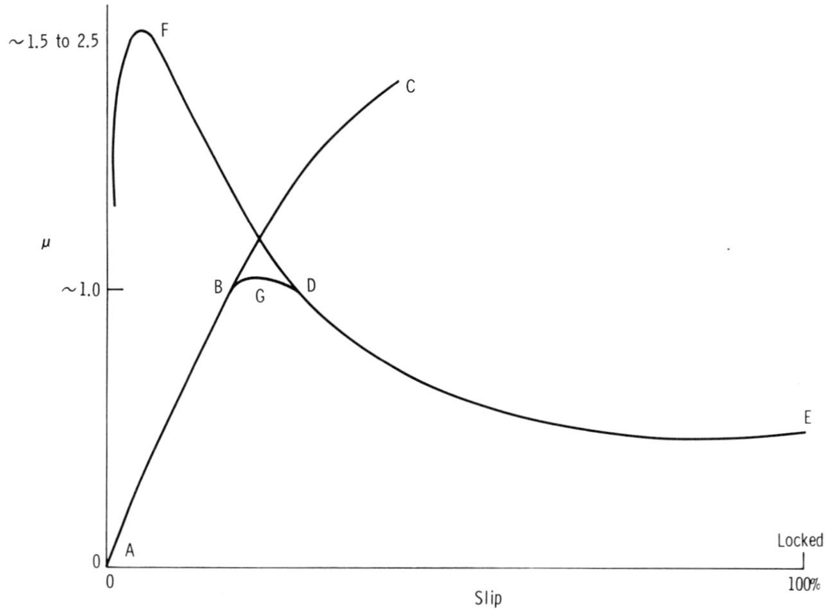

Fig. 1. Retardation of braked wheels.

On the other hand, it is clear that at the locked wheel point E the same piece of tyre tread is sliding bodily over the road. This is a steady state case of sliding, and the retardation value depends on speed together with such features as tyre tread pattern, type of road surface, state of contamination, and temperature. Similar steady state cases are discussed in the first section of this paper.

In the region of medium to high slip, D to E, the retardation values are also speed dependent, and are usually (and reasonably) considered to belong to the same kind of steady state phenomena (such as locked wheel sliding and the cases described in the first section of this paper).

Some physicists consider that the sliding regions are all fully described by the viscoelastic type theory while others consider them to be complex environments in which there are many factors influencing traction of which the aforementioned (rubber viscoelasticity) is only one.

Those who consider these sliding regions to be determined by the viscoelastic factor alone consider the peak, BGD, to be identifiable with data that was observed in the lab tests described in the first section of this paper. I do not consider this to be

References p. 297

the case; and in fact have not for many years, long before my papers in 1958. I believe that the peak G is caused by the intersection of the curves of the two different cases, that of the rising part ABC, which depends primarily on the non-slipping rolling tyre (not on speed) and that of the decreasing part FDE which depends primarily on the slipping tire (on speed). In this interpretation, the peak F of the steady state lab tests is shown in Fig. 1. The peak region BGD is a transition between these two effects involving, amongst other factors, the movement observed before the maximum coefficient and/or the steady state value is achieved (both of these regions have been noted by many if not all workers, e.g., friction meter of Gough). Note that the usual bell shaped steady state sliding curve is usually on a logarithmic speed scale: FDE is on a linear scale.

Just as in the case of the St. Andreas fault or Rift valley, it is the rapidly rising curve of the fault or rift rather than the character of the adjoining plains which specifies this peak region and explains why G is not as high as F.

MICRO-MACRO ROAD TOPOGRAPHICAL FEATURES

It is not possible to set out road surfaces in a monotonic line. There is a dichotomy. Both macro- and micro-features must be stated as can be seen from Fig. 2 (the data used in constructing this figure are listed in Tables 1 and 2). When tyres are tested on a range of road surfaces, it is found that both the macro- and micro-roughness of the roads are important. Four surfaces have been selected to illustrate the findings:

a. Open or coarse macro-texture, sharp micro-texture or harsh or rough aggregate.
b. Open or coarse macro-texture, intrinsically smooth or polished aggregate.
c. Fine macro-texture, sharp micro-texture or harsh or rough surface.
d. Fine macro-texture, intrinsically smooth or polished surface.

A short line (connecting the data points for tests conducted at the same speed) on Fig. 2 means a small dependence on speed – from this it can be seen that coarse or open macro-texture surfaces tend to ensure that the coefficients will be independent of speed. However, sharp or harsh micro-texture of the road surface is essential to ensure high coefficients irrespective of the macro-texture. Harsh micro-texture is particularly necessary on open macro-texture roads.

Tyre tread patterns have the greatest effect on closed or smooth macro-texture roads. Tyre tread patterns do not make up for the deficiencies of polished smooth open macro-texture road surfaces. This is due in part to the fact that, unlike tread pattern grooves, grooves in road surfaces retain and feed water into the contact patch.

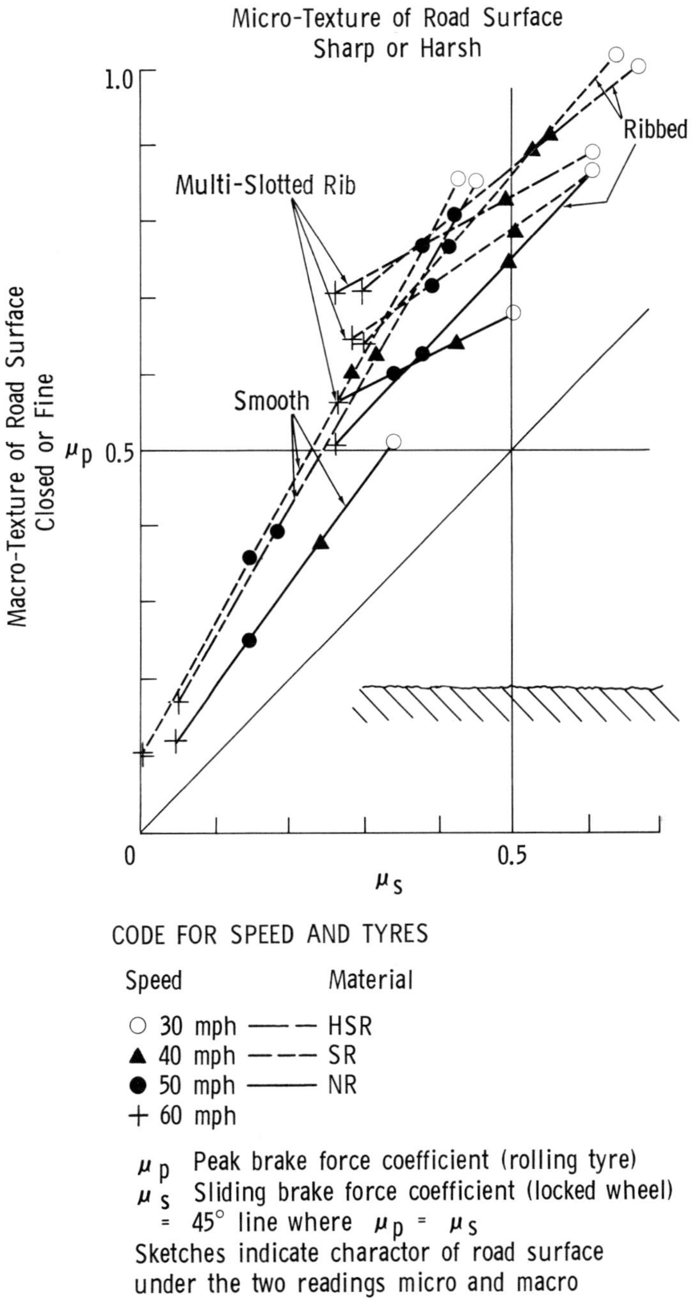

Fig. 2a. Peak (rolling) and slide (locked) brake force coefficients at four speeds of tyres of three types of tread pattern, and three tread materials on four contrasting road surfaces showing the influence of micro and macro texture of road surface and how they interact with tyre variables.

References p. 297

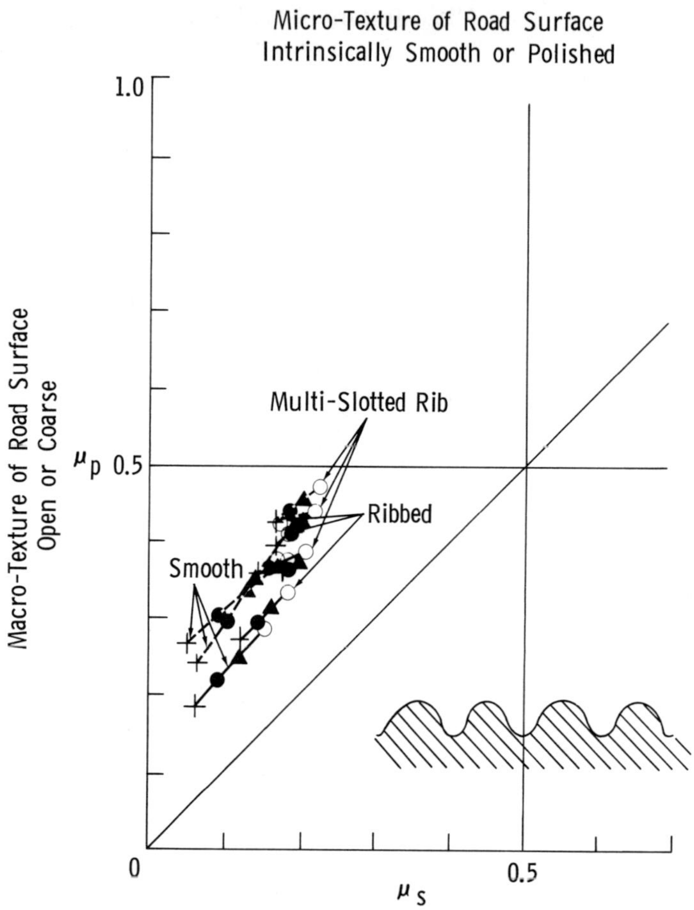

CODE FOR SPEED AND TYRES

Speed	Material
○ 30 mph —— —	HSR
▲ 40 mph — — —	SR
● 50 mph ————	NR
+ 60 mph	

μ_p Peak brake force coefficient (locked tyre)
μ_s Sliding brake force coefficient (locked wheel)
= 45° line where $\mu_p = \mu_s$
Sketches indicate charactor of road surface under the two readings micro and macro

Fig. 2b. Peak (rolling) and slide (locked) brake force coefficients at four speeds of tyres of three types of tread pattern, and three tread materials on four contrasting road surfaces showing the influence of micro and macro texture of road surface and how they interact with tyre variables.

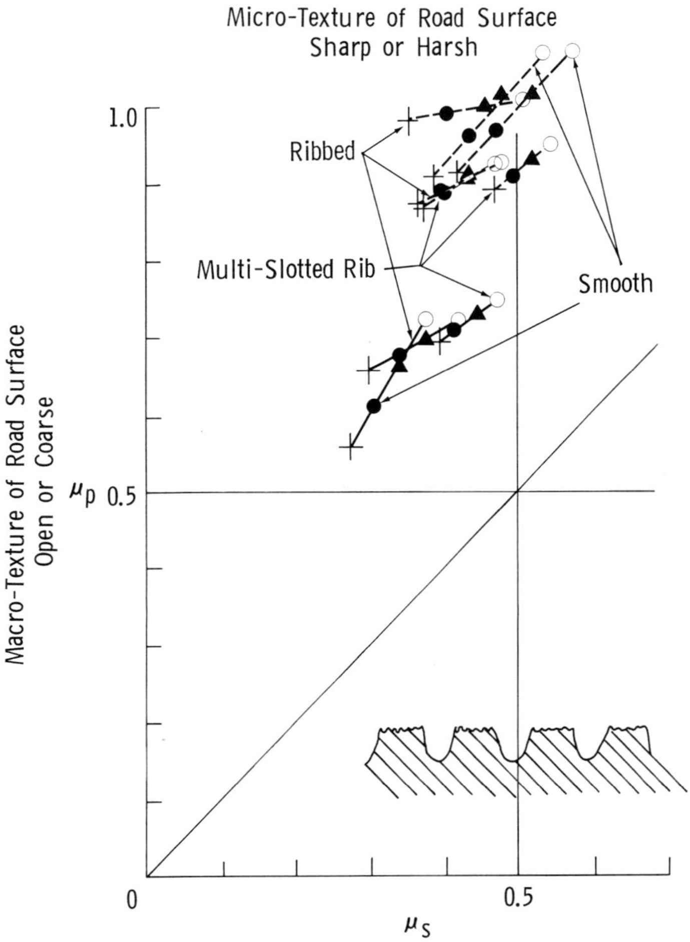

CODE FOR SPEED AND TYRES

Speed	Material
○ 30 mph	—— — HSR
▲ 40 mph	— — — SR
● 50 mph	——— NR
+ 60 mph	

μ_p Peak brake force coefficient (rolling tyre)
μ_s Sliding brake force coefficient (locked wheel)
= 45° line where $\mu_p = \mu_s$
Sketches indicate charactor of road surface under the two readings micro and macro

Fig. 2c. Peak (rolling) and slide (locked) brake force coefficients at four speeds of tyres of three types of tread pattern, and three tread materials on four contrasting road surfaces showing the influence of micro and macro texture of road surface and how they interact with tyre variables.

References p. 297

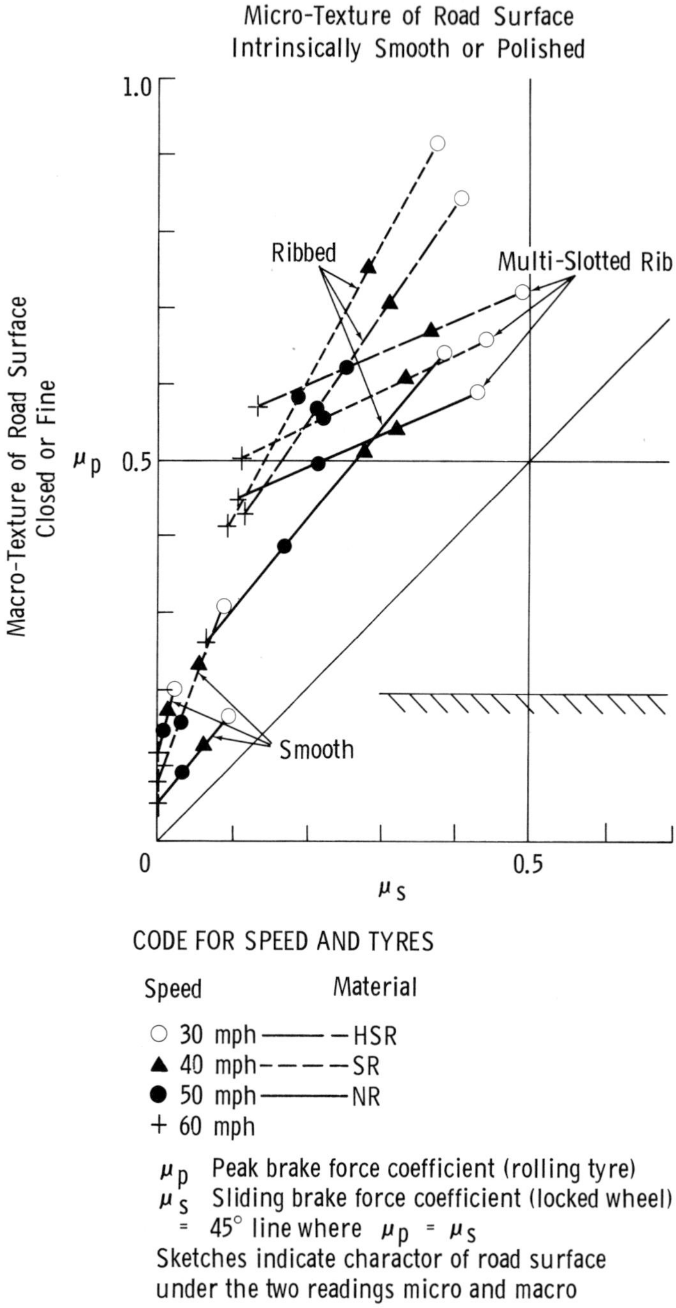

Fig. 2d. Peak (rolling) and slide (locked) brake force coefficients at four speeds of tyres of three types of tread pattern, and three tread materials on four contrasting road surfaces showing the influence of micro and macro texture of road surface and how they interact with tyre variables.

TABLE 1

Influence of tyre and road factors on safety
(based on data in Fig. 2.)

Road / Tyre Tread Rubber / Vehicle Speed Mph / Tyre Pattern or Smooth / Wheel Locking / Road				Micro Texture of Road Surface							
				Sharp or Harsh				Polished or Intrinsically Smooth			
				Best		Worst		Best		Worst	
				30	60	30	60	30	60	30	60
Macro-Texture of Road Surface	Open or Coarse	Locked	Patt	✓	x	✓	x	x	x	x	x
			Smooth	✓	x	x	x	x	x	x	x
		Rolling	Patt	✓	✓	✓	✓	x	x	x	x
			Smooth	✓	✓	✓	✓	x	x	x	x
	Fine or Closed	Locked	Patt	✓	x	✓	x	✓	x	x	x
			Smooth	x	x	x	x	x	x	x	x
		Rolling	Patt	✓	✓	✓	✓	✓	✓	✓	x
			Smooth	✓	x	✓	x	x	x	x	x

✓ Safe 〉 0.5
x Unsafe 〈 0.5

TABLE 2

Summary table showing the effect
of road texture on number of ticks
(indicating safe operation)

re coefficient 〉0.5		Micro Texture: Harsh	Micro Texture: Polished
Macro-Texture	Open	11	0
	Closed	8	4

TREAD DESIGN FEATURES

While the locked wheel state can be reconciled with bearing seizure, what mechanical engineer would accept that the following specification could be met

References p. 297

reliably and repetitively upon instant demand: on a completely flooded surface, "seizure" is to occur within one hundredth of a second, the surface being reflooded and the process repeated ten times a second. This is what happens with a rolling braked tyre at peak coefficient, and is what a tread is designed to ensure.

Furthermore, as the three zone concept (Fig. 4a) shows, the actual seizure contact may last only a few milli-seconds (at high speed conditions), and so the physicists' energy loss concept (described in the first section of this paper) cannot be considered as plausible whereas the transmission of traction force can.

Slots, Cuts, Sipes in Tyre Tread Pattern – On closed or fine macro-surfaces, these being the cases for which patterns have the greatest effect, cuts or sipes should empty the water into the adjacent grooves for highest peak values at low speed. (See Fig. 3 rib a.)

However, closed slots are more efficacious at high speed as the air in the closed slot holds the water back in the early zone but provides a space for the last traces of water in the third zone. (Sec Fig. 3 rib d.) This corresponds in part to the "blotting paper effect". If the slots are close to each other, the water has only a short distance to travel to escape from the interface. This is essential because of the short periods of time that are involved. Cylindrical holes of small cross-section sufficiently close together can be very effective as "blotting paper". Large holes merely ensure that water is not forced throughout the contact patch causing loss of effective contact. A

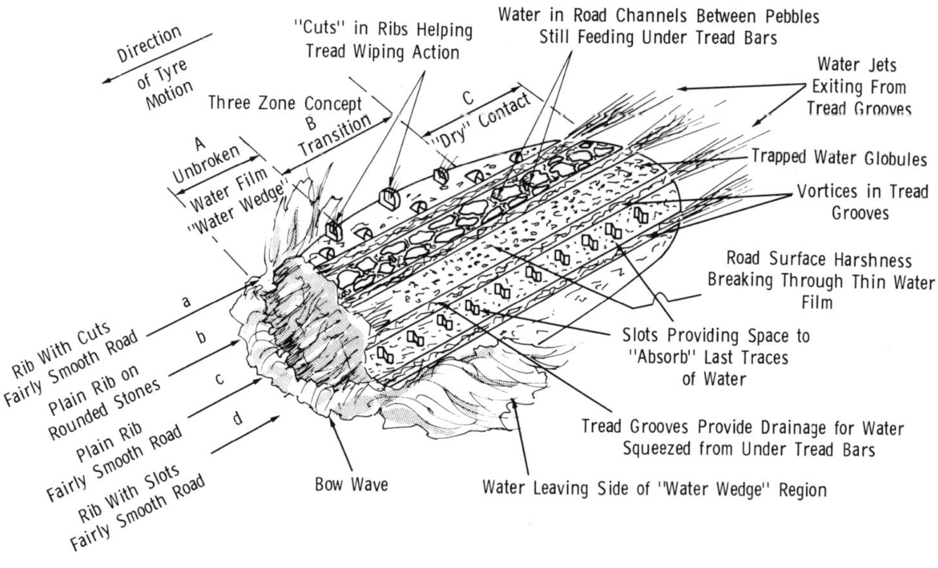

Fig. 3. Schematic composite phantom view of water film between tyre and ground illustrating the wide variety of phenomena which can occur in the tyre to ground contact patch on wet roads for tyres of differing design on roads of various character.

nice analytical problem could be posed in determining spacing, hole diameter, and depth for optimum retardation at various speeds.

Trapped Water Globules – Because the bulk modulus of rubber and that of water are similar in magnitude, globules of water can remain trapped under tread ribs (and certainly on bald treads) causing a loss of effective area of the "dry" third zone. (See Fig. 3 rib c.)

I have known of this for many years, since this can be easily deduced, for example, from an application of St. Venant's principle. Alternatively, the existence of a nearly uniform stress in all directions at the centre of a rib on which this globule entrapment is based can be demonstrated by an analysis of the dynamic relaxation stress based on classical theory of rubber blocks compressed against rigid plates. To put it in a "nut shell", the water does not know which way to go and so remains as a globule. The usual squeeze film theory (Saal etc.) usually assumes infinite rigidity of plate and rubber.

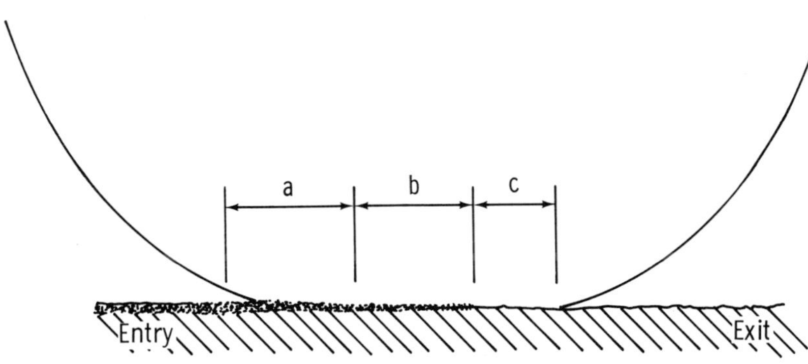

Fig. 4a. Three zone concept of water film in tyre ground contact patch. (a) Unbroken water film. (b) Transition between (a) and (c). (c) Dry Dry or substantially dry contact.

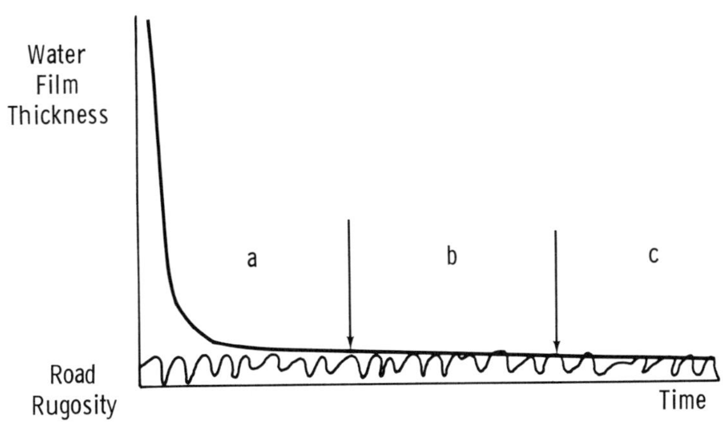

Fig. 4b. Reduction of water film thickness with time. The zones a, b, and c are as in Fig. 4a.

References p. 297

Three Zone Contact – This was the basis of my thinking several years before I disclosed it in public in 1959 (2); it was quite independently discovered by Moore (3) who has developed it in greater mathematical detail. My thinking is indicated in Figs. 4a, b, i.e., the micro-harshness is such that it breaks through the water film in the mid- or transition zone.

This suggests that two types of road surface might be developed for good wet traction: (a) a macro-smooth mastic asphalt which has, however, micro-sharp grit which comes to the surface as the mastic wears but is of sufficient quantity and appropriate size to fade out before it is traffic polished; and (b) a rough macro-texture road made of friable road stone which breaks up to maintain harsh micro-roughness before it traffic polishes.

TYRE LOSSES UNDER TORQUE CONDITIONS AND EFFECTIVE MOMENT ARM

The SAE term "Rolling Resistance Moment" is misleading – nay it is fallacious!

The effective moment arm relating torque and traction force is not equal to the axle height "h" nor is it equal to the effective rolling radius – "γ" (the distance the axle travels per radian of rotation). These statements follow logically from the following:

The equation of moments,

$$F = C/h - Wa/h \tag{1}$$

being basic physics, *must* be satisfied (C is the input torque to the wheel, W is the load, and "a" is the offset of the centre of pressure from the mid-point of contact). The equation of energy.

$$F = \frac{C}{\gamma} - D \tag{2}$$

where D is the effective drag force representing the magnitude of the losses internal in the tyre plus any tyre to ground interface losses, being basic physics *must* be satisfied. In practical terms, D is the force to tow the rolling tyre.

The relation between h and γ *is* empirical and must either be determined experimentally or from a mathematical model which has an empirical assumption *somewhere* within it.

Typically, γ_o (γ for free rolling with no friction loss at the interface) is often taken as

$$R_1 - d/3 = h + 2d/3 \tag{3}$$

where R_1 is the original outer radius of the tyre and d is the tyre deflection. This relationship contains the empirical assumption that the arc of the tyre surface is compressed to the chord length on the ground.

A more elaborate, but still empirical model gives

$$\gamma_o = h + 2D/3 - t\left(1 - \frac{P}{3AE}\right) \tag{4}$$

where t is the thickness of a tread of modulus E, the tread design is such that the ratio, 'land'/total area = A, and P is the tyre inflation pressure. A belted radial casing tyre frequently has γ_o equal to the radius of the undeflected belt because of its high longitudinal modulus. Even this is an empirical assumption. For braked condition $\gamma > \gamma_o$: for traction condition $\gamma < \gamma_o$. Thus, γ is ∞ for the case of locked wheels and zero for spinning wheels that have zero forward movement of the axle.

Equations (1) and (2) lead to

$$D = C(1/\gamma - 1/h) + Wa/h = Fh(1/\gamma - 1/h) + Wa/\gamma . \tag{5}$$

These equations lead to

$$\frac{C}{F} = \frac{Dh - Wa}{D\gamma - Wa}\gamma = \frac{D/W - a/h}{D/W - a/\gamma}h. \tag{6}$$

This shows that the "effective moment arm" relating C and F is neither γ nor h nor is it constant but that it depends on D, h, γ and "a" which all vary with torque under the assumption that W is constant.

Equations 5 and 6 can be written in various forms. One form of equation 5 is

$$\frac{(D/W)}{(a/h)} = 1 + \frac{1 - (h/\gamma)}{\frac{Fh}{C} - 1} \tag{7}$$

which is used as the basis of a chart given in my contribution to Turley's IME paper (4).

Only in the case of C=O does the so-called 'rolling resistance moment' Wa give the correct value for the drag losses D_o. Paradoxically the rolling resistance moment' might be a justifiable name when the vehicle drive shaft does not have to provide a moment or torque!

In all other cases, the loss does not depend directly on the 'rolling resistance moment', Wa. In this connection, it should be noted that during braking "a" decreases and can even change sign. If Wa were the only measure of losses, the tyre would be a source of power – a perpetual motion machine. Tyres are accused of many things but not of that.

References p. 297

TREAD REVERSION

There are many operational conditions and many forms of tread surface reversion which come under this heading and so one unique and simplified explanation of this topic is not possible.

Its most common occurrence is on aircraft tyres but it can be obtained on truck tyres. It is usually associated with high speed and high or moderately high inflation pressures. These conditions result in low effective braking drag.

Certain features are worth placing on record. The first cases that Hardman and I (5) noted were obtained in the early part of World War II with aircraft tyres with 45 psi. inflation pressures. They were on aircraft landing with locked wheels on wet grass airfields with an experimental (and especially long travel) landing strut for ease of landing by injured pilots or damaged aircraft. The tyre damage was observed as a porosity of the tyre surface caused by high temperatures developed by high stressing just below the tread surface. We had several possible theories at the time, but ultimately were convinced that the porosity was caused by the high shear in the tread material caused by the tyre sliding over the ground (or coarse grass) roughness that is predicted by applying the Hertzian theory of contact stresses. However, the most important point to be made here was that the actual tyre tread surface of these tyres was completely undamaged – the moulded surface showed the same scratches as the mould itself proving that the reversion started from within, but close to the tread surface. This statement can be made as a positive fact because previous examples of tyre damage which caused us to investigate the phenomena were claimed to be due to dirty moulds, and we were using tyres from moulds which had been specially cleaned with emery cloth leaving a characteristic finish.

Since that time many and varied types of reverted tread damages have been observed on aircraft tyres used in landing on both wet paved runways and on wet grass airfields – all associated with locked wheels. Some damages have also been observed on high speed truck tyres. In extreme cases, the whole contact patch has reverted, and in most cases the original tyre tread surface is not present or not detectable.

It is of interest to note that in some instances of severe general reversion the edges of the reverted contact patch have the appearance noted in the first observed cases, i.e., subcutaneous blistering without reversion of the actual tread surface.

I do not propose to discuss reversion in much further detail but suggest that many varied forms exist – melted rubber, steam, etc. all of which give low coefficients persisting sometimes down to low speeds, and all of which were started by high speed braking with wheels locked on wet surfaces. I strongly recommend that anyone studying these phenomena "gets his coat off" and examines both tyre damage and road conditions in careful detail. Some of the cases I have seen suggest a real "witches cauldron" in the contact patch; and such factors as road unevenness, patches or puddles of water coupled with axle load variations due to imbalance are necessary factors to get the observed end product. Otherwise how does the water to produce the steam get to the hottest end,which is the exit of the contact; or alternately, why is the exit end of the contact abraded away though it can, in some cases, have the same perverted appearance as the entry end of contact?

Many years ago Obertop (6) suggested that the basic mechanism of skidding was that steam determined the area of the effective "dry" patch, and he got some very good numerical estimates of observed drags. But this general theory "would not hold water" (forgive the pun) – the hottest part of the dry contact region,which is where the steam would be produced, would be at the exit which leaves us with the unanswered question of how did the water get to that region after having been removed by the "dry patch"? Hence, my previous illusion to a witches cauldron implying the presence of non-steady state tyre contact and road characteristics in various if not all regions of the contact patch.

REFERENCES

1. *V. E. Gough, "Discussion of Papers by Allbert and Walker and by Maycock", Journal of Rubber Chemistry and Technology, pp. 816-821, 1968.*
2. *V. E. Gough, "Friction of Rubber on Lubricated Surfaces", discussion of a paper by D. Tabor, Rev. Gen. du Caoutchouc, Vol. 36, No. 10, 1409, 1959.*
3. *D. F. Moore, "Intermin Report on a Study of Tire-Surface Interaction for the Case of Rolling on a Wet Surface", CAL Report No. YD-1969-V-1, Cornell Aeronautical Laboratories, January, 1965.*
4. *J. A. Turley, "Engineering Aspects of Tyre Testing", Institution of Mechanical Engineers, Proceedings 1970-71, Vol. 185 74/71.*
5. *V. E. Gough, J. H. Hardman, and R. J. Mac Claren, "Abraded Tyre Treads," Transactions of the Institution of the Rubber Industry, Vol. 32, 1956.*
6. *D. H. F. Obertop, "Decrease of Skid Resisting Properties of Wet Road Surfaces at High Speeds," Symposium on Skid Resistance, pp. 102-112, ASTM Special Publication No. 326, 1962.*

PANEL DISCUSSION
"TIRE TRACTION – THE ROLE OF THE TIRE"

S. K. Clark (Moderator)
R. N. Kienle
H. B. Pacejka
R. F. Peterson

Clark

Perhaps first of all, I should say just a very brief word about the conduct of this panel session, which may be rather lengthy. We will try to address ourselves to several topics. I will ask for comments from some of my fellow panel members on things designed to generate discussion, I hope not too heated, and after that I would like to ask for participation from the audience with questions, comments, and opinions that you might have, because since we are all in this meeting together to learn, I am sure that we can learn more from you than you can from us.

First, do any of the panel members care to comment on the paper by Eric Gough?

R. F. Peterson

As Mr. Gough has stressed, a regression equation shows only that a correlation exists, and does not necessarily argue for a causative relationship. He has raised many good points and I do not have the data to argue them.

For instance, we measure coefficient of friction either by peak and slide tests, or from stopping distance measurements. Our data obtained using acceleration traction tests did not show large differences between stocks differing greatly in hysteresis; however, the data had a great deal of scatter and I would hesitate to cite this as evidence supporting Mr. Gough's arguments.

R. N. Kienle

I will make some comments. They are going to sound awfully naive, and I guess they are. If you had perfect friction, it seems to me, what you are transferring is energy from the car to the world. I guess the analogy is that if everyone was walking east, the world would be moving west, to a finite extent. You do not really generate friction. You do not generate heat in the brakes to dissipate the energy. This does not help you stop the car. There is the limiting point at which you lose your traction. It is then that heat is generated. What we are looking at here is not a mechanism for transferring momentum or the energy in the car into heat, but we are looking at the borderline conditions. So what we are doing is trying to determine that property of the materials which limits the amount of transfer that you can make from the tire to the world. Well, that, I think, is the point I want to make.

K. A. Grosch

I think Dr. Kienle is right. It is not a question of brakes or road. It is a question of the mechanics of slipping wheels in which the friction coefficient is part, but not the whole story.

Clark

Are there any other comments from the audience? The general subject that we have just discussed here brings to my mind two or three questions. The first one is the differences or rather the explanation for the existence of a peak friction coefficient and a sliding friction coefficient. Would either of you two care to comment on that point?

Kienle

Well, I will venture to mention one thing which I think is involved, and that is that I think you are dealing with a difference. You are dealing with a lateral force. I think that in peak and in slide you are dealing with a difference in the contact area that you are generating or making the transfer of energy through. This is one of the reasons why you have a peak and a slide. Slide, if you wish, occurs more at a right angle to the road, and, therefore you have less area involved (in terms of the surface that you are transferring your energy through). I think that this is one reason why the slide is always lower than the peak.

Peterson

We believe that there are three major factors, and probably a number of smaller ones, which combine to produce differences between peak coefficients and slide coefficients. They are: temperature effects, viscoelastic effects, and drainage effects. First, frictional heating of the rubber raises the temperature substantially during a skid. Since hysteresis above T_g falls with increasing temperature, we see lower friction coefficients at high temperature (during the slide) than at low temperature, when the rubber has not yet had time to be heated. Viscoelastic effects are similar; as speed increases, the rate at which you pass over asperities also increases. Hysteresis, hence friction, passes through a maximum with increasing frequency and speed. The first two factors are interrelated through time-temperature principles. Finally, drainage effects are important on wet roads. At 100% slip, drainage of water to establish contact or interaction between rubber and road is more difficult than in a partial slip. Contact area will be larger, and the tire will be better drained, and coefficients will be higher under conditions of lower percent slip.

G. G. Balmer

Dr. Peterson, the difference between slide and peak braking is very small on dry pavements as contrasted with the difference on wet pavements. Can you relate this to the comments you made a few minutes ago?

Peterson

Your (Kienle's) data indicate to me that a large fraction of the difference between peak and slide coefficients on wet pavement is due to differences in the drainage capacity of the grooves of a rolling (peak) and sliding (slide) tire. Drainage effects, of course, are not operative.

Kienle

I would like to make a comment on that. In recent SAE papers, we made some measurements both wet and dry over a seven-month period. If you look at the low speed ratio of peak to slide, both wet and dry, the ratio is quite constant on any surface, with any tire, even heavy service tires. In fact, in that one slide (projection) where I said neglect that line, it was the average line for all values of 20 mph ratio of peak to slide. You get quite a variation, of course, but you do get quite close to a single ratio of peak to slide on either a wet surface or a dry surface, as long as you keep to low speeds where you can get rid of the water and you do not have this water effect coming into play.

Peterson

Did the tires run in that test include changes in design and changes in tread recipe?

Kienle

Yes, there were radial and bias tires. The tires had different compounds, different tread designs, different aspect ratios, including heavy service as well as passenger tires.

Peterson

And they all had about the same ratio of peak coefficient to slide coefficient at 20 mph, about 5:4?

Kienle

A bit lower than that. I think the final analysis ended up about 1.16.

Clark

I think the first question came from John Tanner. Do you still want to ask it, John?

J. Tanner

I just want to make the comment that the difference between peak and sliding friction coefficients on dry surfaces is highly dependent upon speed, and the higher the speed, the greater the difference between the two. Another thing to consider when you are dealing with a locked-wheel is the risk of getting into a reverted rubber skid which we have observed during aircraft tire testing. This change in the rubber composition at the tire-pavement interface yields an entirely different skid phenomenon which can result in a severe loss of traction.

Peterson

Since reverted rubber is a phenomenon exhibited primarily by natural rubber, it will be a serious problem only in tires made from natural rubber. Since passenger tire tread stocks are now mostly synthetic rubber, reversion is not a major problem here.

K. A. Grosch

I just wanted to make the comment that the peak and slide coefficients on dry surfaces can also be deduced from the master curve. At very low sliding speeds, the temperature rise is low and the friction coefficient tends to increase with increasing speed. But with increasing speed the temperature rises and tends to reduce the $a_T v$ values. The friction passes through a maximum not to be confused with the maximum of the master curve. On the wet surface it is probably a combination of temperature rise and increasing lubrication condition.

A. R. Williams

I believe that Dr. Gough is asking the question, what is the most efficient braking process as determined by the need to dissipate the kinetic energy of the vehicle. From this aspect, the peak rolling condition of friction is the most efficient where over 90% of the energy is dissipated within the brakes.

Under non-rotating wheel conditions, over 90% of the energy has to be dissipated at the tyre to ground interface. For the majority of abrasion conditions initiated during a non-rotating tyre braking cycle the highest temperature is produced at the tread surface with heat flow into the road surface and into the bulk of the tread rubber. Under these conditions, a high rate of rubber abrasion is experienced. The phenomenon known as reversion is somewhat different from the normal forms of tyre abrasion actually occurring under conditions associated with low tyre wear.

Chemical analysis and the measurement of temperatures by the use of thermocouples placed at different depths below the tread surface have shown that the maximum temperature under conditions that lead to reversion is actually below the tread surface, and heat flow is such that the tread surface is a heat sink.

The conditions under which reversion may occur are summarised:

a. non-rotating tyre
b. low friction surface i.e. on wet and highly polished, or ice/frost covered surface
c. open textured surface or rough macro-texture

The porous layer formed below the tread surface originates by the production of volatile materials or by depolymerisation to the monomer which occurs at temperatures in the region of 250°–300°C.

I would like to ask Dr. Peterson if he has considered the different compounding requirements needed to optimise the peak rolling rather than the locked-wheel sliding condition. The latter condition, as stated, is the more difficult from the aspect of energy dissipation, and also for producing adequate drainage within the tyre contact area. It would, therefore, seem that tyre wet traction could be optimised around either condition; however I would suggest the peak rolling condition as being of greater importance.

Clark

I think we have another question.

R. A. Moyer

Over a period of 40 years in which I have been engaged in the measurement and evaluation of coefficients of friction for many different tire and road surface types and conditions, I have relied primarily on the results of braking and cornering tests over a wide range of speeds with a two-wheel trailer to document and postulate explanations for the large differences in the coefficients of friction measured in these tests. I have found it most helpful to explain the large increase in the braking force coefficients under peak rolling-wheel braking versus locked-wheel braking in terms of a rack and pinion gear arrangement or in terms of an analogy of "gearing the tire to the road." The results of tests have consistently shown that under peak rolling-wheel braking with the tire "geared" to the road large braking and traction forces are developed with low percentages of slip. The rubber tire has remarkable properties as a highly elastic and resilient material. It has high tensile and shear strengths which are major factors in the high friction values measured in braking, cornering and traction for the rolling wheel condition when the tire is "geared" to the road. In locked-wheel braking tests on wet road surfaces, the coefficients of friction rarely exceeded 0.70 whereas in peak rolling-wheel braking tests coefficients greater than 1.00 were measured. On certain coarse-textured road surfaces when wet, the tire interlock or "gearing" the tire to the road is clearly evident in tests with high hysteresis butyl rubber tires where peak rolling-wheel coefficients in the range of 1.00 to 1.20 have been measured. In my opinion, the analogy of "gearing the tire to the road" can be used to explain the large variations in the friction values measured for various tire and road surface conditions not only in rolling-wheel braking but also in locked-wheel braking. As far as I know a mathematical model or analysis of friction at the tire-road interface based on the analogy of "gearing the tire to the road" has not been developed but in my opinion such a model or analysis has great possibilities to explain the large differences in the coefficients of friction for various tire and road surface types and conditions.

L. Segel

I find myself a little puzzled by the entire discussion. I thought that perhaps I could try to clarify the issue by calling to your attention the fact that when a tire is side slipping due to the application of a side force at the hub, we observe that the tread deforms in a manner indicating that the tread rubber is adhering to the road and not sliding relative to the road surface. Now, it seems to me that the issue we are debating is whether or not a shear force can exist between tread and road without the existence of a relative velocity between tread and road. It seems to me that all the empirical evidence that exists makes it clear that a differential shear force can exist between tread rubber and road without the existence of a relative sliding velocity. This statement is in support of Dr. Moyer's remarks, of course. Now, if one accepts the fact that a shear force can be generated between an element of tread rubber and the road without the existence of any relative motion at the interface, then one can explain the observed peak in side force and the decrease in side force with increasing lateral slip. Similarly, one can explain the peak in the longitudinal force and the decrease in longitudinal force with additional longitudinal slip, if one assumes (and of course this is a large assumption) that the existence of a relative velocity between tread rubber and pavement (as occurs at large values of lateral and longitudinal slip) will then reduce the shear force that can be generated. I cannot enter into the arguments as to the physics involved in the so-called "frictional coupling." All I can do is point out that if you allow (or assume) that the differential shear force between tread and pavement will decrease (depending on the normal pressures and other factors) with an increase in the relative sliding velocity, then there are many simple mathematical models that have been generated for the tire that will predict a peak longitudinal force to occur in excess of that value produced at the full slide condition.

Clark

Thank you. From my personal association with people working in the traction field, it is my impression that this discussion could continue well into the evening. If I may take the privilege of breaking it off, I would like to make a couple of comments about the other papers that were presented later on here today.

During the intermission period, Don Hays mentioned a question to me that I thought had considerable interest for many people in the audience here, a little bit different from the ordinary question we have been debating, and I wonder, Don, if I could ask you to state that concerning the role of the tread pattern geometry.

D. F. Hays

I do not know to whom I should address the question, and I do not want to put anybody on the spot. I think it goes to the tire manufacturers, though. The general public, when they decide to buy a tire, find that they are faced with a multitude of tread geometries. They go to the store and find that the tread patterns vary markedly in groove width, tread angle, shape, and they then wonder why. We are all interested in traction, especially on wet roads. Consequently, we all must be striving for a tread pattern that will give us that traction. Now if we are getting the required traction with some tread patterns, then this says that not all of these many geometries are really effective, and they are not all contributing to that traction. I come to the conclusion that of our well-designed tires, some must be better well-designed than others. So to the people who do design tread patterns on tires, could you tell us why we have such a variation?

Clark

Thank you very much. That is a well-posed question. Do we have any comments from the tire manufacturers?

Kienle

Obviously, the design is something which you see. It is very hard to sell to a layman what you cannot see. I will certainly take exception to your saying that our designs are not effective. I would rather say they are equally effective rather than say that they are not effective at all. The variations which you see are attempts to appeal to the customer and his aesthetics. Much of the design work is done by artists within certain technical frameworks which are laid down for them by technical people. As tire engineers we get so involved that we think there is a lot of difference in the visual appeal of different tread patterns to laymen. When I talk to my neighbors and find out they just go in and buy a tire and do not even look at the tread, why, I am sure that we are grossly overstating our case. I think there are certain principles to which we do adhere and to which we are trying to get further depth all the time. This is what I was addressing in my talk. Beyond those principles, what differences there are, are for aesthetics and eye appeal, and to catch a customer when he comes in to a store. Just as the outside of a refrigerator looks different and in many cases there is function to it. But there is also an eye-catching appeal.

Clark

Thank you. I would like to call on a couple of people who have not had a chance to respond before. Ken, did you want to say something?

K. L. Campbell

I would like to refer back to the published work from the SAE traction test task force which showed that over a wide range of commercially available tires that had been tested, the difference between the best and the worst on wet slippery surfaces was about 25% in coefficient of friction or effective traction, whatever you want to call it. I explain that partly because the general life of a tread pattern is about 10 years, from the time it is introduced as best and latest until it passes out as being oldest and most obsolete and therefore no longer saleable. You have at least a 10-year period in which you can improve in your technology.

So in the marketplace today there are some old designs still being sold at a low price because the capital equipment is there and available in the tire companies to make those tires. You have a low priced range which is probably not as good as the current higher priced tires *if this property of wet traction is one that is important in the eyes of the tire designer and the consumer,* (and I think it has become that more recently, perhaps more than it had been before.)

And, I will also answer the question by seconding what Bob Kienle has said. If you will tell me why a Pontiac and an Oldsmobile look different, I will tell you why two of our tire designs look different.

Clark

I think Dick Topping had his hand up first, if I could call on him.

R. W. Topping

I would like to add emphasis to that, but also to point out that tires are not designed with particular emphasis on just one performance aspect. Different tread designs reflect this. Some are designed for wear or different types of traction. I would also point out that if the engineering state of the art were where we would like to have it, we would have the best possible compromise between all these. But that is not the case, and in certain instances we emphasize one form of performance over another.

Clark

I think we have time for one more comment on this question.

J. T. Tielking

I do not think we have received an adequate reply to Mr. Hays' inquiry. The contour of the refrigerator and of the automobile are far less important than the pattern of a tire tread. I am certain that more science is employed in tread pattern design. Now, I do not understand how the design life of a tire tread can be 10 years. With what laymen know about tread-road interaction, how are they to know that the 10 year old tread is not better than the new one that is introduced today?

D. F. Moore

I would like to make a brief comment that once you have main channels, feeder channels, and sipes for different reasons, there are a variety of optimized tire designs you can get. These designs are very comparable in performance.

Clark

I would like, if I could, to break that off at this point. I wanted to have a chance to ask Dr. Pacejka a very brief question. I think, that for my part, I had always been led to believe that cornering force was a unique property of slip angle and environmental conditions and I notice that in your paper that you report really a non-unique cornering force situation. Could you elaborate just a little bit on it? Is this the function, strictly, of wet surfaces or are any other factors involved?

H. B. Pacejka

Yes, it has showed up only on very smooth slippery surfaces for speeds in excess of certain critical speeds, and it seemed to be quite reproducible. Maybe it is now a good time to ask if anybody in the audience has ever seen such a phenomenon. I think that we are the only institute which has ever published such kind of news in the steady state characteristic.

Clark

Thank you. That was really the reason for raising the issue. I wonder if anybody else would like to comment on that. Well, I gather you now have complete right of priority to this phenomenon since no one else seems to have seen it before.

SESSION IV

TIRE TRACTION
THE ROLE OF THE PAVEMENT

Session Chairman
K. C. LUDEMA

University of Michigan
Ann Arbor, Michigan

INTRODUCTORY REMARKS (K. C. Ludema)

The fourth session has something to do with the traction of tires, but specifically it is entitled, "The Role of the Pavement," and for this session we have three speakers. I really would like to mention just a little bit about how speakers were selected because I think in the case of our three speakers of this session we find first that they are all from the British Empire. This does not imply that there are no authorities on the subject in the United States or in Germany or elsewhere. The original plan, and the title of the conference is, "The Physics of Tire Traction." When it comes to the physics of tire traction, one has to, of course, define some terms within that bound and we chose to look at mechanisms. Inevitably we got down to the details of the interaction between the tire and the road. The three speakers of this session were selected in particular because they have had some of the newer works available, which in our judgment have high potential for aiding understanding of the complex interaction between the tire and road and for generating a very good discussion here. These papers have breadth and at the same time depth and will make a very fine contribution to the proceedings from this conference. The three speakers are Dr. Yandell, Mr. Schonfeld, and Dr. Williams.

The first paper is by Dr. Yandell, and he is going to be speaking on the general subject of where the friction energy goes and what the features on the road surface do about it, or how they contribute. The second paper by Mr. Schonfeld is on a more empirical way of describing road surfaces, and the third paper by Dr. Williams relates wet skid resistance to wear and noise generation.

THE RELATION BETWEEN THE STRESS SATURATION OF SLIDING RUBBER AND THE LOAD DEPENDENCE OF ROAD TYRE FRICTION

W. O. YANDELL

University of New South Wales, Australia

ABSTRACT

The author's understanding of the relationship between tyre friction and road texture is briefly outlined.

One source of tyre friction dependence on load is shown mathematically, with laboratory confirmation, to be due to the stress saturation of the sliding rubber by the surface texture. It is shown that stress saturation occurs when the stress flux in the rubber fails to keep pace with an increasing normal load. This progressively reduces the coefficient of hysteretic friction.

Losses in the coefficient of friction from this source can be restored by reducing the normal load, by increasing the gross area of contact or by increasing the Young's modulus of the rubber.

INTRODUCTION

The incidence in sliding rubber of stress saturation generated by the road surface texture is thought to cause losses in the coefficient of hysteretic friction (1). Before entering into a detailed discussion of this problem, it may be useful to briefly describe some of the author's ideas on the mechanism of tyre-road friction.

Hysteretic Friction – The surface of the road is composed of a number of scales of texture of pitch sizes ranging from microscopic to about 3/4 inch. The coefficient of hysteretic friction generated by sliding rubber on each scale can be calculated from the scale's average absolute slope in the direction of sliding and the damping factor of the rubber (1). The size of the scale has no direct theoretical effect on hysteretic

References p. 321

friction. The coefficient of hysteretic friction is the sum of the coefficients generated on each scale. In, say, an unconfined compression test, the damping factor is defined as the energy dissipated divided by the energy applied to a rubber sample during one cycle of load-unload. There are some modifying factors to the above idealization.

Frequency Effect – Since rubber is a viscoelastic material, there is a particular number of asperities that a point in the rubber will pass per second to give a maximum damping factor to the rubber. This peak damping factor will obviously occur at a higher sliding speed when the pitch of the asperities increases. Fig. 1 shows the additive effect of the coefficients of friction and the influences of friction peaks on each scale (see (2) for a description of these scales of texture). When the largest scale of texture (scale 1), consisting of the aggregate pieces, is absent, the loss of the coefficient of hysteretic friction with sliding speed is more rapid.

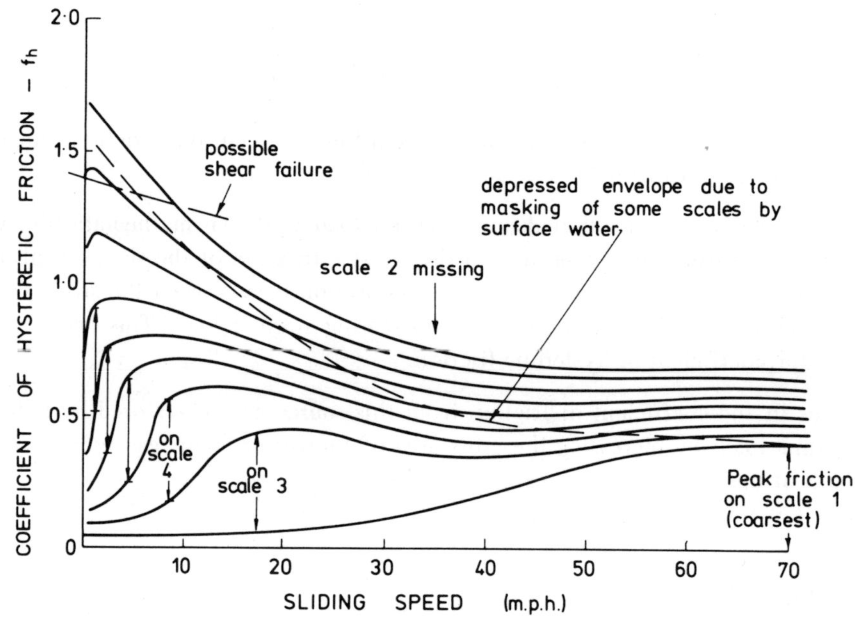

Fig. 1. The additive effect of the coefficients of hysteretic friction generated on each scale of texture of a hypothetical road surface versus the sliding speed of a rubber tyre generating it.

Asperity Shape and Heat Effects – When the asperities become rounded due to polishing (see Fig. 2) or when heat generated on, at first, the finest scales at high speeds alters the properties of the rubber (causing the envelope of Fig. 1 to become less horizontal), the coefficient of friction generated by each scale will be less than that calculated from the average absolute slope in the direction of sliding.

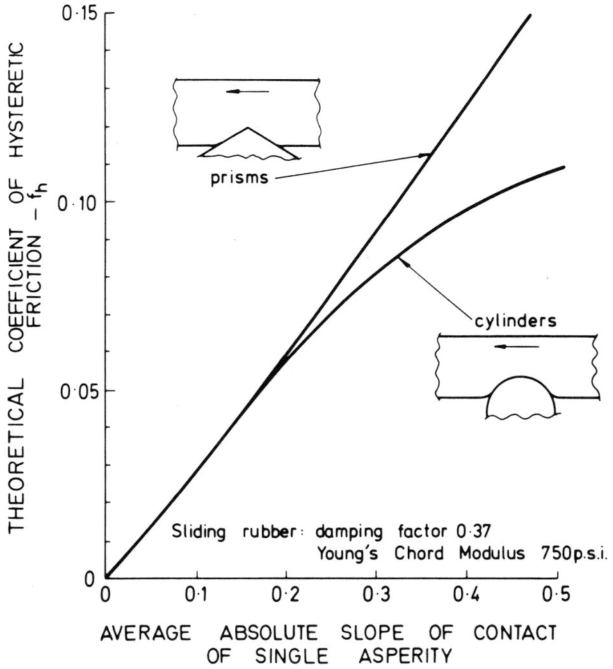

Fig. 2. The theoretical coefficients of hysteretic friction generated by rubber sliding on a triangular prism and on a cylindrical asperity respectively, versus the average absolute slope of contact.

Masking Effect – Starting with the finer scales and increasing in size with sliding speed, components of the texture can be rendered ineffective by the masking effect of surface lubricant films. This reduces friction.

Load Dependence – When the normal force holding the sliding rubber against a certain scale of texture reaches a particular value and a further increase produces a change in the coefficient of hysteretic friction on that scale, we can say that the broad phenomenon of load dependence is having an effect.

The Load Dependence of Hysteretic Friction – Variation of the coefficient of hysteretic friction with changes in the normal force between sliding rubber and a particular scale of texture can have a number of causes.

Shape of Asperities – The shape of the individual asperities can account for the change in the coefficient of hysteretic friction with load in sliding rubber. The coefficient of hysteretic friction, in the absence of other effects, is constant on triangular prismatic asperities no matter what the load, since the average absolute slope of contact remains constant. Increases in the normal load on rubber sliding

References p. 321

across cylindrical asperities results in increases in penetration and, hence, increases in the average absolute slope of contact, producing a higher coefficient of hysteretic friction (see Fig. 3).

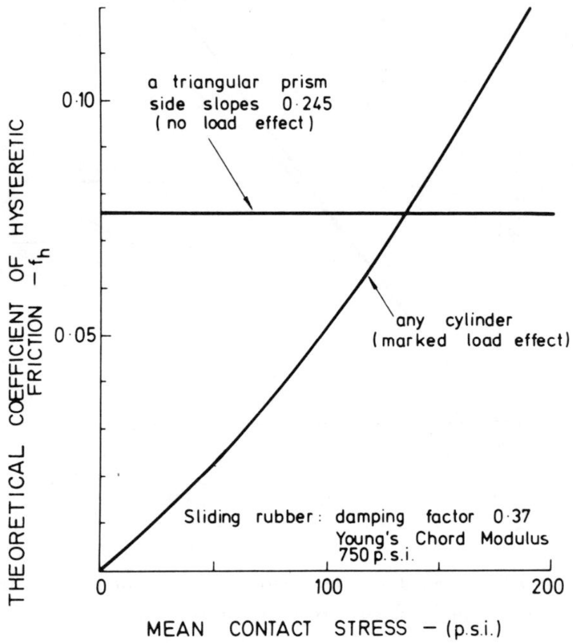

Fig. 3. The theoretical load effects on the coefficient of hysteretic friction of rubber sliding across a prism and across a cylinder.

Change in Film Thickness – An increase in the normal force between sliding rubber and a particular scale of texture that is masked by water can decrease the film thickness and partially unmask that scale causing it to generate hysteretic friction. This will correspondingly increase the total coefficient of hysteretic friction.

Stress Saturation of the Rubber – Another cause for loss in the coefficient of hysteretic friction with increased load is that attributed to the texture's penetration into the rubber being great enough to prevent, what the author called (1), the stress flux in the rubber – from increasing as rapidly as the normal load. Since the stress flux generates hysteretic frictional force the coefficient of hysteretic friction will then decrease as the normal load increases. This phenomenon which was predicted by the author (1) we call "the stress saturation of rubber by the texture".

The object of this paper is to demonstrate, by rigorous stress analysis, that this phenomenon of stress saturation exists. Some laboratory confirmation is also presented. The phenomenon is investigated theoretically and in the laboratory as a two dimensional problem instead of the more general three dimensional case only for reasons of computer time economy.

THEORETICAL

The author's mechano lattice analogy (1) is used for calculating the stresses and deflections in a rubber slider of infinite length, unit width, and finite thickness fastened to a rigid backing plate. It is assumed to be sliding with plane stress across a texture at such a speed as to give it a damping factor of 0.37 and a "Young's Chord Modulus" of about 750 p.s.i.. The texture consists of an infinite number of triangular prismatic asperities of unit length arranged side by side. The slope of these smooth prisms is 0.245.

The stresses, deflections and coefficients of hysteretic friction generated in the rubber are calculated for a number of specific normal stress levels so that the effect of stress saturation can be studied. The specific normal stress is defined here as the normal load transmitted to an asperity divided by the width of the asperity measured from trough to trough. When the rubber is contacting the whole of the texture surface, the specific normal stress will equal the average normal contact stress.

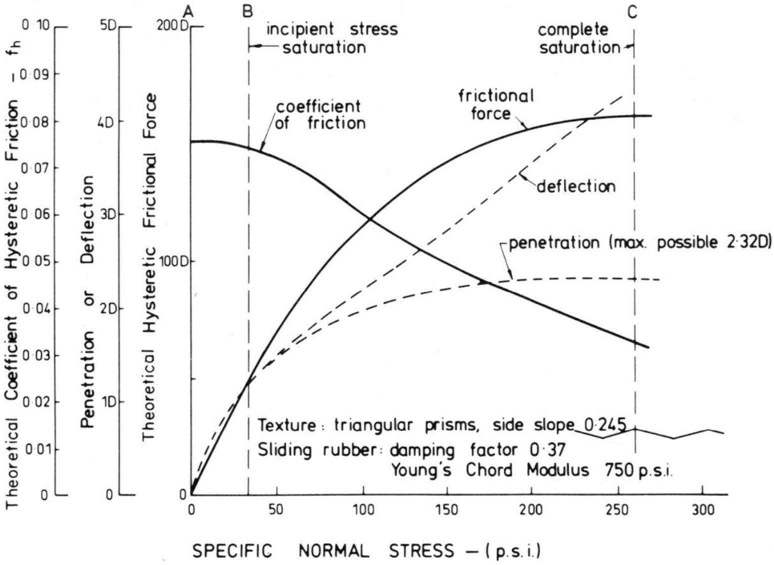

Fig. 4. The theoretical variation in the coefficient of hysteretic friction, the hysteretic frictional force, the absolute deflection of the rubber and the penetration of the rubber into the texture as the specific normal stress increases.

References p. 321

Change in Hysteretic Friction – As the load and, hence, the specific normal stress increases from zero (Fig. 4), it can be seen that at first the theoretical coefficient of hysteretic friction remains constant; but then it decreases in a nonlinear fashion. At low specific normal stresses, the hysteretic frictional force increases proportionally to the specific normal stress. The coefficient of friction, thus, remains constant and at a maximum value from condition 'A' to condition 'B'. As the specific normal stress continues to increase past 'B', the frictional force fails to keep pace with it, and the coefficient of friction decreases at an initially increasing rate until complete stress saturation occurs, at which point 'C' the maximum frictional force possible is reached. Further load increases result in an exponential drop in the coefficient of friction.

It can be said that up to a specific normal stress represented by 'B' in Fig. 4 the rubber is virtually free from stress saturation by the texture. 'B' to 'C' represents a transition stage, and from 'C' onwards, the rubber is fully stress saturated by the texture.

Penetration of Rubber into the Texture – Fig. 5 shows the shape of the sliding surface of the rubber at the various critical specific normal stresses indicated by 'A', 'B', and 'C' in Fig. 4. At condition 'A', the rubber is only touching the peaks of the asperities and generating the maximum coefficient of hysteretic friction of 0.076. With increased specific normal stress when the effects of stress saturation are incipient, condition 'B' is reached. The coefficient of hysteretic friction has dropped slightly, and the rubber has penetrated to less than half the peak to trough depth of the texture. At near full saturation, the rubber has almost filled the texture as shown by surface 'C'. Any further increases in specific normal stress will cause little increase in hysteretic frictional force, and obviously, no further penetration can occur. The coefficient of hysteretic friction will then decrease exponentially (although never reach zero) as the specific normal stress increases.

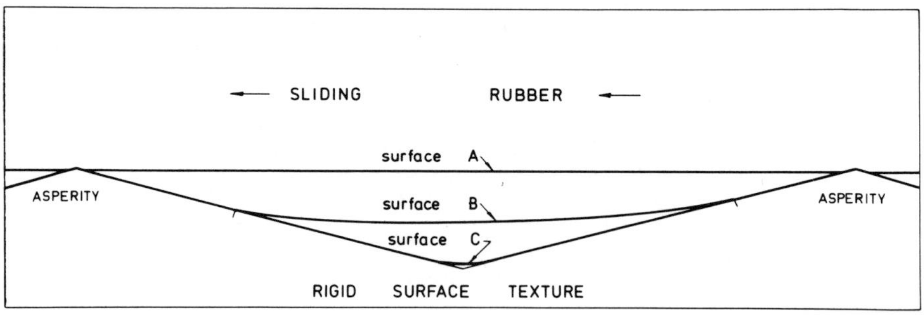

Fig. 5. The theoretical penetration of the sliding rubber into the texture at low specific normal stress, A, at incipient stress saturation, B, and at near complete saturation, C.

It will be noted that the penetration curve in Fig. 4 departs from the absolute deflection relationship near a point representing the commencement of the stress saturation of the rubber. Thus, from this specific normal stress upwards, all points on the rubber surface are in a continuous state of some upward vertical deflection during sliding.

Stress Behaviour – The horizontal axes of Figs. 6, 7 and 8 can be viewed as time axes increasing to the left as the rubber slides to the left across the infinitely long texture. They can also be viewed as the stress or strain conditions at all points in the sliding rubber at one point in time. Therefore, it can be seen that at any particular depth, an element of rubber experiences cycles of stress and strain as sliding proceeds.

A comparison of the patterns of major principal stresses in the rubber at incipient saturation 'B' and at full saturation 'C' is shown in Fig. 6. The lack of symmetry of the stresses is evident in both cases. The coefficients of hysteretic friction can be calculated from the resolved reactions at the smooth interface. Since the depth of rubber is not infinite, the major principal stresses increase as the specific normal stress increases.

The minor principal stresses are shown in Fig. 7. When the rubber is suffering from incipient saturation 'B', the stress pattern shows that the whole volume of rubber is subjected to periodic reversals of stress as it slides. On the other hand, the minor

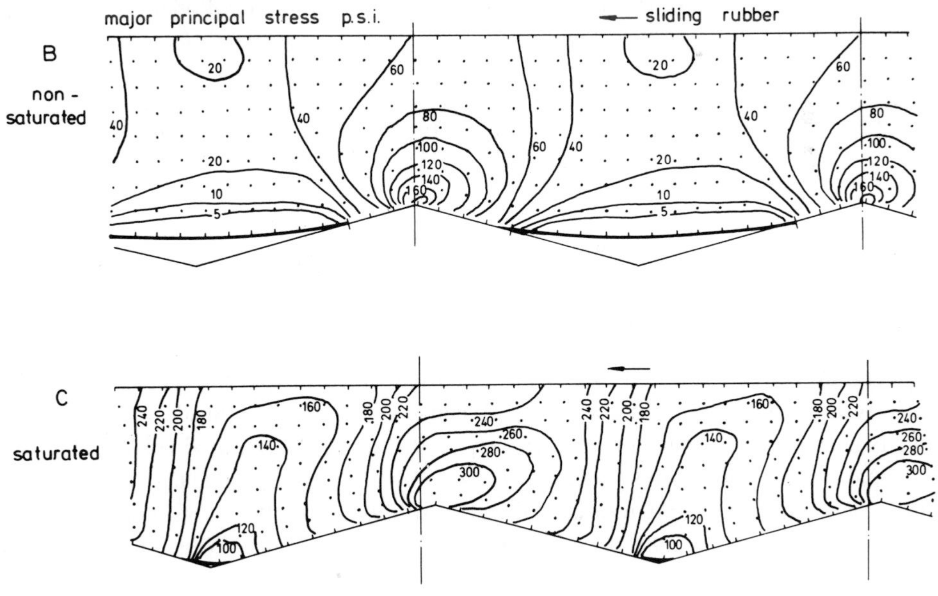

Fig. 6. The comparison of major principal stress patterns in the saturated rubber, C, and in the unsaturated rubber, B.

References p. 321

principal stress pattern of the fully saturated rubber, 'C', shows that no part of the rubber is at any time subjected to tensile stress.

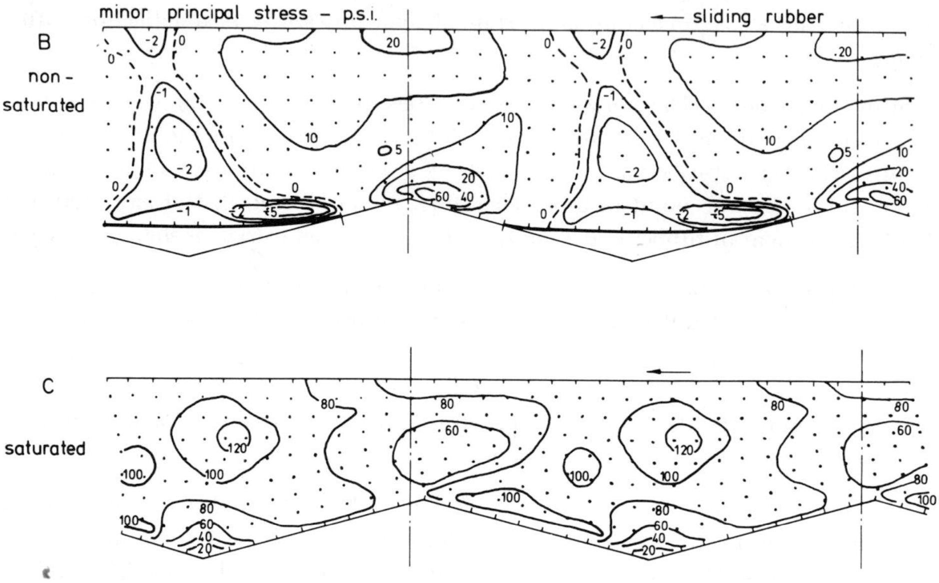

Fig. 7. The comparison of minor principal stress patterns in the saturated rubber, C, and in the unsaturated rubber, B.

It has been mentioned earlier, (1,2) that the maximum hysteretic frictional force occurs in the rubber only when the stress flux is at a maximum. Figure 8 gives a pictorial appreciation of the stress flux, but we will not quantify it here. The major and minor principal stresses, divided by the specific normal stress – to scale them to the coefficient of friction – have been plotted vectorially at intervals and at two depths.

In the non-stress saturated rubber, 'B', the planes of principal stress make complete revolutions both near the surface and at depth while the magnitudes of the scaled principal stresses fluctuate greatly. However, in the stress saturated rubber, 'C', the principal planes merely rock back and forth, and the scaled principal stresses do not fluctuate as much as in the unsaturated sliding rubber. Thus, the stress flux emanating from the rotation of planes of principal stress decreases as the specific normal stress increases. Also, the stress flux emanating from the fluctuation of the magnitude of the scaled principal stresses decreases as the specific normal stress increases beyond that present at the incipient saturated stress condition 'B'. This causes the observed loss in the coefficient of hysteretic friction.

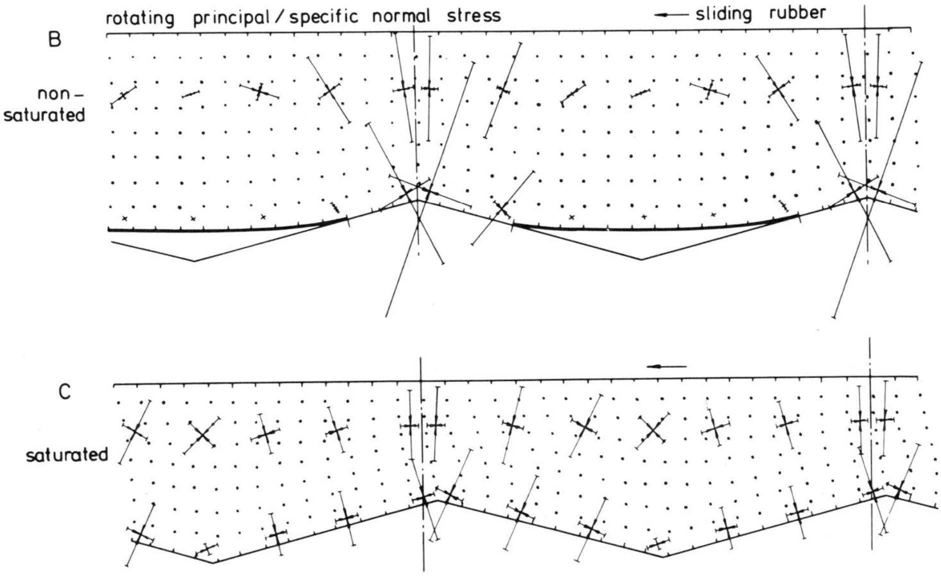

Fig. 8. The comparison of the rotation of planes of principal stress and the ratio of principal stresses to the specific normal stress in the saturated rubber, C, and the unsaturated rubber, B.

Relation Between the Degree of Stress Saturation and the Cycling Rate of the Stress Flux – It is inaccurate to assume in all situations that the frequency of loading which controls the damping properties of a rubber is equal to that obtained by dividing the velocity of sliding by the pitch of asperities. All this gives is a lower limit to the frequency. The frequency of loading that controls damping is the inverse of the time taken for the stress effect of the asperity to pass; or more precisely, the cycling rate of stress flux controls damping.

Fig. 8 shows that in the non-stress saturated case 'B', the cycling rate of stress flux near the surface is about twice that occurring at depth. Hence, because the damping factor of the rubber is frequency dependent, the damping factor may vary with depth in cases where the rubber is not fully stress saturated. However, the cycling rate of stress flux in the stress saturated rubber, 'C', is equal to the velocity of sliding divided by the pitch of the texture in all parts of the rubber.

LABORATORY INVESTIGATION

Fig. 9 compares the theoretical loss of the coefficient of hysteretic friction due to stress saturation with that measured in the laboratory. In both theoretical and practical cases, the rubber has a damping factor of 0.45 and a Young's Chord Modulus of about 750 p.s.i.. For the laboratory tests, the prismatic textures with the two

References p. 321

pitches, 0.012 and 0.006 inch, and with side slopes of about 0.6 and 1.0 respectively, were rolled onto aluminum plates. Due to the fineness of the texture, it was impossible to avoid rounded crests which reduce the peak coefficients of hysteretic friction on both textures. The cylindrical crests also caused the coefficient of friction to rise at first since the average slope of contact increased rapidly during the initial penetration into the rubber per unit increase in specific normal stress. At approximate incipient saturation penetration, the average slopes of contact were about 0.4 and 0.35 on the 0.012 inch pitch and on the 0.006 inch pitch, respectively.

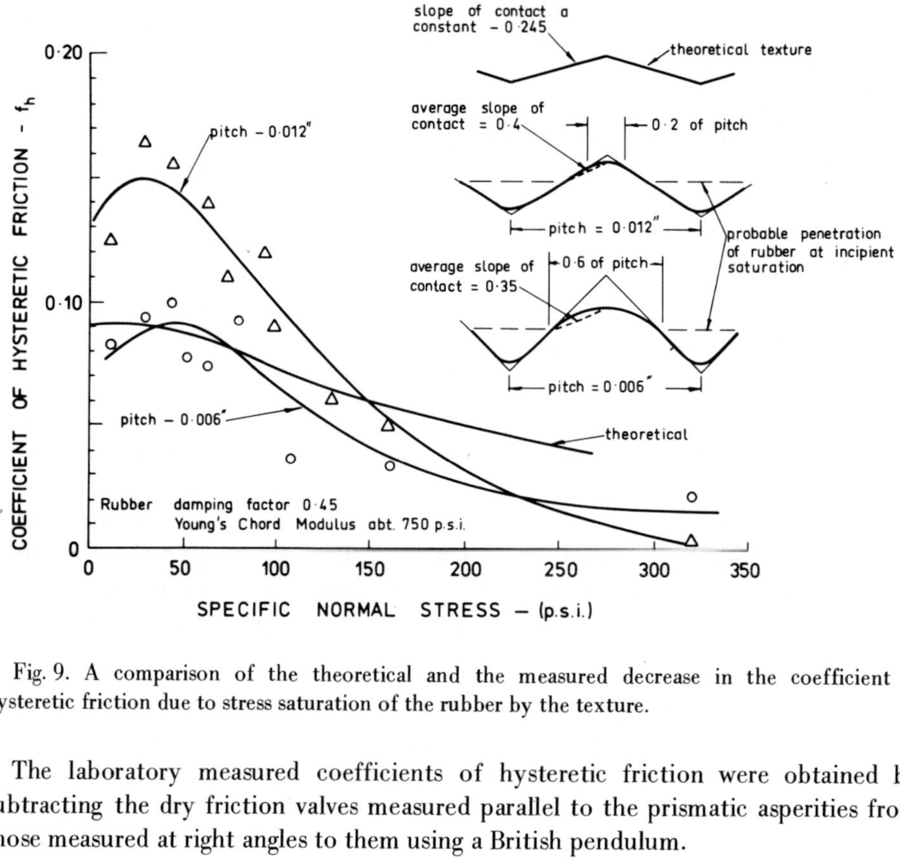

Fig. 9. A comparison of the theoretical and the measured decrease in the coefficient of hysteretic friction due to stress saturation of the rubber by the texture.

The laboratory measured coefficients of hysteretic friction were obtained by subtracting the dry friction valves measured parallel to the prismatic asperities from those measured at right angles to them using a British pendulum.

It can be seen that, in the theoretical case and on the 0.012 inch pitched texture, the coefficient of hysteretic friction began to drop after the specific normal stress passed 40 p.s.i.. However, this point was not reached until 60 p.s.i. on the 0.006 inch pitched texture. This may have been due to the fact that the effect of the increase in the average slope of contact at first counteracted the initial stress saturation effects since the cylindrical peaks constituted a larger proportion of the asperities in the finer texture. There may also have been a frequency effect in the rubber. The shapes of the textures were measured by the light cut method.

CONCLUSION

The so-called load dependence of road-tyre friction has been explained in terms of the hysteretic phenomenon – the stress saturation of the rubber by the surface texture. This was supported by laboratory tests. It was shown theoretically that the drop that occurs in the coefficient of hysteretic friction when the sliding rubber becomes stress saturated is associated with the failure of the stress flux to increase at least at the same rate as the increase in specific normal stress. The load effect attributed to the stress saturation of the rubber is separated from that caused by the shape of the surface texture asperities.

There are two ways in which the condition of loss of friction due to stress saturation can be relieved.

a. The specific normal stress can be reduced either by decreasing the load or by increasing the gross area of contact, or
b. The "Young's Chord Modulus" of the rubber can be increased, causing saturation to initiate at a higher normal stress.

It is likely, due to the high normal stresses there, that the fine scales of texture near the peaks of the coarser texture will cause local stress saturation in the rubber with a consequent reduction in the expected coefficient of friction.

ACKNOWLEDGEMENTS

The author wishes to thank his student, Bill Wild, for carrying out the laboratory friction tests and Professor D. F. Orchard for his encouragement.

REFERENCES

1. *W. O. Yandell, "A New Theory of Hysteretic Sliding Friction," WEAR, 17, pp. 229-244, 1971.*
2. *W. O. Yandell, "The Simulated Traffic Polishing of Roadstones", WEAR, 21, pp. 313-338, 1972.*

DISCUSSION

Ludema

Well, are there questions for Dr. Yandell? Dr. Moore has one.

D. F. Moore

The first question, Dr. Yandell, is that you assume a lubricated condition, is that correct? I do not think that you stated that it was for a lubricated contact condition.

Yandell

Yes, I must apologize for that. Also this time I neglected any lubrication effect. The surface of the asperities are assumed to be frictionless in this example.

Moore

Well, the second thing was the superposition of effects as demonstrated on the blackboard. I gather from these curves that you superimposed the roughness to get the overall effect.

Yandell

Superimpose the friction generated on each scale of texture.

Moore

I thought these were profiles of roughness effects.

Yandell

They are components of the total profile.

Moore

Yes. We are in agreement there. In getting the hysteresis-velocity curve we see that, depending on the roughness, the peak occurs at different velocities. But I am wondering about the magnitude of each roughness effect. I would have thought that had you superimposed these effects the individual peaks would be lower at lower frequencies. Therefore, the net effect would be a gradual rise rather than a drop.

Yandell

Why should the peaks of friction be lower on the finer scale if the scales are of similar shape?

Moore

Because hysteresis contributes less in energy dissipation on finer scales of roughness.

Yandell

I contend that this is a false supposition. I am assuming that Fig. 1 is apt providing each peak damping factor is the same height, and each scale of texture is of a similar shape but of different size, so mathematically speaking, you should get the same coefficient of friction, although each peak friction occurs at a different speed of sliding.

Ludema

It is a difficult point, and we have probably only two minutes left. Perhaps we can bring it up in our discussion, as we should not let it drop. Have we another short question?

A. R. Williams

The contours of principal stress which you showed in the illustrations, were they derived theoretically, and if so, have you attempted to substantiate them from practical experiments? With regard to the depth or rubber considered in the theoretical analysis, was this taken as a finite value?

Yandell

The depth of rubber was finite. The contours of principal stress were calculated by a mechano-lattice analogy. I have not attempted to measure those stresses inside the rubber. I would be interested in contacting anyone who has.

R. F. Peterson, Jr.

In your diagram, it looks like you are summing the coefficient of hysteretic friction as due to roughness of many sizes. If you take large-scale roughness as contributing 0.1 to the coefficient of friction, and the next smaller size roughness as contributing another 0.1, and the next smaller size again adding 0.1 to the coefficient, I do not understand where you stop. If you keep taking smaller and smaller asperities, each contributing 0.1 to the coefficient of friction, your sum will be infinity. Obviously it is not, but where is the flaw in this argument?

Yandell

Before that happens of course, the rubber fails doesn't it? You have cohesive failure in the rubber, and that is delayed in the presence of water films and the melting of rubber. The finer scales of texture generate the same amount of heat in a smaller volume of rubber, therefore they are more likely to provoke melting providing there is no heat transfer.

PAVEMENT SURFACE TEXTURE CLASSIFICATION AND SKID RESISTANCE PHOTO-INTERPRETATION

R. SCHONFELD

Ontario Ministry of Transportation and Communications, Downsview, Ontario, Canada

ABSTRACT

The pavement surface texture classification method presented in this paper is based on the concept that the pavement surface is a geometrical structure which can be expressed by six parameters: height, width, angularity, distribution, harshness of projections above the matrix from which they arise and harshness of the matrix itself.

The need for surface texture classification stems from the fact that the highway engineer must manipulate the surface texture in order to control the pavement's skid resistance. Texture classification is a means by which skid resistance measurements, tire wear and noise measurements can be related to identifiable pavement surfaces. Similarly, the mechanical engineer who is concerned with vehicle performance needs pavement surface data. Without this information, he cannot properly relate performance data of different vehicles or vehicle components to each other, nor can he relate the performance data of the vehicle on different surfaces.

Photo-interpretation of skid resistance is discussed. Like the texture analysis to which it is linked, photo-interpretation is the outcome of an empirical approach which resulted in a two-tier model of tire – pavement contact: on the upper level, supporting the tire directly, are the pavement surface projections; on the lower level, is the matrix which provides additional tire support to a degree dependent on height, distance between the asperities, and on tire construction.

Finally, the role of pavement texture classification in the surveillance of pavement slipperiness is discussed.

References p. 328

PRINCIPLE OF PAVEMENT SURFACE TEXTURE CLASSIFICATION AND SKID RESISTANCE PHOTO-INTERPRETATION

Pavement surface textures have been stereo-photographed and studied on both macroscopic and microscopic scales by several researchers (1) (2).

The present texture classification system (as described in the Appendix) deals with surface textures as perceived by the human eye at a magnification of approximately six times natural scale and expressed in terms of six texture parameters. The function of pavement classification is to classify pavement textures by identifying six geometric features of the pavement surface which are known to influence skid resistance and to use the texture classification for correlating the pavement surface textures with the accumulated test results of skid testing devices.

EFFECTIVE SURFACE TEXTURE

The effective pavement surface texture, as distinct from the surface texture of the pavement structure itself, includes sediments of pavement material, tire and windblown debris, and other contamination. There is evidence that the effective surface texture is of a more volatile nature, i.e. less durable than the surface texture of the pavement structure itself and that it is responsible for the seasonal fluctuations of skid resistance (3) which coincide with precipitation and with the change of seasons. The probable effect of debris on the surface is that it submerges asperities and fills in crevices which, when cleansed, contribute to skid resistance and surface drainage.

COMPUTATION OF PHOTO-INTERPRETED SKID NUMBERS

The photo-interpreted skid number of a pavement surface is the sum of the texture parameters' friction weights. The friction weight is the friction increment assigned to the parameter according to its contribution to the total skid resistance of the pavement (4). The friction weight of a parameter varies according to the other parameters present in the complete set of six texture parameters (referred to as the Texture Code Number). Friction weights are obtained from photo-interpretation charts (described in the Appendix), which are the result of multiple regression analyses of texture code numbers and skid test results. Figs. 7-10 in the Appendix apply to random textured asphalt pavements. Charts have also been prepared for random textured Portland Cement concrete pavements, but they are not shown in the Appendix. Charts for orientated textures, e.g. longitudinal and transverse grooving, have not been prepared to date.

The photo-interpreted skid numbers can be abstracted from the charts in the ordinary way, or they can be calculated by electronic computer.

TWO-TIER CONCEPT OF TIRE-PAVEMENT CONTACT

The tire tread may be compared with a suspension bridge as shown in Fig. 1. The tread between asperities is the suspended structure between the towers; the tire locus pressing on the asperity is the supporting structure; the stone projection is the bridge foundation; and, the matrix is the topography between supports which, in the case of the tire, provides additional support for the suspended structure. The bridge load is unevenly distributed on main and additional supports, and the load distribution depends on height, distance between asperities and on tire construction.

The two-tier concept is inherent in the skid resistance photo-interpretation charts. This concept can explain the different roles in pavement skid resistance played by the coarse aggregate (projections) and by the fine aggregate and the binder (matrix) between the coarse aggregate. It can be seen that hard and angular fine aggregate is important if stone projections are low and less important if the coarse aggregate is strong.

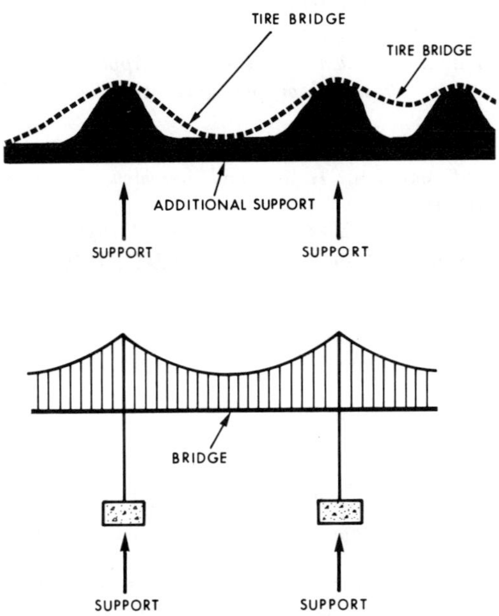

Fig. 1. Two-tier tire-pavement contact.

EXAMINATION OF CAUSE OF PAVEMENT SLIPPERINESS

Photo-interpretation can reveal the reasons for the existence of a slippery pavement condition. An analysis of the pavement surface with respect to the six texture parameters can reveal possible defects. These defects could include such items as polished aggregate (texture parameter E1*), smooth matrix in the form of an excess

References p. 328

of binder or an excess of dust size particles in the fine aggregate (texture parameter F1 or F2*), an excessive wear of pavement aggregate (angularity parameter C1*), and stripping of aggregate (distribution parameter D1 or D2*).

CONCLUSION

Photo-interpretation has been used in Ontario during the past two years for evaluating the skid resistance of pavements and laboratory samples. A knowledge of the surface geometry of pavements combined with skid resistance experience enables the engineer to better understand the cause, the prevention, and the remedy of a particular skidding problem.

REFERENCES

1. *B. E. Sabey G. N. Lupton, "Measurement of Road Surface Texturing Using Photogrammetry," Road Research Laboratory, RRL Report LR57, Berkshire, England, 1967.*
2. *T. Holmes, G. Lees, and A. R. Williams, "A Combined Approach to the Optimization of Tyre and Pavement Interaction," Presented at the American Chemical Society Meeting, Miami, Florida, 1971.*
3. *M. A. Furbush and K. E. Styers, "The Relationship of Skid Resistance to Petrography of Aggregates," Bureau of Materials, Testing and Research, Department of Transportation, Harrisburg, Pennsylvania, 1972.*
4. *R. Schonfeld, "Photo-Interpretation of Skid Resistance", Highway Research Record Number 311, pp. 11-35, Highway Research Board, Washington D.C., 1970.*

APPENDIX

Method for Classifying Pavement Surface Textures Suitable for Skid Resistance Photo-Interpretation

Scope – This method is used for the visual analysis and classification of pavement surface textures by means of stereo-photography.

The surface texture of a pavement is defined in terms of a texture code number which consists of six texture parameters, viz. height A of projections, width B of projections, angularity C of projections, density of distribution D of projections, harshness E of the projections' surfaces, and harshness F of the pavement surface between projections.

Texture code numbers may be correlated with the skid resistance properties of the pavement surface texture as recorded by skid resistance testing devices.

**See Surface Texture Classification Table in the Appendix.*

Summary of Method – The equipment used for pavement texture analysis consists of a camera mounted on a box with a light source, a sliding seat for taking stereo-photographs, and a horizontal and vertical reference scale.

Stereo-photographic transparencies of the pavement surface are viewed over a transparent centimeter square grid on a light table. The six texture parameter numbers of each of ten random centimeter squares are noted. The texture parameter number of the pavement photograph is the average of the parameter numbers of the ten squares. The interpreted skid number of the pavement photograph is the average of the interpreted skid numbers of the analyzed centimeter squares.

A texture sample deemed to be representative of a pavement section consists of not less than three suitably spaced pavement photographs. The parameter number of the pavement section is the average of the parameter numbers of the pavement photographs. The interpreted skid number of the pavement section is the average of the interpreted skid numbers of the pavement photographs.

Equipment – A 35 mm single lens reflex camera, focal length 55 mm, is used for taking pairs of stereo-photographs. The camera is mounted on a box at a height of 18 inches above the pavement. The box is equipped with an electronic flash unit which illuminates the photographed area at an angle of approximately 45 degrees. The camera is attached to a sliding seat and takes pavement photographs from two positions 3-3/4 inches apart. The stereo-photographs cover a pavement surface area of approximately 4 inches square (see Fig. 2). The camera box is placed in the middle of the wheel track with the light-source side of the box facing to the right of the traffic direction.

Pavement stereo-photographs show reference scales for estimating texture dimensions. A horizontal reference scale with one-millimeter divisions is attached to the light box at approximately pavement surface level. A vertical reference scale is attached to the light box at approximately pavement surface level. The vertical scale is wedge shaped, 1/4 mm high at one end and 6 mm high at the other end, with 45 degree side slopes. Heights are indicated at increments of the texture height parameter (see Fig. 3).

Pavement stereo-photographs are viewed under a mirror-stereoscope or under a micro-stereoscope. If a mirror-stereoscope is used, the 35 mm pavement photographs are first enlarged into natural-scale prints. These are viewed at 6 x magnifications. A micro-stereoscope with 25 x magnification, mounted on a light table, may be used for viewing 35 mm colour transparencies (see Fig. 4.).

Texture Classification – The surface texture of a pavement is defined in terms of six texture parameters, viz. height A of projections, width B of projections, angularity C of projections, distribution D of projections, harshness E of the projections' surfaces, and harshness F of the pavement surface between projections.

Parameter A denotes the height above the matrix of projections on the pavement surface.

Parameter B denotes the width of the surface projections measured at the level of the top of the matrix.

Camera Box

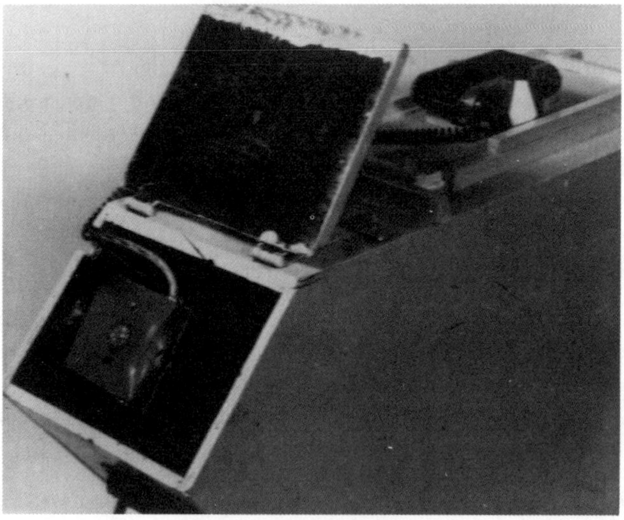

Camera Box Showing Flash

Fig. 2. Views of the camera box.

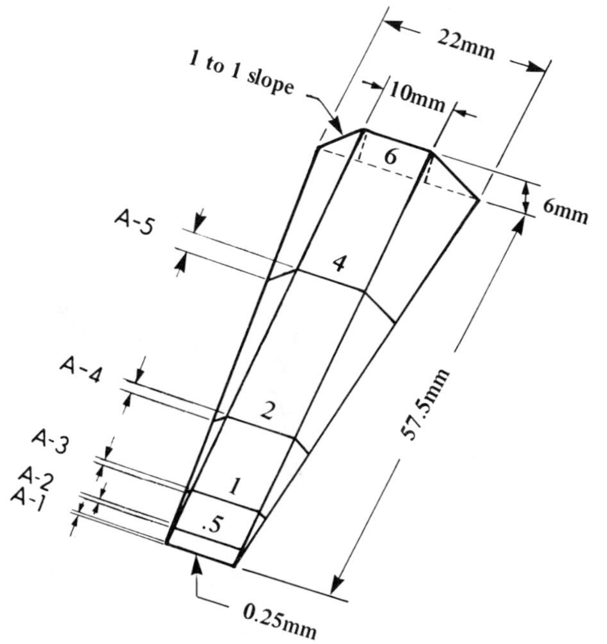

Fig. 3. Vertical scale wedge for pavement surface texture analysis.

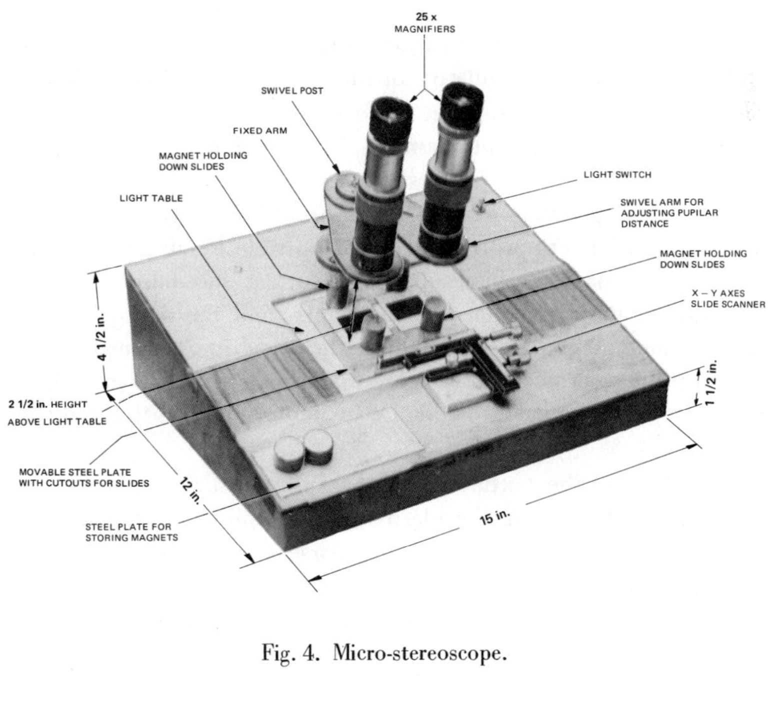

Fig. 4. Micro-stereoscope.

Parameter C denotes the shape of the projection: round, subangular or angular (see Fig. 5). Round projections have curved sides and no edges or nearly plane sides and well rounded corners and rounded edges. Subangular projections have somewhat blunted edges. Angular projections have sharp edges and pointed corners.

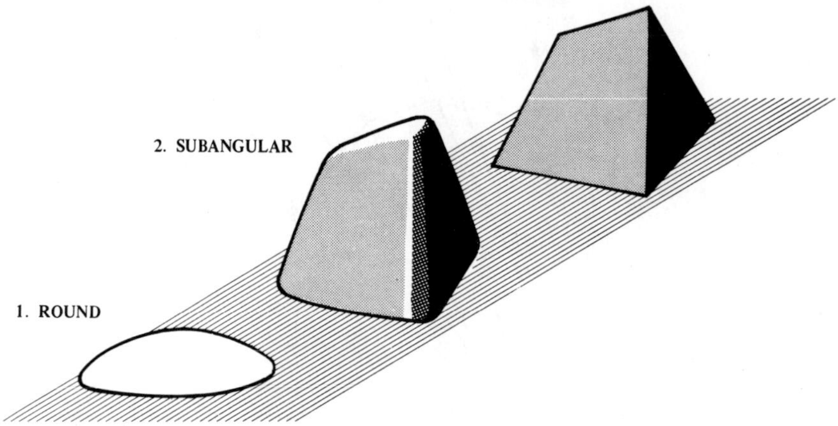

Fig. 5. Angularity of projections.

Parameter D denotes the density of distribution of the projections as the proportion of the whole surface area occupied by projections.

Parameter E denotes the harshness (micro-texture) of the projections' surfaces in terms of apparent height and angularity of the micro-projections: polished, smooth, fine grained, coarse grained-subangular, or coarse grained-angular. A polished surface has no visible texture. A smooth surface has a visible texture, but the micro-projections are too small for a visual estimate of height with the aid of the vertical reference scale. A fine grained surface has micro-projections which are approximately 1/4 mm high, and the micro-projections must protrude by an amount not less than half their width. A coarse grained-subangular surface has blunt micro-projections which are approximately 1/2 mm high or higher, and the micro-projections must protrude by an amount not less than half their width. A coarse grained-angular surface has sharp micro-projections which are approximately 1/2 mm high or higher, and the micro-projections must protrude by an amount not less than half their width.

Parameter F denotes the texture of the background surface between projections in terms of apparent height and angularity of micro-projections: polished, smooth, fine grained, coarse grained-subangular, or coarse grained-angular. A cavity is an undrained area below the surface of the matrix. The five other categories for background surface texture are the same as those five that have just been applied to parameter E.

For the scale of the texture parameters refer to the pavement surface texture classification table (Table 1).

TABLE 1

Surface Texture Classification

Pavement Surface Texture Classification

<table>
<tr><th colspan="5">Macro-Texture Of Projections</th><th colspan="2">Micro-Texture Of</th></tr>
<tr><th>Parameter Number</th><th>Approximate Height A*</th><th>Approximate Width B</th><th>Angularity C**</th><th>Density Of Distribution Of Projections, As A Percentage Of Total Area D</th><th>Projections Harshness E**</th><th>Background Harshness F</th></tr>
<tr><td>0</td><td>0 mm</td><td>16 mm</td><td></td><td>0-12 Percent</td><td></td><td>Cavity in Surface in Matrix</td></tr>
<tr><td>1</td><td>1/4 mm</td><td>8 mm</td><td>Round</td><td>13-37 Percent</td><td colspan="2">POLISHED (No Texture Visible)</td></tr>
<tr><td>2</td><td>1/2 mm</td><td>4 mm</td><td>Subang.</td><td>38-62 Percent</td><td colspan="2">SMOOTH (Texture Visible But Micro-Projections Too Small For Visual Estimate Of Height)</td></tr>
<tr><td>3</td><td>1 mm</td><td>2 mm</td><td>Angular</td><td>63-87 Percent</td><td colspan="2">FINE GRAINED (Micro-Projections Appr. 1/4 mm High) †</td></tr>
<tr><td>4</td><td>2 mm</td><td rowspan="3">Particles Less Than 2 mm Wide Are Regarded As Background</td><td></td><td>88-100 Percent</td><td colspan="2">COARSE GRAINED-SUBANGULAR (Micro-Projections Appr. 1/2 mm High Or More) †</td></tr>
<tr><td>5</td><td>4 mm</td><td></td><td></td><td colspan="2">COARSE GRAINED-ANGULAR (Micro-Projections Appr. 1/2 mm High Or More) †</td></tr>
<tr><td>6</td><td>8 mm</td><td></td><td></td><td colspan="2"></td></tr>
</table>

**A texture element has a height dimension only if the surrounding area below its peak is drained.*

***For the use of the Photo-Interpretation Chart for Asphalt Pavements the following adjustment is made: if the C-Parameter Number is 2, then the E-Parameter Number is raised by 1; if the C-Parameter Number is 3, then the E-Parameter Number is raised by 2.*

†Micro-projections must protrude by an amount not less than half their width.

The surface texture of a pavement is classified by the Texture Code Number which is the complete set of the six texture parameter numbers, e.g. 2.7-1.4-1.8-55-2.1-3.0. In the case of heterogeneous surface textures, where part of the photographed surface has been expressed in terms of one single micro-texture parameter and the other part by the full set of six parameters, two texture code numbers and their proportion are stated, e.g. 30% x 0-0-0-100-3.4 + 70% x 2.7-1.4-1.8-55-2.1-3.0.

Texture Analysis – A transparent grid, equivalent to one-centimeter squares, is placed over one stereo-photographic print or under one colour transparency, depending on whether a mirror-stereoscope or micro-stereoscope is used. On the grid, ten random 1.0 cm-squares are marked and numbered (Fig. 6).

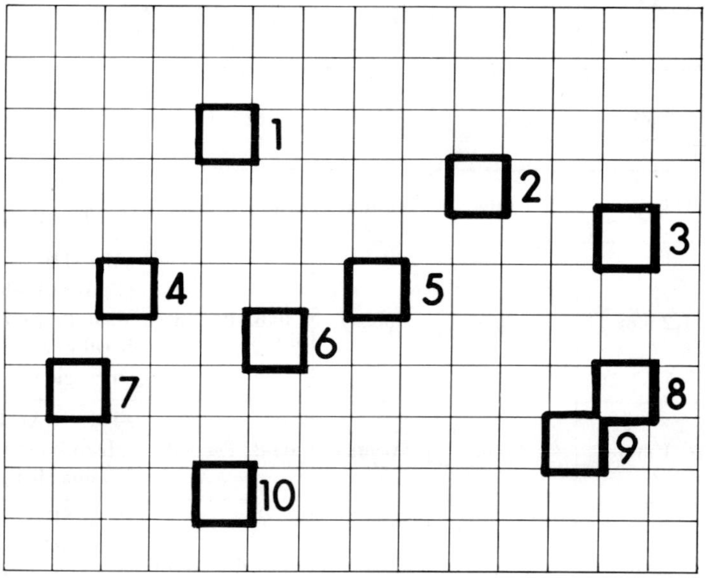

Fig. 6. Transparent grid.

Each of the numbered cm-squares of pavement surface is examined in turn under the stereoscope, and the number of each of the six texture parameters A, B, C, D, E and F is assessed according to the Pavement Surface Texture Classification Table. The numerical average of the parameter numbers of the centimeter squares is the parameter number of the photographed surface area. In the case of parameter C only, the average parameter number is rounded to the nearest whole number. The set of the six average parameter numbers is the texture code number of the photographed surface area.

Sampling the Pavement Texture in Test Sections – At least three pavement stereo-photographs shall be taken in each test section. A test section shall be defined as a section of pavement of uniform age and general composition which has been subjected to essentially identical wear throughout. Sharp curves and steep grades should not be included in the same test section with level tangents, nor should passing lanes be included with driving lanes.

Skid Resistance Photo-Interpretation – Correlation charts of texture code numbers and skid test results of an ASTM type two-wheel trailer have been prepared (see Figs. 7-10). Similar charts based on regression analyses of texture code numbers and test results of other skid testing devices may be prepared.

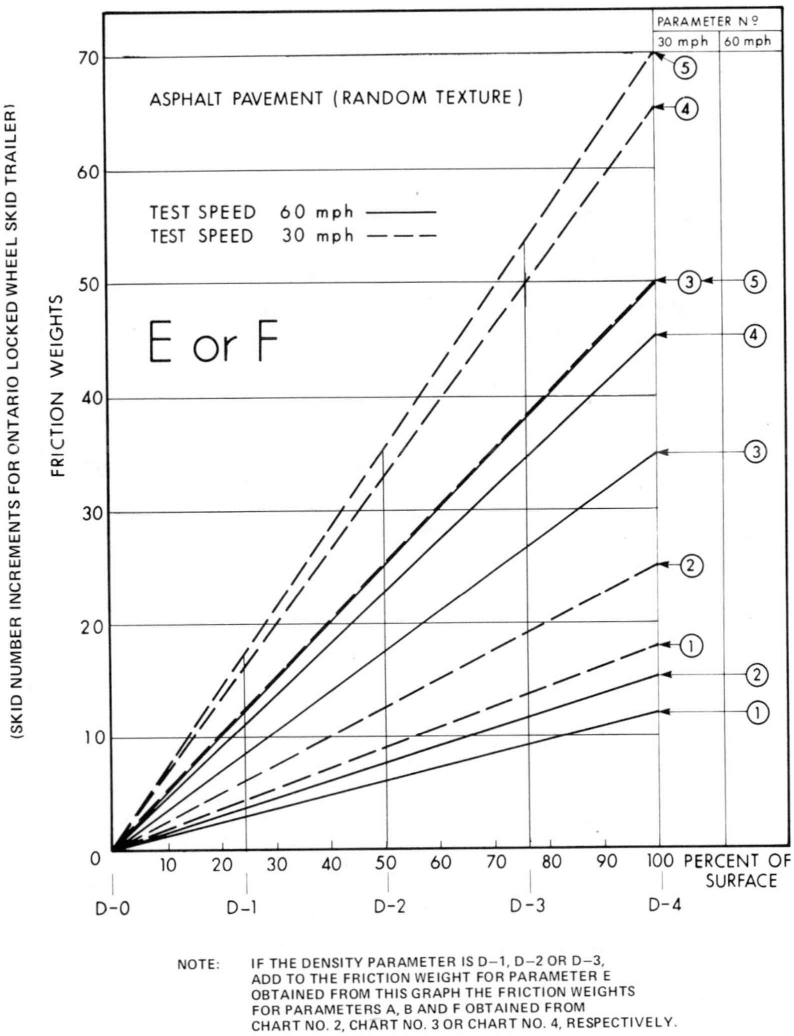

Fig. 7. Skid resistance photo-interpretation (friction weights) of pavement texture parameter E or, if macro-projections are absent (D zero), of parameter F.

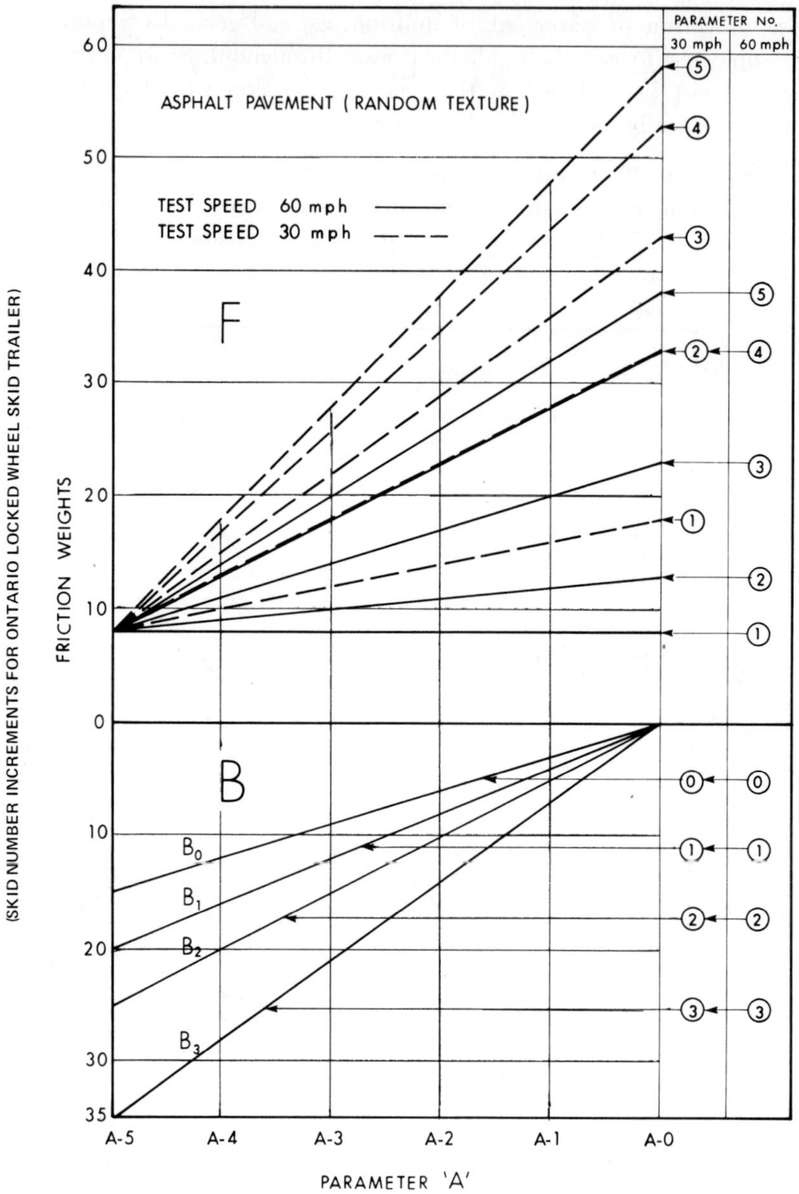

Fig. 8. Skid resistance photo-interpretation (friction weights) of pavement texture parameters A, B and F, for D1.

Report Data – The report shall include such data (as are given in the following list) as are pertinent to the variables or combination of variables studied in the texture examinations:

- Data photographed

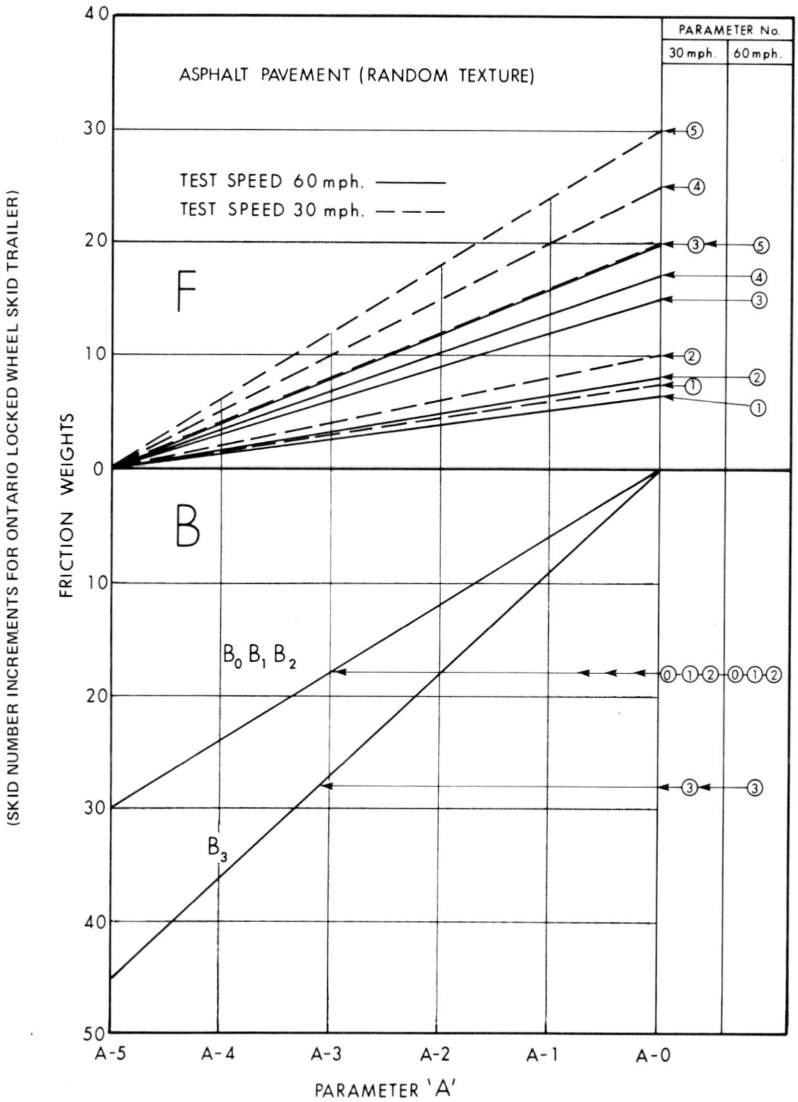

Fig. 9. Skid resistance photo-interpretation (friction weights) of pavement texture parameters A, B and F, for D2.

- Location of test sites
- Age of pavement surface at time of test
- Pavement type and mix design
- Number of traffic lanes
- Lane and wheel path tested
- Grade

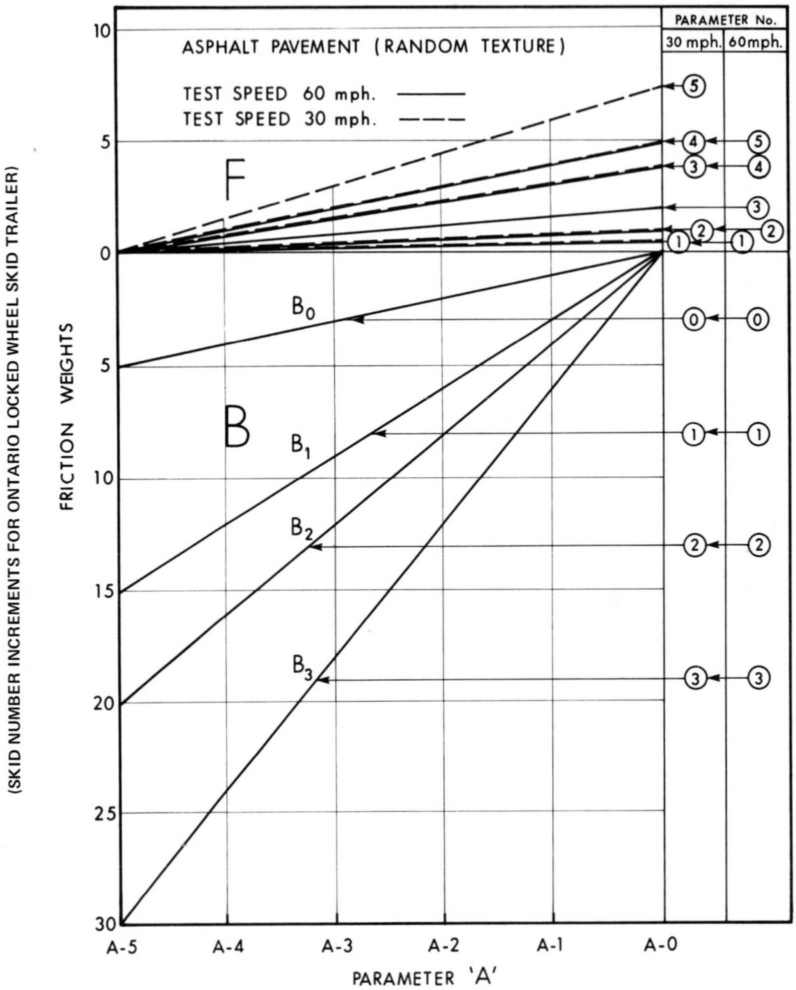

Fig. 10. Skid resistance photo-interpretation (friction weights) of pavement texture parameters A, B and F for D3.

- Average daily traffic
- Texture code number
- Photo-interpreted skid numbers at respective speeds
- Remark concerning cause of low or high skid resistance level in the light of the pavement texture analysis.

AN APPROACH TOWARDS THE UNDERSTANDING AND DESIGN OF THE PAVEMENT'S TEXTURAL CHARACTERISTICS REQUIRED FOR OPTIMUM PERFORMANCE OF THE TYRE

R. BOND, G. LEES and A. R. WILLIAMS

Dunlop Tyre and Rubber Company, Birmingham, England

ABSTRACT

Pavement wet skid resistance is characterized and governed by its macro-texture required to dissipate bulk water, and its micro-texture, essential for the creation of real areas of contact with the tread rubber.

Pavement macro-texture is also a major factor controlling noise generation, spray generation, and ride performance of tyres; and the micro-texture helps control the wear rate of tyres.

A theoretical approach to optimise the potential real area of contact between the rolling tyre and the pavement whilst maintaining adequate water drainage is discussed. The resulting macro-texture recommendation is shown to produce lower tyre noise and superior ride characteristics to other surfaces with similar levels of water dissipation.

The minimum level of micro-texture is related to the thickness of the water film; increasing levels of micro-texture are shown to increase tyre wear without a proportional increase in the wet friction. Direct observation of micro-texture characteristics of various aggregates has led to levels of maximum and minimum levels of micro-texture being suggested. The concepts of micro- and macro-texture optimisation have been applied to a new form of wearing course mix design currently undergoing road trials. The results from these experiements are used to support the theoretical approach.

Improved pavement surfaces will influence tyre design, some aspects of which are discussed.

References p. 360

INTRODUCTION

In the U.K., the wet friction of tyres as measured on low friction pavements has increased three-fold during the period from 1955 to 1970. Increases in traffic density and average traffic speed over the same period of time may explain why there has been no decrease in the percentage of accidents involving skidding under wet conditions which remains at over 30%. It is considered, by the present authors, that it is not possible by tyre technology alone to appreciably reduce this figure. Further improvements are required not only in the design of the tyres and braking systems, but also in the pavement wearing course.

The importance of the pavement in obtaining safer levels of wet friction has been demonstrated by the Transport and Road Research Laboratory's four year study of an urban motorway in London (1). Accident statistics showed that by increasing the Sideway Force Coefficient of the pavement from 0.35/0.40 to 0.50/0.60 measured at 50 km/h the total number of accidents fell by 45% and accidents occurring in wet conditions fell by 63%.

There is a need to specify the textural characteristics of the pavement surface since these affect all aspects of the interaction with the tyre. Such specification may alter the design, materials used, and the economics of the pavements laid, and in turn, may also lead to modifications in tyre design. The pavement textural characteristics governing the wet skid resistance are the micro-texture and macro-texture. The former is required to penetrate thin water films and, thus, create areas of effective contact with the tread rubber, and the latter is required to aid the removal of bulk water. The tyre alone cannot produce an adequate level of wet skid resistance on pavements that possess inadequate micro- or macro-textural characteristics. From the aspect of tyre technology, these two textural characteristics of the pavement govern other tyre performance factors in addition to wet skid resistance.

Other conditions being equal, the micro-texture controls the rate and type of tyre abrasion, and the macro-texture controls the amount of spray generation, the degree of tyre to pavement noise, and the tyre ride characteristics.

By identification of the individual requirements needed of the pavement in order to optimise the tyre performance, it is possible to design pavement surfaces not only with high values of wet skid resistance but also with reduced noise generation, a controlled level of tyre wear, and of improved riding comfort relative to more highly textured pavements.

This is made possible through the rational method of mix design developed by one of the authors (2, 3, 4). The method of mix design is equally applicable to asphalt or concrete mixes and has been developed in order to relate more exactly the design of mixes to the packing properties of the aggregate particles comprising them. These packing properties are affected by the shape and surface texture of the particles, the lubrication effects, the compactive effort, and the boundary conditions.

This paper discusses how, within the concepts of the rational method of mix design, it is possible to produce pavement surfaces capable of enhancing the performance and safety of the tyre.

THE MICRO-TEXTURE

Wet Friction – Micro-texture of aggregates is that property which governs the level of wet friction between tyre and pavement and is essential for the generation of safe values of friction over the whole range of traffic speeds. The importance of the micro-texture is shown in Fig. 1. The seasonal change in wet skidding resistance, which correlates with the seasonal change in accidents involving wet skidding (5), has been associated with the polishing and, hence, removal of the aggregate micro-texture during the summer months and the roughening of the micro-texture during the winter period as a result of frost and other forms of weathering action (6). The photomicrographs in Fig. 1 illustrate changes in micro-texture topography of the order of 5-20x10^{-4} cm that occur as a result of the natural weathering actions which are dominant during winter and the polishing action of traffic which is dominant during the summer. Under extremes of traffic density or at areas of heavy braking or cornering, the polishing effect of slippage between the tyre and the pavement will mask the effect of natural weathering processes.

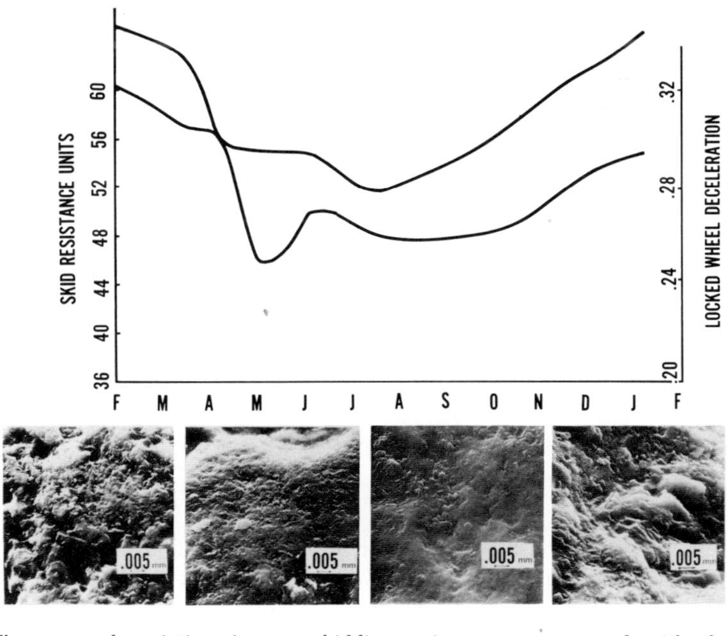

Fig. 1. The seasonal variation in wet skidding resistance as measured with the T.R.R.L. pendulum and an instrumented test vehicle within a braking area of a medium traffic density road. Photomicrographs of aggregates are also shown through the same time period.

References p. 360

The value of wet friction obtainable with aggregates remains low until the surface topography reaches a level of texture depth that is greater than a thin film of water and which has sufficient angularity to penetrate the water film to produce real areas of contact with the tread rubber.

The minimum level of micro-texture to achieve this penetration of thin water films is of the order of $5x10^{-4}$ cm, being limited by the film thickness reported to be 1 to $3x10^{-4}$cm (7). Fig. 2 is a graph illustrating the trend of the increase in wet friction with harshness of micro-texture. The values of wet friction are shown to increase sharply when the harshness of the aggregate topography is above $5x10^{-4}$cm. The rate of increase however, reduces with further increases in harshness. This lowering of the rate of increase is associated with the reduction in the real contact area between surface and rubber. These results are supported by the results from pavement skid resistance tests made with an instrumented test vehicle (8).

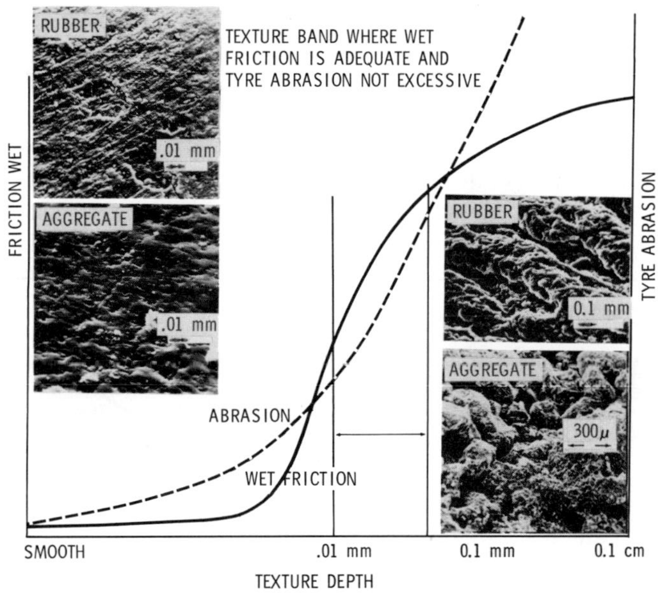

Fig. 2. The increase in wet friction and tyre tread compound abrasion with increase in aggregate micro-texture.

Tyre Wear – Seasonal changes in tyre wear have also been observed, during the present study, during routine tyre wear testing on a country road circuit. The estimated pattern life of tyres tested during the winter period may be only half that forecasted by tests during the summer.

The changes have been associated with the dependence of the wear rate of the rubber compound on the ambient temperature (9), and may also be a function of driver behavior. The change in micro-texture from polished during the summer to harsh during the winter is an additional factor which tends to increase the tyre wear rate in winter. Fig. 3 illustrates seasonally induced changes in aggregate micro-texture in the range of 10 - 50 x 10^{-4}cm. The samples A-E were selected from the same positions around the circuit during the summer and winter periods. The winter sample from site B clearly shows the freshly produced cleaved surfaces, the edges of which have not become polished by the action of traffic. The effect of changes in surface micro-texture on the rate of tyre wear reduces with an increase in severity of use, an effect which is also implied by the work of Brenner, Kondo and Cohen (10).

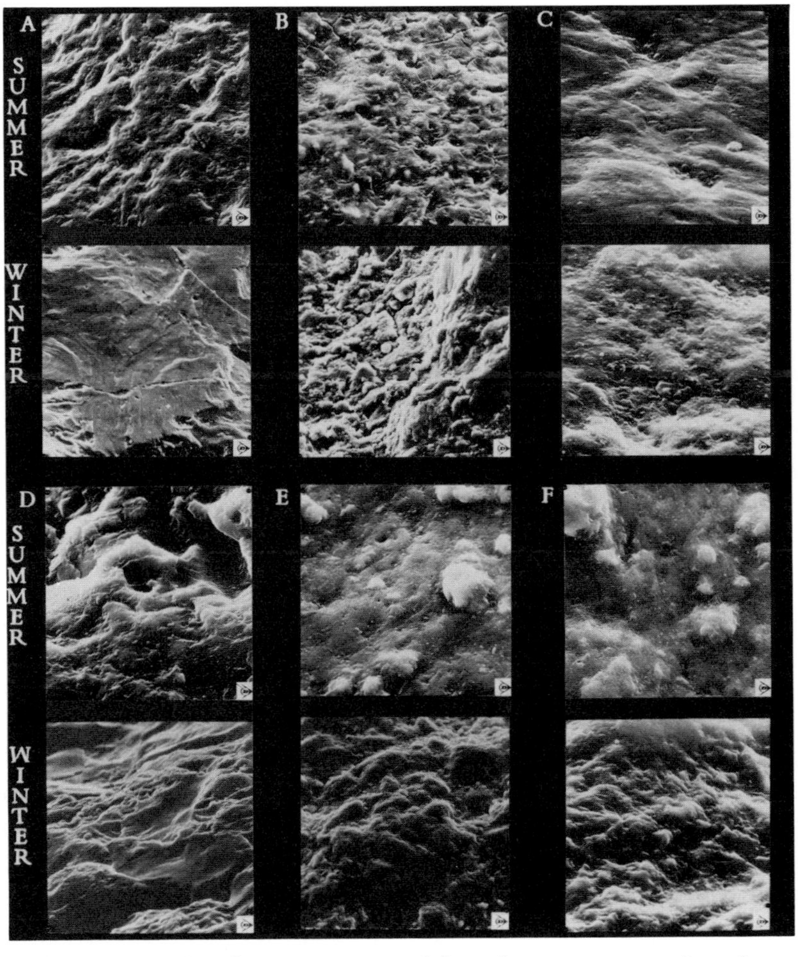

Fig. 3. Photomicrographs of aggregates removed from the pavement at various places around a tyre wear test circuit during the summer and winter periods showing the seasonal change in micro-texture.

References p. 360

Laboratory tests using a recently constructed variable speed internal drum machine (11) have shown how the scale of micro-texture influences the rate of tyre wear under various degrees of wear rate severity. The general form of the relationship between abrasion rate and increase in pavement micro-texture is shown in Fig. 2.

The greater influence of pavement micro-texture as compared to that of tread compound formulation on the wear rate of tyres may suggest a change in the grouping of these two factors in the exponential term 'n' of the wear equations (12).

From the data in Fig. 2, it is possible to suggest limits of micro-texture harshness between 10 and 100×10^{-4}cm within which an adequate level of wet friction may be obtained without resulting in excessive tyre abrasion. By the careful selection of aggregates of all sizes that are used in the wearing course of pavements, it is not only possible to produce a micro-texture within the suggested range but also to maintain the desired harshness level throughout the design life of the material. Such aggregates would include those with sufficient hardness to resist traffic polishing and those that possess a degree of friability so that their texture is renewed with wear and, hence, do not reach a polished condition.

Concerning natural aggregates, it is those in the latter category (friable) that are the more plentiful and which have been chosen as the preferred aggregate type for use in the rational method of mix design, whose goal is the maintenance of high skid resistance.

MACRO-TEXTURE

Bulk Water Removal – The second basic requirement for a pavement surface which is to have good wet friction properties is the need for adequate macro-texture. This is the property which gives the road surface its bulk water drainage, a more open texture being required on pavements where traffic speeds are high.

A theory concerning pavement drainage has been developed for the rolling tyre condition, where it is assumed that the relative speed between the road surface and tyre contact patch is zero. The theory is not suitable for the sliding tyre condition where there is a high relative speed between the pavement and the tyre contact patch. The road surface gives its best performance in terms of drainage when the relative speed between the road surface and the tyre contact patch is zero. This leads to the suggestion that the most important parameter in the road surface tyre contact patch interaction is the rolling friction value experienced when the tyre is subjected to braking, since the relative speed between the road surface and the tyre contact patch is low in this condition, and the value of friction is high when compared with that obtained in the locked-wheel condition. Fig. 4 illustrates the dependence of the degree of water penetration within the tyre contact patch on speed, pavement macro-texture and tread pattern depth. The data for part of this figure was obtained by taking a photograph of the tyre contact patch through a glass plate and measuring the area of water penetration within the contact patch by use of a Quantimet image analysing computer.

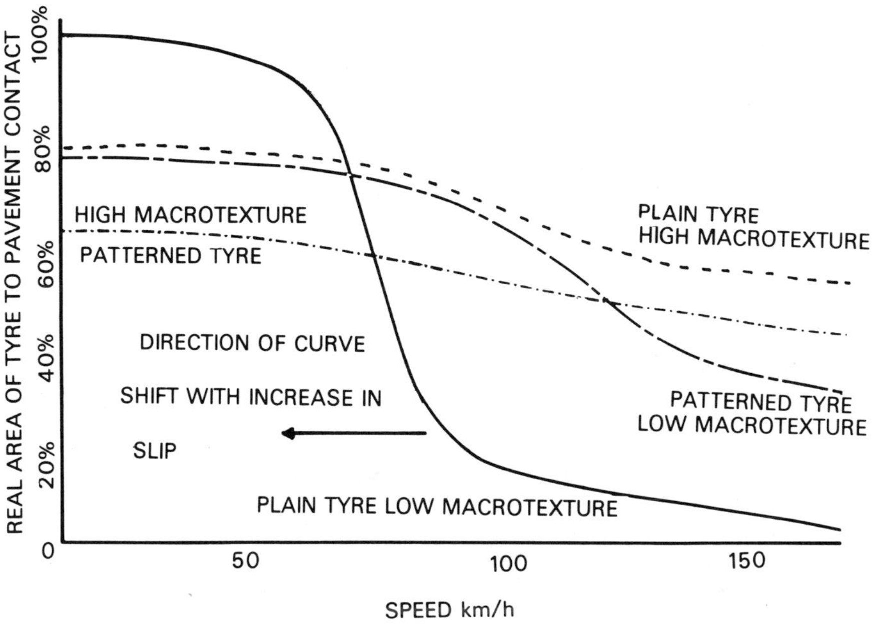

Fig. 4. The influence of tyre pattern wear, vehicle speed, and mode of operation on the ratio of real contact area to apparent contact area for a range of speeds.

The theory which has been developed predicts the bulk water drainage properties of road surfaces made up of various parameters, e.g., size of aggregate (aggregates assumed to be spherical), separation between aggregates, and texture depth, together with various tyre parameters, e.g., pressure in the contact patch, length of the contact patch, width of the contact patch, and modulus of the tread rubber.

The theory has been developed with the assumption that the friction between the tyre and pavement is proportional to the real area of contact and by considering the road surface to be made up of hexagonally packed spheres with the only outlets for the water being in the plane of the surface, the notation used being shown in Fig. 5.

Fig. 6 shows how the ratio of the actual area of contact in the contact patch to the apparent area of contact varies with the radius of the spheres making up the idealised road surface for different separations between the spheres. It can be seen that when the separation between the spheres is zero, the ratio (A/Ao) is independent of the radius of the spheres and gives the maximum possible value, this value being governed by the modulus of deformation of the rubber and wheel load. Thus, it would appear

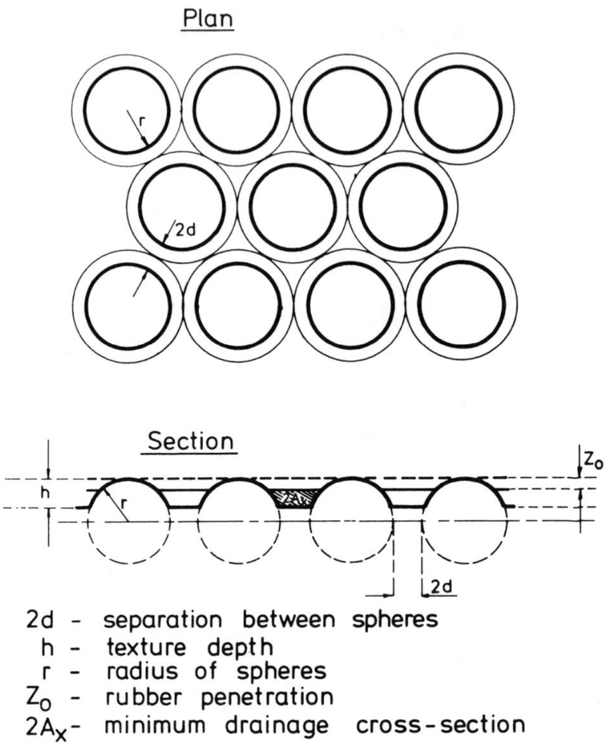

Fig. 5. Notation used in the development of the drainage theory, theoretical road surface model for macro-texture.

that if a road surface can be designed to give good bulk drainage properties whilst maintaining zero separation between the aggregates, the optimum condition would be obtained. It can be shown that the time taken to remove the water from the contact patch (t_a) is given by:

$$t_a = \frac{2\,\ell\, w\,(SPL_{eff})\,Z_o \left\{1 - \dfrac{0.908}{(r+d)^2}\left(rZ_o - \dfrac{Z_o^2}{3}\right)\right\}}{\dfrac{(P-P_{atmos})}{2\pi\eta}\,A_x^2\,n_{eff}\,\dfrac{(PC)}{(PD)}}\,.$$

$$\text{where } Z_o = \frac{(r+d)^2}{1.815r}\,\frac{(A)}{(A_o)}\,.$$

The symbols used in the preceding equation are defined in the following manner:

Z_o represents the rubber penetration;
ℓ represents the length of the contact patch;
w represents the width of the contact patch;
r represents the radius of the spheres;

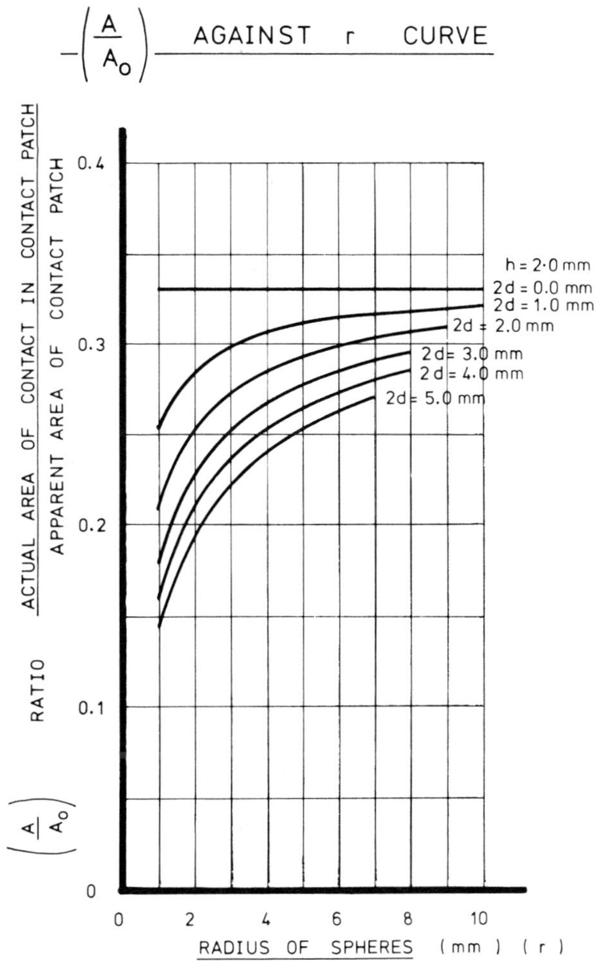

Fig. 6. Graphs of the ratio of actual area of contact to apparent area of contact for a rubber compound on spheres of different radii and spacing. (From theory)

d	represents half the separation between the spheres;
P	represents the uniform pressure in the contact patch;
A_x	represents half the minimum drainage cross-section;
η	represents the coefficient of viscosity of water;
SPL_{eff}	represents the shortest path length;
n_{eff}	represents the number of outlets for the water;
A	represents the actual areas of contact in the contact patch;
A_o	represents the apparent area of contact patch;
PC	represents the perimeter of an equivalent circular cross-section for the minimum drainage cross-section; and
PD	represents the perimeter of the minimum drainage cross-section.

References p. 360

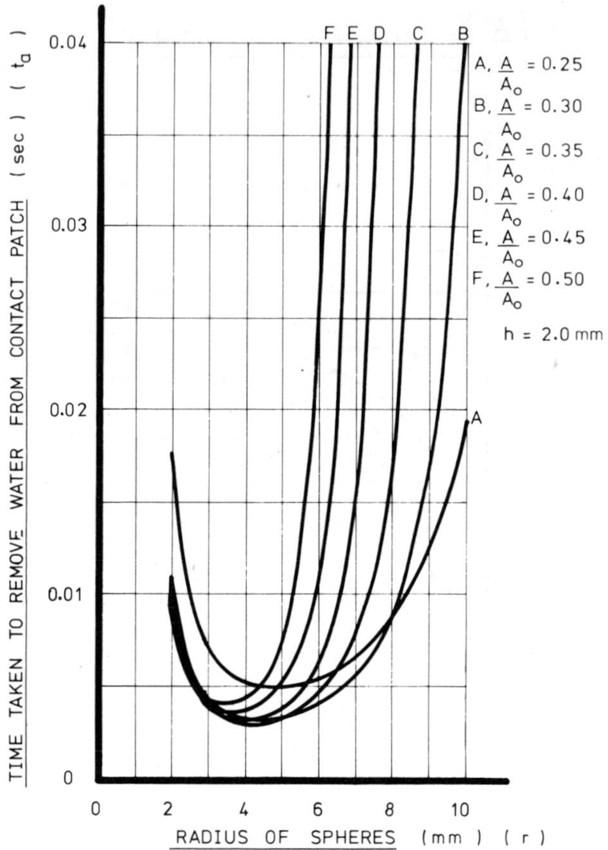

Fig. 7. Graph of the time taken to remove water from the contact patch area for different sphere radii and contact pressures. (From theory)

Fig. 7 shows how the time taken to remove the water from the contact patch varies with the radius of the spheres making up the idealised surface for different values of the ratio (A/Ao) when the separation between the spheres is taken to be zero. The curves A to F are shown for a range of contact pressures from 1.31 to 3.71 kg/cm^2 which is equivalent to a range of A/Ao values from 0.25 to 0.50. It can be seen that the most efficient surface for bulk water drainage, given a texture depth of 2 mm and maintaining the maximum possible contact between the tyre and road surface, will be made from aggregates between 6.0 mm and 12.0 mm diameter.

The theory has been extended to predict the variations that will occur in the road surface drainage chacteristics with changes in speed, and to include transverse grooved pavements. The theoretical curves for the drainage characteristics against speed for:

1. Grooved concrete – Groove size 6 mm x 4 mm, e.g. Motorway M40.

2. Hot rolled asphalt with 19 mm chippings (British Standard 594 (13) e.g. Motorway M40, (prior to any future loss or embeddment of chippings).
3. Development surface produced to the rational method of mix design using the size of aggregates as predicted by the theory.
4. Smooth concrete.

can be seen in Fig. 8, where A_W represents the water free area under the contact patch and A_o represents the apparent area of the contact patch.

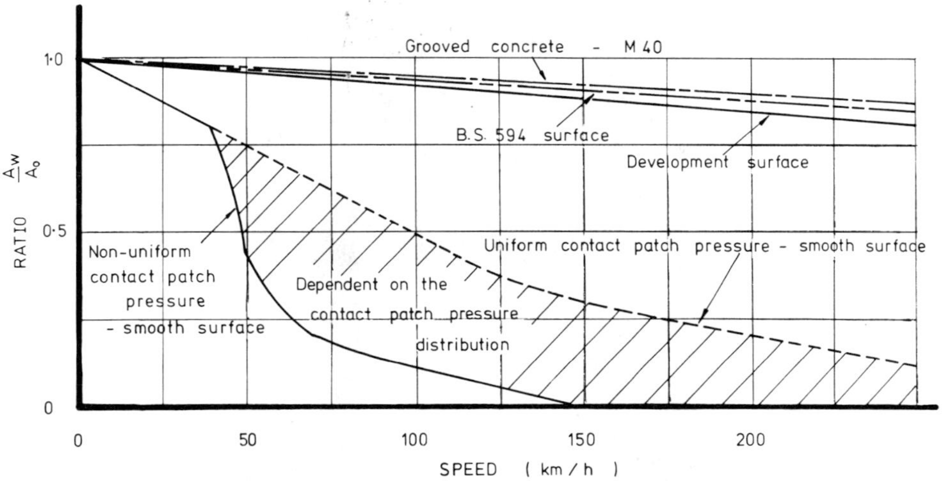

Fig. 8. The drainage characteristics as determined from plots of the ratio of real area of contact to apparent area of contact (rolling conditions) for surfaces of dissimilar macro-texture as a function of speed (from theory).

An important feature of Fig. 8 is the dependence of the drainage characteristics on the pressure distribution within the contact patch. For tyres possessing a higher contact pressure at the shoulder region, the theory predicts and direct measurement confirms the tendency for inferior drainage of bulk water from within the contact patch relative to the drainage from tyres with an even pressure distribution. The flow of water beneath the contact patch of a worn tyre possessing a higher contact pressure at the shoulder than at the centre is shown in Fig. 9.

In an intial attempt to verify the theory, two sets of results will be considered: –

1. Experimental results showing how the water free area in the contact patch of a 5.60-13 smooth cross-ply tyre actually varies with increasing speed are compared in Fig. 10 with the variation predicted by the theory.
2. Experimental results for peak deceleration obtained for the four surfaces previously described plus,

i. Hot rolled asphalt with 12 mm precoated chippings. British Standard Specification 594 – A4, London;

ii. Smooth polished concrete;

References p. 360

Fig. 9. Photograph of water penetration down the centre of the contact patch for a worn tyre under free rolling conditions at 100 km/h in 1.5 mm of water.

are plotted in Fig. 11, and the results show that the surfaces are rated in the same order as was proposed by the theory (shown in Fig. 8).

The rational method of design produces an initial level of macro-texture for bulk water drainage through the creation of a system of surface interconnected voids. The degree of macro-texture is designed to be related to the traffic speed of the pavement area to be laid. The initial macro-texture is maintained during the design life of the material by the incorporation into the aggregate grading of two or more different types of aggregate. Each must have the micro-texture characteristics discussed in the second section of this paper with the aggregates making up the smaller sizes of the

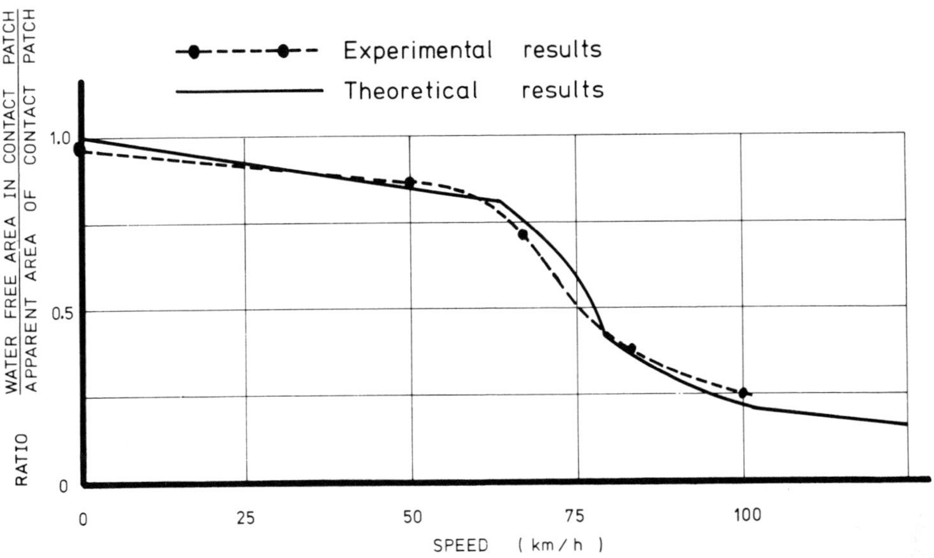

Fig. 10. Comparison of theoretical and experimental results (rolling conditions) for the ratio of the water free area in the contact patch to the apparent area of the contact patch against speed.

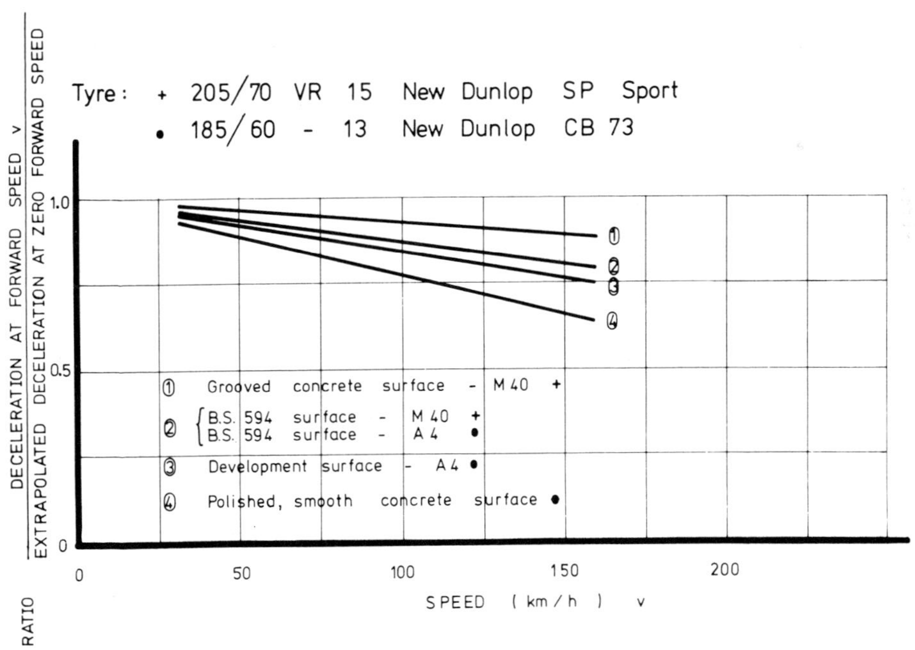

Fig. 11. Experimental results (peak friction conditions) plotted as the ratio of deceleration at a given forward speed to the extrapolated deceleration at zero forward speed against speed for various surfaces of dissimilar macro-texture.

References p. 360

grading being chosen with a higher value of friability, as determined by the aggregate abrasion test (British Standard 812 (14).

Noise Levels Generated at the Tyre to Pavement Interface – The Department of the Environment in the U.K. has proposed that the maximum value of car traffic noise in 1975/76 be 79 dB(A) and that this value be further reduced in 1980/81 to 76 dB(A). Similar reductions are proposed for commercial vehicles. It has been reported that relatively high noise levels are experienced on smooth surfaces and also on those with an extremely high macro-texture (such as textured concrete) when measured under dry conditions with patterned tyres (15). Surfaces with the type of macro-texture discussed in the third section of the paper, as an optimum for water drainage have also been found to result in a lower value of external noise than that present with these two extremes of macro-texture. The increase in noise generation with speed for two of these three surfaces is shown in Fig. 12 for the cases of plain and patterned automobile tyres. A reduction of between 3-4 dB(A) has been

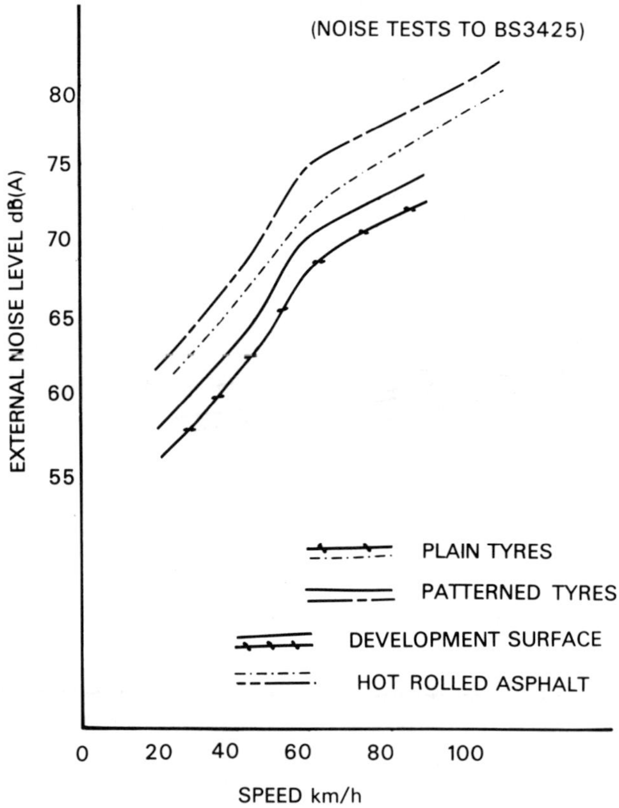

Fig. 12. The external noise generated between tyre and pavement, measured with the engine off on two dissimilar surfaces – the development surface and a hot rolled asphalt with 19 mm chippings. Results are given for plain tread and patterned tyres.

measured on the development surface relative to a surface coated with a hot asphalt material with 19 mm chippings that is extensively used as a wearing course in the U.K..

The interior noise spectra generated on the development pavement and on the hot rolled asphalt for both plain and patterned automobile tyres and a ribbed truck tyre are compared in Fig. 13 (a and b). For both vehicles, a positive reduction in the

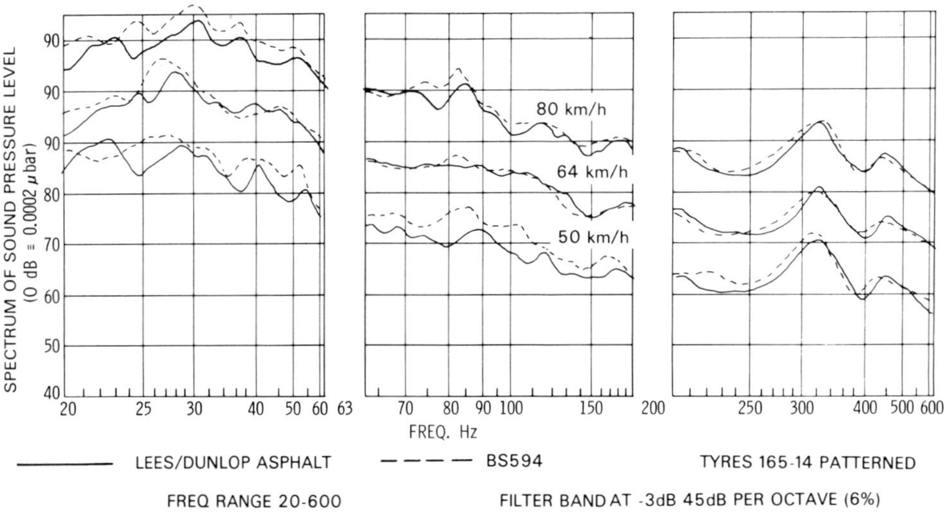

(a)

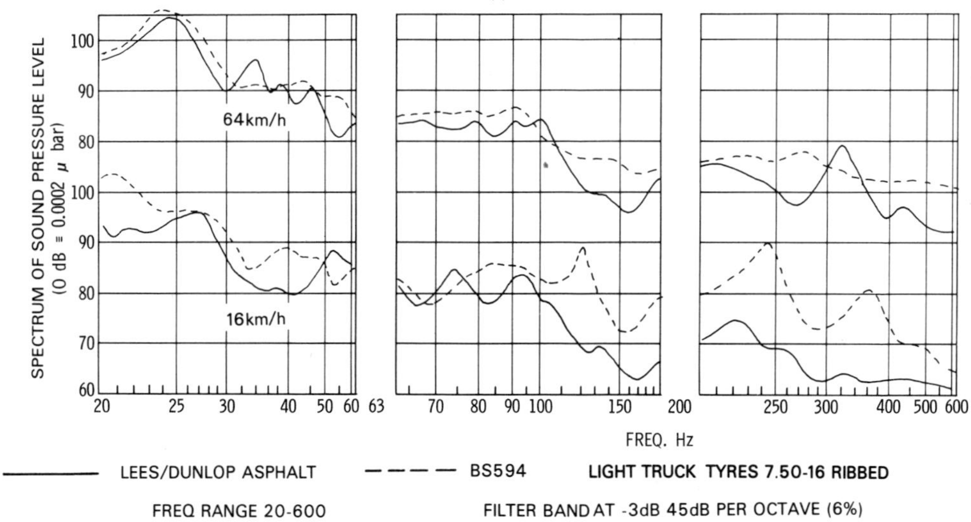

(b)

Fig. 13. (a) Internal noise spectrums for patterned automobile tyres on dissimilar surfaces and at various speeds. (b) Internal noise spectrums for a light truck tyre (7.50-16) with a ribbed pattern at speeds of 16 km/h and 64 km/h.

References p. 360

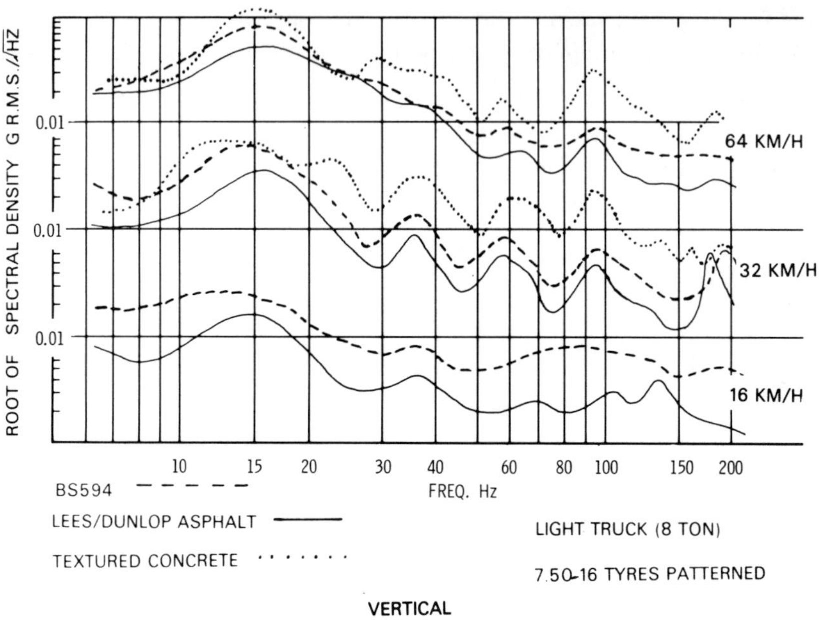

(a)

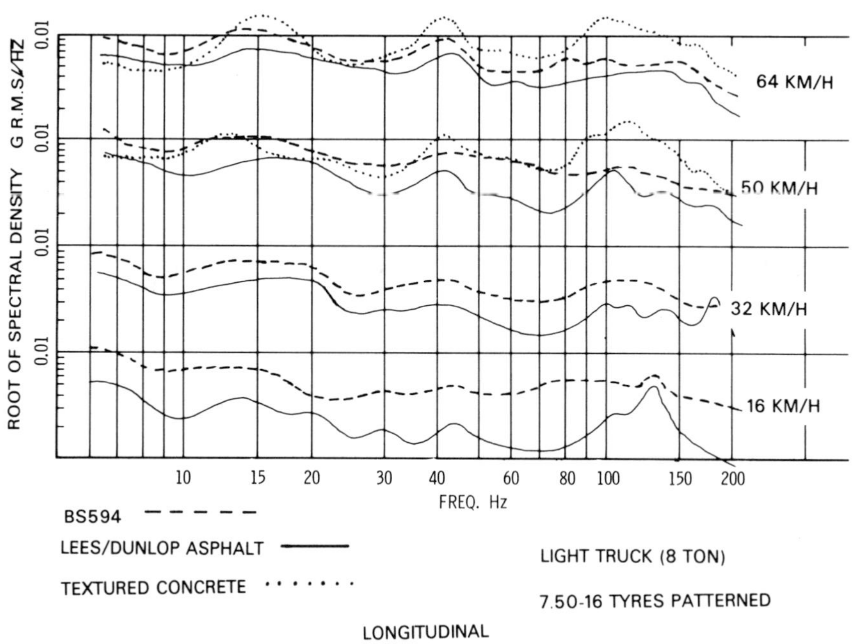

(b)

Fig. 14. (a) (b) Unsprung mass acceleration in vertical and longitudinal modes for light truck tyres on dissimilar surfaces.

internal noise spectra produced on the development surface is found when compared with the other pavements. The design of a tyre tread pattern for minimum noise generation will depend on the pavement macro-texture. This, in turn, needs to be developed in conjunction with the tyre industry to ensure that such resonances associated with the pavement texture do not reinforce those present with the tyre.

Ride Characteristics of Pavements – Analysis of the vertical and longitudinal accelerations at the unsprung mass of both an automobile and a light truck show a positive reduction in the value of the power spectral density (16) over a range from 2 to 200 Hz. This will result in a positive improvement in the riding comfort detected by passengers, (see Fig. 14 a and b), when travelling on the development surface.

The Influence of Pavement Performance on Tyre Performance – The values of sideway force coefficient plotted against speed as measured using a Sideway Force

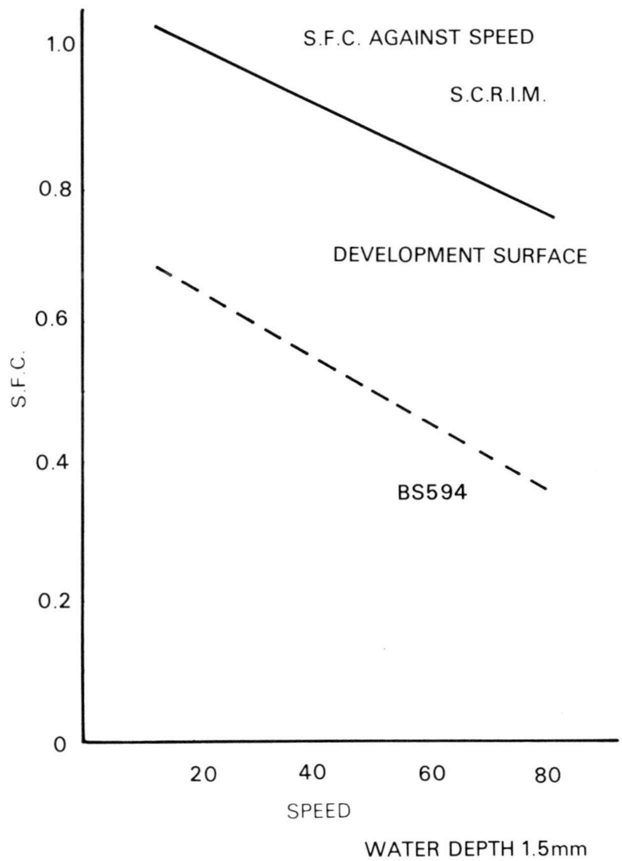

Fig. 15. Sideway force coefficients at various speeds for the development surface and for a hot rolled asphalt with 12 mm precoated chippings.

References p. 360

Coefficient Routine Investigation Machine (S.C.R.I.M.) (17) are shown in Fig. 15 for the development surface that was produced to the rational method of design and incorporating the concepts of pavement textural characteristics discussed in this paper. The results indicate a significantly higher level of wet skid resistance compared to that for a hot rolled asphalt standard surface with 12 mm precoated chippings. Both surfaces have been subjected to the passage of 50,000 vehicles per day for an equal time period. The location of this site is on the A4 in London. A comparison of the macro-texture characteristics of these two types of surface is shown in Fig. 16, the photographs being taken from another test site in Staffordshire. The interconnecting voids between the upstanding aggregates are seen on the development surface.

A Standard Asphaltic Mix

The New Road Surface

Fig. 16. Comparison of surface macro-texture between a hot rolled asphalt with 19 mm precoated chippings and the development surface.

Fig. 17 illustrates the improvement in peak braking force coefficient that can be achieved on the development surface after trafficking when compared to that for a Bridport pebble macadam proving ground test strip. The improvement in braking force coefficients is such that the peak braking force coefficient on the Bridport pebble surface, obtained with a high styrene S.B.R. tread compound, is only equal to the locked-wheel braking force coefficient obtained with an 80/20 SBR/BR blend on the development surface.

The further advantages that can be achieved by use of high friction tread compounds have been shown to be significantly greater on surfaces possessing the specified levels of micro-texture.

The introduction of pavements of improved wet skid resistance will, to some extent, increase the rate of tyre wear. The general trend toward the increased use of steel breaker radial tyres is reducing the dependence of the mileage obtained on the

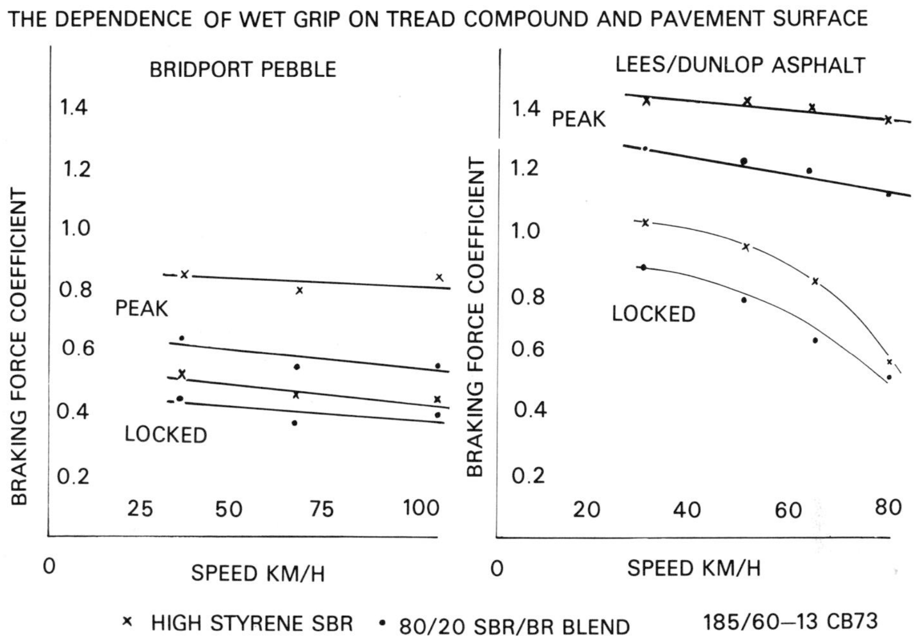

Fig. 17. Comparison of the peak braking force coefficients measured on a Bridport pebble Macadam and on the development surface for dissimilar tread compounds over a range of speeds.

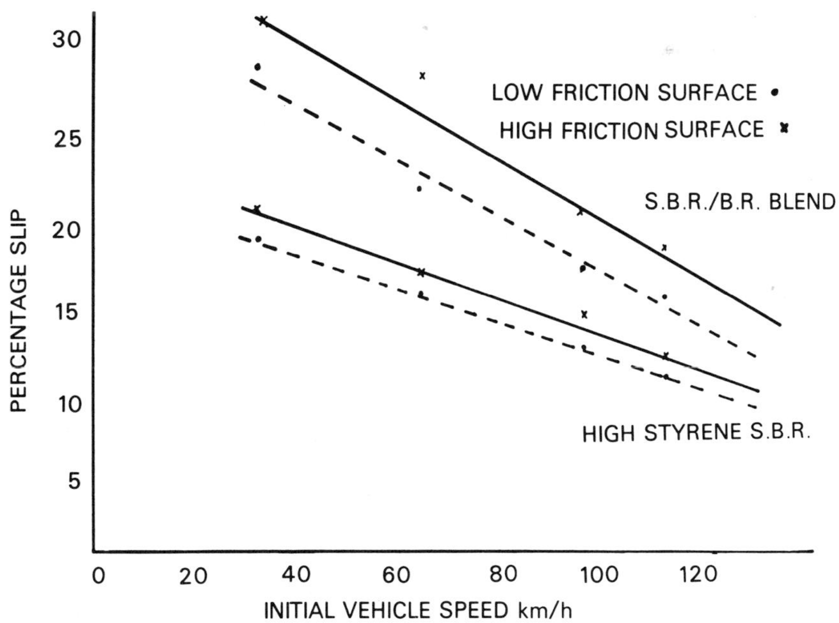

Fig. 18. The influence of compound and pavement on percentage slip at peak friction.

References p. 360

tyre tread compound formulation and possibly presents an opportunity for further improvements in compound wet friction characteristics. However, these changes in tyre construction may alter tyre performance under the straight ahead braking condition. At a given speed and rate of brake application, the time that it takes and the percentage of slip that is necessary to generate the maximum braking force are dependent on the pavement friction and the elastic stiffness of the tyre in the longitudinal direction. The longitudinal stiffness of a tyre is increased by a change in construction from a cross-ply tire to a textile radial tire and further increased by a change to a steel radial tire. These also increase with tread wear due to an increase in the tread rubber profile stiffness per unit area (18). The longitudinal stiffness is also altered by changes in tyre aspect ratio. This may be partly due to changes in contact patch shape.

The influence of tread compound and pavement friction characteristics on the peak braking force coefficient of patterned tyres at various speeds is shown in Fig. 18. The high friction compound produces a high value of peak braking force coefficient at a lower percentage of slip than a compound of lower friction but of equal modulus. A

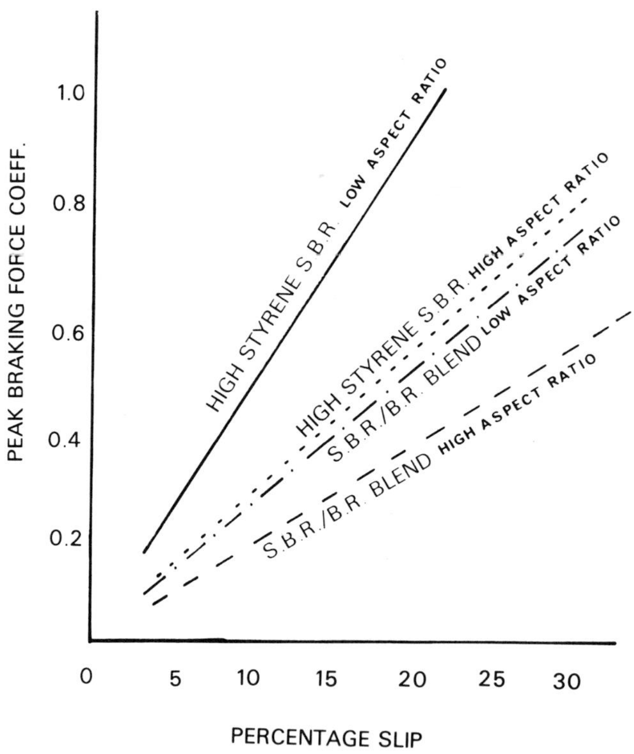

Fig. 19. The influence of longitudinal stiffness, tread compound, and speed on percentage slip at peak braking force.

reduction in the pavement friction characteristics reduces the percentage slip at peak as does an increase in vehicle speed. The reductions in percentage slip and in the associated sliding velocity at peak braking force may be explained in terms of the dependence of polymer friction on the sliding velocity and second order transition temperature of the polymer (19).

The changes in values of the peak braking force coefficient and of the percentage slip at which this peak occurs are shown in Fig. 19 for tyres of different aspect ratios having an initial velocity of 50 km/h.

The trend of tyre development is to reduce the percentage slip at peak which will influence vehicle control under the straight ahead braking condition. However, this trend will be countered by improvements in pavement wet friction characteristics.

CONCLUSION

The optimum performance of the tyre cannot be achieved without a specification for the pavement's textural characteristics. Table 1 summarizes these requirements for all of the aspects of tyre performance that were considered in this paper: wet skidding resistance (of primary importance), tyre wear, tyre/pavement noise generation, and ride comfort. The joint programme of research between the Department of Transportation and Environmental Planning, University of Birmingham and Dunlop Limited has led to the combining of the pavement surface parameters for optimum tyre performance with a rational method of mix design to produce a wearing course material currently undergoing trials and entering commercial use in the U.K..

TABLE 1

Pavement Textural Characteristics Required for Optimum Tyre Performance

Tyre Property	Pavement Property	Specification
Abrasion, Wet Friction	Microtexture (All Aggregates in Mix)	10-100x10-4cm
Wet Friction, Ride/Comfort, Noise	Macrotexture	Aggregate Particle Size 6-12mm DIA
		Inter Particle Spacing 1-4mm
		Absolute Texture Depth 1-3mm (dependent on aggregate size)

References p. 360

ACKNOWLEDGEMENTS

The information in this paper resulted from a research programme between the Department of Transportation and Environmental Planning, University of Birmingham (Head of Department, Professor J. Kolbuszewski) and Dunlop Limited to whom thanks are due for their joint interest and encouragement in the research and also for permission to publish this paper. The authors also thank their colleagues for their help in obtaining the information recorded in this paper. The theoretical approach to water drainage forms a part of a thesis for the degree of PhD to be submitted to the University of Birmingham by one of the authors (R Bond).

REFERENCES

1. *M. M. Miller, H. D. Johnson, "Effect of Resistance to Skidding on Accidents Surface Dressing on Elected Surface of M4 Motorway," Transport and Road Research Laboratory, Report LR542, 1973.*
2. *G. Lees, R. L. Sharif, "The Design and Field and Laboratory Testing of High Friction Dense Asphalts," The Highway Engineer, pp. 4-20 June 1973.*
3. *G. Lees, "The Rational Design of Aggregate Gradings for Dense Asphaltic Compositions," Proceedings Association of Asphalt Paving Technology, 39, pp. 30-69, 1970.*
4. *G. Lees, "The Rational Design of Continuous and Intermittent Aggregate Gradings for Concrete," International Symposium on the Grading of Concrete Aggregates H.R.B., January 1973.*
5. *B. E. Sabey, "Road Surface Characteristics and Skidding Resistance," J. British Granite and Whinstone Federation Vol. 5, No. 2, 1965.*
6. *A. R. Williams, T. Holmes, G. Lees, "Toward the Unified Design of Tyre and Pavement for the Reduction of Skidding Accidents," SAE Paper No. 720162, Detroit Michigan, January 1972.*
7. *D. V. Deyaguin, Z. Physik, 84, p. 657, 1933.*
8. *G. Lees, A. R. Williams, "The Relation Between Tyre and Road Surface Design," I.R.F. meeting, Munich, October 1973.*
9. *D. Bulgin, M. H. Walters, "The Abrasion of Elastomer Under Laboratory and Service Conditions," Inst. Rubber Conference, Brighton, England, 1967.*
10. *F. C. Brenner, A. Kondo, G. B. Cohen, "Research for a Uniform Quality Grading System for Tyres V. Effect of Environment on Tread Wear Rate," Rubber Chemistry and Technology 44, p. 4, 1971.*
11. *A. R. Williams, G. Lees, "A Machine For Friction and Wear Testing of Pavement Surfacing Materials and Tyre Tread Compounds," Journal of the IRI, October 1973.*
12. *A. Schallamach, Rubber Chemistry Technology, 41, p. 209, 1968.*
13. *British Standards Institution BS594 Rolled Asphalt (Hot Process), 1961.*
14. *British Standards Institution BS812 Mineral aggregates, sands and fillers, 1969.*
15. *M. C. P. Underwood, "A Preliminary Investigation into Lorry Tyre Noise," Transport and Road Research Laboratory Report to be published 1973.*
16. *V. E. Gough, C. W. Barson, J. C. Hutchinson, D. H. James, "Tyre and Vehicle Vibration," Institute of Mechanical Engineers, 179, 2A, 1964-65.*
17. *Sideway Force Routine Investigation Machine, T.R.R.L., Crowthorne England.*
18. *H. C. A. van Eldik Thieme, H. B. Pacejka, "The Tyre as a Vehicle Component," Chapter 7, Mechanics of Pneumatic Tyres, edited by S. K. Clark N.B.S., Monograph 122, 1971.*
19. *D. Bulgin, D. G. Hubbard, M. H. Walters, "Road and Laboratory Studies of Friction of Elastomers," 4th International Rubber Conference, London, 1962.*

PANEL DISCUSSION

TIRE TRACTION – THE ROLE OF THE PAVEMENT

K. C. Ludema (Moderator)
D. F. Moore
R. Schonfeld
A. R. Williams
W. O. Yandell

Ludema

We have now covered a large number of topics. Alan is going to pull it all together later. I should like to invite discussion with the panel and with audience interaction on some of the subjects we have covered. In particular, I think the future directions for work in this area is worthy of some exploration. We have had some discussion of the relative effect of tires and the relative effect of road surfaces, on skid resistance. Some claim that the potential for higher friction lies in the tire and others say the answer lies in the pavement. Others say the answer lies in getting rid of the water and everybody is doubtless correct. It has become clear in the last two days that there is no ready model – no simple expression – that includes all of the variables we discussed. You could almost say that we hardly got off our starting blocks because, after all, the one term we do not have yet defined is adhesion. What is adhesion? How does one justify a viscous film on top of little stones (where angels fear to tread) or on these little asperities. In the face of the fact that cars do not slide downhill when parked for a month or so, there must be more than a viscous film present, there has to be adhesion. Roberts has showed adhesion, or some surface interaction at least, to be acting. In essence, we do not even have a way of expressing some of the basic quantities involved, but I would say we have to go one step further back than that. Are we sure that we have cataloged all of the quantities that we wish to define? From that point of view then, perhaps we ought to take a look at the prospects for some analytical expressions forthcoming from the work we have seen in the last two days. Consider now how we have danced around that central prospect of an analytical expression uniting or combining all of the variables. We have had an equation by Pottinger on the first day. In the denominator of that equation is an expression for the effective drainage contributed by the tire. We have had pictures by Yeager in which we watched drainage in action. We have seen vortices, not described mathematically. We have seen other phenomena described in terms of viscoelastic transforms. There has been some discussion of mechanical properties but no useful general expressions for them. But even apart from that, there may be some structural factors, such as the ripples of Schallamach. Do they or do they not exist under tires, i.e. real tires on real roads? They exist in the laboratory and more work should be done there. We have the correlations discussed by Peterson involving the old familiar properties, rebound and hardness. Neither of the two properties are compatible with each other. We must admit that describing rubber in terms of two disconnected properties and correlating those terms with performance has limitations. We had water explusion equations of Kienle this morning and the equations of Yandell and Williams. Much work is required to pull it all together. On the other hand, we have the very obvious and graphic method of Mr. Schonfeld showing what road surfaces actually look like. His is a refreshing point of view, isn't it. Rather than simply sitting and calculating things, he first looks at the road before sitting back to express what each little feature means. Back to the point, we are making good progress. What now is the prospect of coming up in the next two or three years with some analytical expressions for the purpose of optimizing wet skid resistance, noise generation, tire wear, and all these other important variables? This panel of experts on either side of me should be able to answer this question. Who is ready to take it on?

D. F. Moore

The question of adhesion I will take on first. I understand there is general agreement that adhesion exists. In fact, from what I understand, adhesion contributes at least 80% of the friction under practically any conditions of the tire-road situation. Now, as regards the question of where adhesion begins and hysteresis leaves off, there is a certain area of gray there. And it is not entirely clear whether in fact they originate from the same mechanism. But most of us here, I think, agree that adhesion exists, and there is one relevant question I will pose to Dr. Yandell. I do not think he likes the word "adhesion", but we will come to that later. If, in fact, it does not exist, Dr. Yandell, how do you explain the friction of two smooth surfaces where there is no texture saturation. Can you answer that for us?

W. O. Yandell

I did not say that adhesion does not exist. I say that in our experiments and theoretical work, we have been able to ignore it so far. It probably does exist but no one knows much about it or what the mechanism is. Since it is so vague, I prefer to ignore it until I have to take notice of it.

Moore

Well, in your paper which I have read very carefully, you specified lubricated conditions and in that case you can get away with it, partly.

If however I plot the friction coefficient against the logarithm of velocity, we have the hysteresis coefficient which rises to a maximum at fairly high speeds and the adhesion at lower speeds. If we combine them, we get a curve having two peaks. Now the question is, if we have a lubricated surface, the adhesion peak virtually disappears as we have seen from other speakers, and if we go to very high speeds, we may approach the hysteresis peak. But the recent thinking is that hysteresis is, in fact, quite a small contribution. If I can go a little further to the question of the roughness on a surface where we can break it up into a macro-roughness with superimposed layers of finer and finer micro-roughness, then by accepting the superposition principle as Dr. Yandell mentions, we get different individual hysteresis peaks. By further accepting that the friction factor versus frequency curve has a peak at some frequency for a given temperature, and by writing the equation $V = \omega\lambda$ so that the sliding speed is proportional both to frequency and spacing then for a given ω which defines peak friction the sliding speed corresponding to these decreasing roughnesses actually causes these hysteresis peaks to move to the left. You've already said that essentially, but in superimposing the roughness effects, I cannot see how they can grow in magnitude with decreasing frequency. I would surmise they would decrease and that therefore the net effect would be an envelope which would rise with speed. Now, if you, in fact, include somehow what we all call adhesion in this breakdown, then perhaps you may get a total coefficient of friction which decreases with increasing V.

Could you elaborate on that?

Yandell

Yes. Why do you suppose that the individual peak coefficients of friction for the finer texture are smaller than the peak for the large scale? There is no reason why they should be.

Moore

Well, I think within a small range, you may be correct because as the size decreases, the number increases in fact. But I think if you carry that to the limit, the friction contribution must disappear which your model does not allow. I guarantee that. And if you go to a perfectly smooth case where we have no hysteresis, then your coefficient should be due to adhesion.

Yandell

Is there such a thing as a perfectly smooth case?

Moore

No, there is no perfectly smooth case, but our terms of reference would be different. We can approach the smooth case, and if we are speaking of hysteresis, I will maintain that in so doing the coefficient will drop. If you include adhesion in your terms, then of course it may continue to rise and eventually drop.

Yandell

No, I do not agree. A lot of people have the misconception that it takes a large scale of texture to generate significant hysteretic friction, but mathematically at any rate, that is not the case.

Moore

I think this cannot be true because hysteresis is defined as energy dissipation in bulk, and the larger the asperities the more the bulk of the rubber which has this energy dissipation, the more is involved. Now there may be a range, and this is where I yield somewhat to you, where you can vary the texture, in fact, the macro-roughness, and perhaps have a fairly constant hysteresis. But if you carry it to the limit, your hysteresis must decrease.

Yandell

You are quite wrong.

Ludema

Well, you fellows live half way around the world from each other. You will have to do some communication by mail on that, I guess.

R. Hegmon

The friction process, by definition, is an energy dissipating process and rubber hysteresis is the mechanism responsible for the dissipation of the friction energy. The question really is how do we get the rubber to do the job, I mean, to dissipate energy. And here another mechanism comes into effect and this is the excitation of the rubber. Many of us agree that there are two mechanisms to excite the rubber to dissipate energy. One is by repeated deformations and the other is by adhesion. If we accept this, we do not have any contradiction, because even a perfectly smooth surface, which may not exist in reality, would cause, through adhesion, dissipation of energy in a friction process. Now the question of why small asperities do generate less friction; I think it is not just the loss factor, of the rubber, but also the amount of deformation. As the asperities become smaller, the deformation becomes smaller, and the total friction generated will decrease. This is as far as this discussion is concerned. Now I have one question to Dr. Yandell. If I understand your graphs, the coefficient of friction seems to decrease with pressure. If so, this is in contradiction to my experience and any experimental evidence we have on textured surfaces. Could you answer this?

Yandell

Yes, after the pressure has reached a certain stage and the rubber fills the troughs of the texture and the phenomena which I have called "stress saturation" exists, I do predict that the friction drops. We have conducted a number of experiments to confirm this. As far as the first part of your question is concerned, for a given sliding speed, the hysteresis loops generated by the finer scales of texture are moving at a much higher frequency. Even though deformation may be "smaller," the frequency per unit length of rubber is higher to compensate.

Moore

The fact, is, it does not compensate.

Hegmon

But for a given asperity size there is just one speed which corresponds to the frequency of peak loss. And vice versa, at one particular speed, only one particular size of asperities will produce high friction peaks, but it certainly would not hold for all sizes of roughness.

Yandell

What evidence have you to suggest that?

Hegmon

Well, in your own words, the deformational loss is a function of frequency, and it has a peak at one particular frequency. This frequency can be related to the spacing of asperities by one speed only. You cannot, at one given sliding speed, excite the rubber at its frequency of maximum deformation losses for all sizes of asperities.

Yandell

I do not know that I understand you correctly. The peak friction occurs at a progressively lower speed as the size of the texture decreases, and mathematically, there is no reason why a fine style of texture should not generate the same coefficient of hysteretic friction as a coarse scale of texture.

W. E. Meyer

Dr. Hegmon has already expressed what I meant to comment on, but I did want to add that Hartwig Kummer tried some ten years ago to find a bridge between these opposing views by postulating a molecular roughness, so to speak, involving the stressing and breaking of bonds on the molecular scale and comparing these processes to the bulk hysteretic behavior of the friction. Kummer's approach might well provide the bridge between the adhesion and the hysteresis theories.

A. R. Williams

I would suggest that a solution to this problem may be found from consideration of the practical conditions. I have already stated that, in my belief, the peak rolling condition of friction is of greater practical value than the locked-wheel state, where under the former condition the tyre-ground interface sliding velocity will be greatly reduced. The condition of reversion, or more correctly the existence of thermally degraded rubber under the tyre tread surface may be considered as the physical manifestation of bulk hysteresis. The process has been shown to be one of very rapid tyre wear, and a theoretical study (not published) has indicated that the maximum amount of energy dissipated by this process is less than 10% of the kinetic energy of the vehicle.

For an efficient energy dissipation via the higher thermal conductivity of road surface aggregates, the energy loss processes must be at the tyre/ground interface.

Due to the need to penetrate the thin films of water that remain between the tyre and road surface under car tyre contact pressures, a minimum level of microtexture harshness and angularity can be specified as discussed in the paper presented. The penetration of the thin water film is essential for the production of effective areas of contact with the tread rubber. Increasing the harshness of the microtexture from the suggested limits has the effect of rapidly increasing the rate of rubber abrasion.

The presented paper indicates that the macrotexture be only considered for aiding the removal of the bulk water from within the contact patch, and suggested limits to the practical geometry of this texture are given.

It is possible with the understanding gained over the last twenty years of intensive research effort by the tyre and road industries that we are now in a position to redefine the problem. This is illustrated by two examples. Bob Schonfeld's excellent ideas regarding the characteristics of textures is limited when one considers the friction course porous textured surfaces. At the other end of the texture scale, we have considered microtexture on the micron scale. This is far below that which can be detected by the technique currently used for characterising the surface.

The role of tread patterns could well be altered if the road industry adopted such surfaces as the friction course material or the wearing course described in the presented paper.

We have heard that for optimum wet traction the tyre compounders require low modulus tread compounds, and we have heard Hans Pacejka stating that from the aspect of stability a hard tread compound is required. I believe that only by closer communication and joint studies between the tyre and road industries will the correct weighting be given to the various compromises of tyre and road design.

G. G. Balmer

There are a number of the state highway departments that have been examining the Schonfeld method, and it appears to be a promising procedure. It has two disadvantages. One of them is that the method is tedious. The other disadvantage is that it is subjective. The Maryland State Highway Administration has plans to automate the system. Maryland expects to investigate the possibility of using electron scanning of these stereo photographs and digitize the results so as to computerize them. This will remove the subjective aspect and a great deal of the tediousness. The research has not been initiated, but it is a plan for automating the method.

In a different research project, the Federal Highway Administration is examining the possibility of evaluating pavement texture from a moving vehicle at traffic speeds. In this procedure a laser is being used to measure or evaluate the texture. This research is not completed.

Moore

Mr. Schonfeld has six parameters to characterize surfaces and some years ago I came up with five. A lot of other people came up with five as well, but there is one that appears to be missing as far as I understand it. And that is the distribution of asperity heights. Would you care to comment?

Schonfeld

There is, of course, ample room for refining the method which I am using now and which, I believe, is at present in its rudimentary stage. I think both your suggestions may turn out to be practical and I hope that the efforts that Mr. Balmer mentioned will be successful. All this will help. In the meantime, I wanted to say that the method is being used in practice in its rudimentary form and that it makes one rather optimistic about its efficacy for clearing out substandard surfaces over large areas, i.e. on a national scale.

Moore

I would like to make one other comment. I think if you remove the background texture from Mr. Schonfeld's list which is perhaps the least important from an engineering point of view, then we would be largely in agreement. His angularity parameter, I think it is C, I call shape, his density of distribution I call spacing, his microtexture I call micro-roughness, and I have included the distribution of asperity heights which will compliment your background texture. Admittedly they are different, so that we are in agreement on the fact that you need about five to characterize texture.

Schonfeld

I think that proof of the pudding is in the correlation. If you find that you are able to predict the friction condition by five parameters, fine. It is not a matter of principle whether four, five, six, seven or eight parameters are used, nor is it a matter of convenience, but it is a matter of expediency.

Moore

Well, years ago people were content with one, and then jumped to three, and three can be adequate in certain cases, but I think five or six is about what you really need to define surfaces.

W. E. Meyer

I wanted to say, that eventually though, as Dr. Williams suggested, we may have to take advantage of the fact that pavements can be constructed that permit the water to escape through the pavement. This, incidentally, may perhaps be the only way in which we can effectively suppress the spray generated by trucks. I cannot possibly see any practical way in which we can control water that once has been picked up by the tires, thrown on the truck structure and become entrained in the turbulence field around the truck, from being atomized and blown into the path of passing vehicles. The only effective way of dealing with this problem is to make the water go through the pavement, so the ultimate pavement of the future will surely be a porous one.

J. R. Watson

I would like to make two points. First, I wish to concur, with the point that Prof. Meyer has just put in and second, I wish to amplify upon the point that Dr. Williams just put in about his "ultimate pavement". The Federal Highway Administration has been advocating for approximately 9 or 10 years now, the use of an "open graded plant-mix seal" or an "open graded asphalt friction course." This is the use of a hard durable aggregate that is practically, for all intents and purposes, a single aggregate size, held together with an asphaltic binder or virtually any other suitable type of binder. When mixed and placed on a highway surface, this material has interconnecting voids that have the capability of absorbing some of this particular water that is causing the problem in wet weather traction. We have been advocating this concept, open graded friction courses, for some time now but we have been unable to get them into widespread practical use. Now, it is probably the eminence of this particular panel that could be just the thing necessary to promote an increased acceptance of open graded friction courses or porous textured materials. There is now available an excellent report on this subject (Report No. FH WA-RD-74-2), "Design of Open Graded Asphalt Friction Courses."

Williams

I am very pleased to hear the comments of J. R. Watson. Under the rolling tyre condition, tyre tread pattern designs are such that the bulk water is removed from within the contact patch; however this water is often air borne and creates the problem of reduced visability due to spray. If the majority of water has been removed from the road surface prior to the arrival of the tyre (through the wearing course design) then this problem is eliminated.

E. A. Whitehurst

One question and one comment. The question is for Mr. Williams. Could he tell us what aggregate he found to use in his developmental pavement that met the rather rigid specifications he outlined for us?

Williams

I mentioned that there are two possible forms of aggregate that meet the microtextural requirements (discussed in my paper), one being of sufficient hardness to resist polishing and the other being of a more friable nature which would not polish but wear. Artificial aggregates e.g. calcined bauxite have been produced of sufficient hardness to resist polishing; however, of those aggregates that exist in nature, it is the friable types which are more abundant.

Within the U.K. a geographical survey has shown that there are adequate resources of this type which in general fall in the gritstone trade group classification. However, to take full advantage of the properties of such materials a new method of mix design is required, the rational method of mix design developed by my colleague Dr. G. Lees being one such approach which in trials in the U.K. is showing encouraging potential.

Whitehurst

Is a British grit stone an American sand stone?

Williams

I believe that what we call in the U.K. a gritstone is termed a sandstone in the U.S.A.. The degree of friability of such materials in the U.K. is termed the aggregate abrasion value. At present there is a limit to the A.A.V. i.e. degree of friability of aggregates, used in wearing course designs of 12 and we usually use materials of A.A.V. of up to 6 or 8. Using the new method of mix design and employing the concepts of a high wet skid resistent surface I believe we could use aggregates of a higher value than 12 and still produce a long life surface that will maintain its skid resistance throughout its service life.

Whitehurst

Thank you. My comment is in response to John Watson's statement about the porous mix. While I agree with him completely as to some of the desirable attributes of this mix, he states he has been working hard for nine years to get it accepted somewhere. I think his real problem is that his organization and many others have worked equally hard for at least 35 years before that to persuade highway people that the worst possible thing you can do is to let water get into the pavement. It is, perhaps, not really surprising that this is not a highly saleable feature right now.

I. D. Wilken

I have a question for Mr. Williams. In your presentation you discussed a developmental road surface. You explained the logic and thinking in regard to developing wet skid resistance, and you did get an improved surface. You indicated that it was also a quieter surface. Did the developmental surface just happen to be quieter, or was there also some logic to developing the surface for minimum noise generation? I would also like to ask Mr. Schonfeld if there have been any studies done to correlate his texture classification parameters with noise generation.

Williams

The trial areas of wearing course material that have been laid have been studied regarding both external and internal noise generation. These results (and those from friction course material) have shown the surface to be significantly quieter than the traditional hot rolled asphalt with 3/4 inch (19mm) precoated chippings used in the U.K. and randomly grooved concrete (lateral grooving).

Whilst a complete theoretical explanation of noise generation on different surfaces is still awaited, I believe that high wet friction surfaces need not be any noisier in the dry or wet condition.

Schonfeld

I would add that noise measurements are being done. We have no correlation between texture and noise yet, but they are being done systematically and perhaps in a year we shall know more about how to correlate these two.

Ludema

I would like to get back to the second question. It really was, to some extent, discussed here. And perhaps the need for what I mentioned earlier was brought out again – the subject of communication between the tire half of the problem and the road half of the problem. If you did not have two days to discuss these things with people on the other side of the water boundary, would you know what they are talking about. The tire people talk about rebound, as I mentioned, and hardness; the road people talk about 19mm. things and hot asphalt mixes and so on and so on. Their terminology are incompatible with each other. What this produces is doubtless, in a conference of this sort, an intuitive feel for the problems of the other half. Now, the tire and the road do not meet each other intuitively. Should not then our communication between the two halves of this problem be on a more formal basis? Are we moving most efficiently in the direction of a total analytical expression of the events at the road surface? If so, do you feel it is necessary? If you feel it is necessary, how much progress do you see? Who will use such results? What terminology should we discipline ourselves to accept and use in order to communicate with each other? For example, going back to the tire side of the problem, I think in every analysis one needs at least three viscoelastic properties to express tire-rubber behavior. What you find in the literature generally is two – hardness and rebound. Mutually exclusive properties, by the way they are measured. What about the road side? Can you use the parameters of Schonfeld in an equation anywhere and calculate a number? The Schonfeld method needs a correlation chart, correlation with the skid trailers and tires of today. Is there predictive value in the Schonfeld method? Can this method be used in 1980 tires and tread designs? I do not think so. Can we move in a direction of analytical expression so that we can forever, as it were, communicate with each other in a rational manner. Are there any feelings on that subject?

M. G. Pottinger

I think you highly over-emphasize the theoretical in a practical sense, because I think that it is the type of problem where everything that can be done in a practical sense, will be done, long before we have a thorough theoretical analysis. I believe work like Dr. Williams has done is very valuable in designing a better pavement. You gentlemen refer to a porous pavement. Do you mean a pavement that has nice water drainage? When you first said porous, I was thinking of something made of sintered ceramic balls, all the water falls right through it, to a drainage channel.

Williams

That is exactly what happens, the term friction course is reserved for a porous wearing course with vertical porosity. The surface referred to in my paper is designed to increase the horizontal drainage of water.

Pottinger

This is a very porous sponge, sort of a ceramic sponge, for example. This would be a very desirable improvement. Does it exist?

Williams

Within certain areas of the U.K. the friction course material has been laid in substantial quantities on all classes of roads and trial areas have been laid on motorways. It has been in use on aircraft runways for up to 12 years.

Pottinger

Let me ask another question related to this. What happens when you have water in the porous mixture? As an example suppose it's 32°F, you have a little salt in the water, and it suddenly drops to 24°F over night, what happens?

Williams

I think this is a very fair comment. We are probably lucky in the U.K. in that it only rains.

Ludema

And rains, and rains, and rains. Any further comments on the analytical and other such methods?

R. M. Gallaway

I have a comment reinforcing Mr. Pottinger's statement. In spite of what Mr. Williams said about the lack of our ability to build roads economically out of synthetic aggregates, or manufactured aggregates, we have built over 10,000 lane miles in Texas under competitive bids, and many of these include, or at least some of them include, the free-draining type. We have had no qualms about building these where we have freeze-thaw cycles. The reason being, generally speaking, that the surface moisture will freeze from the surface down. In the meantime, the water has drained from the lower part of that one-inch mat which prevents any stress development to speak of; or it, in other words, leaves an avenue of release for the expansion of the water which has now turned to ice. So we do not experience any appreciable problems with the freezing of water. In the spring when a reversal may take place, where you might have ice completely through by successive free-thaw cycles, then you begin to melt from the top of the mat down, and there has been readjustment over a period of months for the binder which is there in very thick films and the aggregate which is so located that the void content is on the order of 25%. Then the problem does not exist. Our policy has been not to look for a problem. We do not ever look for the problem; we simply look for solutions and let the problem find us, if it can.

D. F. Hays

I am one who has rather mixed feelings. In the first place, from the standpoint of being engineers, if we did not do anything until we had a complete theory available for it, then I am afraid that we would probably still be walking rather than driving. We would not have constructed a great many things. We cannot wait for the theory to arrive before we make a move. On the other hand, there is another side of the coin. We want to know why things happen as they do. We want to understand the physics of the situation and this takes us back to the fundamental theme of the symposium. I think that there is in each of us a desire to understand; and for those of us who do analysis, we cannot do the analytical modelling until we begin to get a feel for the physics of what is really occurring. You can, of course, do many things from an engineering standpoint and still not be able to predict how to make something more efficient or more effective. I think that when we move into Ken's ballpark, then we say, "I do not have to run 1,000 tests. I can do some analytical work, and I can obtain some guidelines; and from these I can narrow down what I want to do experimentally." From this approach, I too think that we will be farther ahead in terms of expenditures of money, manpower, and time, and we will move faster in achieving that which we desire.

I am saying that we are not going to stop doing things just because we do not have an equation that is appropriate. But, on the other hand, we are not going to stop trying to get the appropriate equation. I think there are two sides represented here. Hopefully, we would like to move together in our work. Some people have a bent toward the experimental, some toward the analytical. Both are important. Ultimately, I think that if we are going to optimize our designs and have a basic knowledge of what is happening, we will have to be moving in the direction of understanding the physics of tire traction. And I think this is one of the thoughts that we might leave with you – we need to understand the physics of tire traction. And we still have both sides on the coin to consider: the experimental and the analytical.

Ludema

Thank you for those few kind words.

H. A. Smith

My comments fit in quite well with Dr. Pottinger's and Bob Gallaway's; I have been jotting them down and this is the most appropriate time to jump in. There is empirical work that indicates that braking performance of vehicles, particularly on wet pavements is quite variable. But, the worst case estimates seem to give a pretty good handle on the kind of marrying that occurs of pavement characteristics to the tire-vehicle. But, in the cornering mode there is an entirely different problem. Many miles of pavement for which surface characteristics should be determined are not intersections or braking sites but places where the most critical maneuver may be passing, and that involves cornering, or combinations of cornering and braking and acceleration. Empirical data seems to indicate a much greater influence of the tire or tire-vehicle than the pavement surface characteristics on the critical speed for cornering on wet pavement. I am a little disappointed that there has not been much emphasis on cornering. Some people stated that we do not know much about that yet. So I would ask the question, when are we going to have the academic, or at least, additional academic information that will help us predict critical cornering speeds for various combinations of vehicles, tires, and pavements?

R. W. Topping

One quick comment. There is considerable knowledge of vehicle cornering performance. It just seems that this symposium is not addressed to that; we should not conclude that the knowledge does not exist.

Ludema

Right, yes.

Pottinger

I want to state that I am not against a theoretical approach to the problem. There are some people that I work for that think I am too much for a theoretical approach. But the point is this – a sound balance of the experiment and theory is needed. It should also be recognized that 90% or more of the final solution may be found before we are able to do the problem in a proper theoretical sense. The study of the physics of the problem is a combined study. It is nice to talk about a complete theoretical expression, but I believe partial models such as Dr. Williams' are likely to give help to people and guidance long before we have a complete model.

Ludema

I agree. Could we have one comment further from Dr. Schallamach, and I think we ought to close it up very soon.

A. Schallamach

It has been asked in the course of these discussions if cornering requires frictional properties different from those operating in other maneuvers. I do not think so but it should be remembered that accident black spots can be created in places where hard cornering is frequent, and the concomitant wear leaves debris on the road. These contaminations can reduce the coefficients of friction, specifically in wet weather.

Ludema

A good point, but that requires another half day of conference, I gather. We will have to close it up. Our time is about ended. Do we have, in the 30 seconds remaining, any further comments here?

K. A. Grosch

May I, as a tire man, make one suggestion to a road engineer? I think the classification of Mr. Schonfeld of the road surfaces and texture is an excellent one, but perhaps one should add a seventh parameter, that is a drainage factor. Because when it rains, it is the rate of application that matters and the speed with which one can get rid of the water in any convenient way. Whether it goes down or sideways is probably not even all that important. But if one specified a drainage factor for a road, it might move road engineers toward a direction in which perhaps they have not been motivated too much in the past.

Schonfeld

What you have said is one of the aspects that can be investigated once you know what pavement you are talking about, and if you add a parameter for, say a porous pavement, or for any properties of a new pavement type which are not included in the present six parameters, one is free to experiment and regress their effect on skid resistance.

Grosch

I meant in the development of specifications for the road surface; it is a legal factor.

Schonfeld

It is quite possible.

Ludema

Well, we have come to the end of the session. This part of the session is closed.

Ludema

The next speaker is really not on the program. He is one of the patron saints, though, of skid resistance and he has a few things to say, Wolf Meyer.

W. E. Meyer

Thank you, Ken. As Prof. Ludema has said, I am not listed in the program, but various people have expressed the need for someone to thank our hosts for a most worthwhile experience. Not only was the slate of speakers most judiciously selected, but the invited audience represented a most useful cross-section of people from the tire, vehicle, and highway fields. It was repeatedly expressed to me, that one of the merits of this meeting was that these variegated groups were brought together under one roof, were, so to speak, locked up for two days and have had to communicate with each other. It took a little time to really work up some steam; but when it was, the pressure got pretty high, as the last panel discussion indicated. It certainly reflects great credit on the symposium organizers and on the Advisory Committee to have selected people in that fashion.

The program that the audience was offered was an absolutely excellent one. Not only was there representation amongst the speakers from all over the world, but I think, most of the important researchers in the tire traction field were here; and, hence, the audience not only had the benefit of listening to them and hearing their views, but also of meeting them and talking to them in person. Unquestionably, the largest share of the credit for this successful symposium belongs to Mr. Hays and Dr. Browne for having organized it in the first place. For having assembled the lists of speakers and participants they must share, however, some credit with the Advisory Committee consisting of Drs. Clark, Ludema, and Smithson. I cannot express adequately how impressed I am about how well this symposium has worked out, not just in terms of program content and the selection of the speakers, the sequence in which the various subjects were presented, but also in overall organization. Not least of all, of course, we all are obligated to General Motors as our host, having received treatment beyond the call of duty.

I am sure that all of us are going to leave here with the good feeling that we have been spending two consistently worthwhile days at a meeting, something which does not always happen when one spends two days at a technical meeting. I wish, in the name of the audience and of all attendees, to express to General Motors Research our heartfelt thanks for a very memorable experience. We appreciate having had the privilege of being here.

AN ANALYSIS OF THE LITERATURE ON TIRE-ROAD SKID RESISTANCE

K. C. LUDEMA and B. D. GUJRATI

University of Michigan

ABSTRACT

Over the last 20 years there has been an exponential rise in the number of reports in the area of Tire-Road Skid Resistance. In a recent publication*, the authors surveyed the major findings and conclusions on this subject up to 1970. The present paper is a summary of that publication: it is designed to assist one in quickly becoming acquainted with the significant findings and landmark publications on the many different aspects of Tire-Road Skid Resistance research.

INTRODUCTION

Tire-road skid resistance is a complex subject which is far from being thoroughly understood. This lack of knowledge as well as the importance of the topic have stimulated an exponential increase in the number of publications in this area over the past twenty years.

This proliferation of the literature has made it nearly impossible for those working on tire-road skid resistance to keep abreast of the numerous advances that are being reported; it has also made it an arduous task for a researcher just entering the field to determine what is known.

STP-541, "An Analysis of the Literature on Tire-Road Skid Resistance", by K. C. Ludema and B. D. Gujrati, prepared at the request of Committee E-17 of ASTM and supported by R. L. Hess, Director of the Highway Research Institute, The University of Michigan. Available from the ASTM, 1916, Race Street, Philadelphia, Pa. 19103.

References pp. 380–384

The present paper is an attempt to summarize the major findings and conclusions reported in the literature up to 1970. Even though this paper was not presented at the Symposium, the advisory committee felt that its inclusion here would contribute greatly toward making this Proceedings a complete reference volume on the subject of tire traction.

SKID RESISTANCE

It is often found that skidding is not a simple event. A number of transient forces appear during vehicle motion and, at some point, one transient force is too large, or acts too quickly for the driver to detect and/or correct and skidding occurs (47)*. Kern (173) lists in his review paper all of the primary forces which influence tire skid and traction, including the driver, the vehicle, the tire and the road. His goal is optimization of the system for safety. Unfortunately, at this time in history it is not possible to characterize many of the variables Kern has listed. In lieu of listing all of the variables of Kern, in this review we will categorize and briefly discuss the various factors that are recognized in the literature to have some influence on skid resistance.

The first category is "Vehicle Factors". Tire load seems to have no simple effect except perhaps where the tire load changes due to the pitch, roll, etc., that can be present during acceleration, cornering, or braking of a vehicle (147, 167, 200). In such cases the load distribution between tires is different from that for which brakes were designed so that the total skid resistance and wheel spin during acceleration will be strongly influenced (66). The influence of speed alone is observed in 30 references, some of which are given here (2, 36, 150, 157, 172, 173, 191, 196, 222, 245, 262, 286, 338). The precise influence of speed on skid resistance depends upon the road surface texture (150, 247). It appears that the different types of carcass construction influence skid resistance (150, 172, 222, 225).

The influence of tread materials on skid resistance is apparently correlatable with the hardness and rebound properties of the rubber that is used in the tread (16). It appears that the full range of skid resistance force available from different rubber mixtures is approximately 20% (222). The chief effects of varying amounts of carbon black, curing agents and other chemicals is to influence wearing ability, tread groove cracking and other properties not particularly related to skid resistance.

Tread design appears to have very little effect on dry skid resistance. The role of grooves in the tire is to aid in the removal of water from the contact patch. An obvious improvement in the wet skidding and braking characteristic is observed for grooved (ribbed) tires as compared with that for smooth tires (150, 152, 172, 222, 278); however, no formal method is given for the design of such grooves. It appears, however, that tread pattern is not particularly effective on rough and harsh textured surfaces (99).

**The reference numbers used in this short summary are the same as those in STP-541.*

Road surface design strongly influences skid resistance. The curvature, grade, tangent, crown, and superelevation are important from the point of view of traction requirements and water drainage. The binder materials and aggregate type in the wear course of roads have some influence on skid resistance. It appears that the mortar in Portland Cement concrete may contribute to high skid resistance by virtue of the fine aggregates in the mortar (230). Bituminous binders may also influence the anti-skid value depending upon the method of construction (247, 320). On the average, however, identical results are usually recorded for roadways built with either Portland Cement or bituminous binder (195, 315). Surface geometry is generally accepted to be a far more important factor than binder composition (78, 96). The aggregate features which have a direct influence on skid resistance are the type, size, shape, wear state, and resistance to polishing wear. Some studies indicate material type more explicitly by giving chemical composition (34, 250). It is now generally agreed that carbonaceous road stone, namely all types of lime stone, is highly polishable and results in extremely slippery roads when wet, although new roads with such surfaces have high skid resistance (118, 245, 339, 364). Sandstone and quartzite aggregates possess the highest permanent anti-skid properties and are resistant to polish and wear (320, 339). Slag obtained from furnaces and other industrial operations also provides good skid resistance, in spite of the fact that the crushing strength and abrasion resistance are low (78).

In an attempt to define road surface texture in an objective manner, the terms macro-texture and micro-texture are often used. Some authors describe texture in terms of size and geometric distribution of protruberances or asperities (2, 189, 211). Reference (165) discusses in detail how frictional characteristics at the contact interface are related to the micro-topography of surfaces. Other investigators have found good correlation between road texture parameters like texture depth and profile ratio and the speed dependence of braking force (286). Various investigators have considered the variations of skid resistance directly in terms of surface texture (2, 109, 150, 196, 222, 245, 325, 368). Though it is intuitively known that rough and coarse road surfaces are desirable for good anti-skid properties, reference (245) was the first to point out the importance of the sand paper type fine scale finish in obtaining better wet skid resistance. However as speeds increase, the role of the coarse texture becomes more important. Thus, it is now widely contended that large scale texture decides the speed dependence of the surface skid resistance while the fine scale texture is responsible for the general skid resistance level (at low speeds) (2, 189).

The subject of speed dependence is often treated under the subject of hydroplaning. Hydroplaning has been studied theoretically (64) and in many tests (72, 98, 153). It has been pointed out (2) that on intrinsically smooth surfaces complete hydroplaning of a smooth tire will occur at 40 miles per hour with an effective water film ahead of the tire not less than 0.006 in. thick. From the skid

References pp. 380–384

resistance results for various depths of water on three different road surfaces, it has been shown that tread pattern effects dominate at low and moderate speeds while the road surface texture becomes important at higher speeds (278).

Traction on ice and snow covered pavements has also been studied (75).

SKID RESISTANCE MEASUREMENT TECHNIQUES

Mention is frequently made of the manner by which most investigators restrict their attention to a small part of the total system. Investigators usually single out one type of test for examining their narrow topic. Thus, since total skid resistance has very many facets, very many test devices have been developed. A good review of the historical development of towed vehicle techniques for measuring the skid resistance is given in (105, 189). The wide use of the towed trailor technique for measuring skid resistance has lead to the drafting and adoption of a tentative ASTM standard E274-65T. Basically, the test consists of measuring the force needed to drag a locked wheel and tire over the road surface in question at some standard speed, usually 40 miles per hour. Water is supplied to the sliding tires to simulate wet skid conditions. Slave wheels attached to a larger vehicle have also been tried but no standard has been adopted for such configurations. Several indoor devices have been constructed as well which involve running a tire against such surfaces as a drum, a flat plank, or a high speed belt. Each has its limitation. Considerable effort has gone into correlation of all devices with real vehicle behavior. Fortuitous correlations appear between certain of the devices, but the reasons for these are usually not known. A much greater confidence in results from different test techniques is needed, however, if these are to be used for routine testing and maintaining a prescribed anti-skid level for improved road safety.

FUNDAMENTAL STUDIES

Specific points were covered in the symposium, and they will not be repeated here. Rather, a list of prime references in fundamental studies is given (11, 25, 26, 79, 117, 123, 128, 143, 164, 188, 198, 203, 204, 235, 238, 239, 275, 283, 284, 295, 306, 342, 386).

REFERENCES

2. *B. J. Allbert and J. C. Walker, "Tire to Road Wet Friction at High Speeds," Proceedings, Automotive Division, Institute of Mechanical Engineers, Vol. 180, Part 2A, No. 4, pp. 105-121, 1965-66.*

11. *G. M. Bartenev and A. I. Elkin, "The Friction of Polymers at the Initial Stage of Sliding in Various Temperature Ranges," Wear, Vol. 11, pp. 393-403, June 1968.*

16. *E. M. Bevilacqua and E. P. Percarpio, "Lubricated Friction of Rubber," Rubber Chemistry and Technology, Vol. 41, No. 4, pp. 832-894, September 1968.*

25. *B. V. S. T. Boonstra and E. M. Dannenberg, "Abrasion and Friction of Vulcanized Rubber," Rubber Chemistry and Technology, Vol. 29, p. 774, 1956.*
26. *F. P. Bowden and D. Tabor, Friction and Lubrication of Solids, Clarendon Press, Oxford, Part I: 1954, Part II 1964.*
34. *James L. Burchett, Jr. and L. Ronalds Rizenburgs, "Pavement Slipperiness Studies," Kentucky Department of Highways, Research Report KYHPR-64-24, HPR-1(5), Part II, March 1970.*
36. *R. A. Byrdsong, "Investigation of the Effect of Wheel Braking on Side-Force Capability of a Pneumatic Tire," National Aeronautics and Space Administration, NASA TN D-4602, 1968.*
47. *S. K. Clark (Editor), Mechanics of Pneumatic Tires, August 1971, National Bureau of Standards (also Chapters on "Properties of Cord-Rubber Laminates," and "The Contact Between Tire and Roadways,").*
64. *H. Daughaday and G. G. Balmer, "A Theoretical Analysis of Hydroplaning Phenomenon," National Research Council, Highway Research Record 311, pp. 1-10, 1970.*
66. *W. DeVinney, "Factors Affecting Tire Traction," SAE Paper No. 670461, May 1967.*
72. *R. C. Dreher and W. B. Horne, "Ground-Run Test with a Bogie Landing Gear in Water and Slush," National Aeronautics and Space Administration, NASA TN D-3515, July 1966.*
75. *A. H. Easton, "Skid Prevention on Ice and Snow," Proceedings of the First International Skid Prevention Conference, Virginia Council of Highway Investigation and Research, Charlottesville, Virginia, p. 549, August 1958.*
78. *E. A. Finney and M. G. Brown, "Relative Skid Resistance of Pavement Surfaces Based on Michigan Experience," Proceedings of the First International Skid Prevention Conference, Virginia Council of Highway Investigation and Research, Charlottesville, Virginia, p. 435, August 1958.*
79. *C. A. Foord, L. D. Wedeven, F. J. Westlake and A. Cameron, "Optical Elastohydrodynamics," Proceedings Institute of Mechanical Engineers, Vol. 184, Part I, 1969-70.*
96. *C. G. Giles, "Some European Methods for the Measurement of Skidding Resistance," Proceedings of the First International Skid Prevention Conference, Virginia Council of Highway Investigation and Research, Charlottesville, Virginia, pp. 267-298, August 1958.*
98. *C. G. Giles and B. E. Sabey, "Skidding as a Factor in Accidents on the Roads of Great Britian," Proceedings of the First International Skid Prevention Conference, Virginia Council of Highway Investigation and Research, Charlottesville, Virginia, pp. 27-44, August 1958.*
99. *C. G. Giles, "The Skidding Resistance of Roads and the Requirements of Modern Traffic," The Institution of Civil Engineers (London), pp. 216-249, February 1957.*
105. *W. A. Goodwin and E. A. Whitehurst, "A Review of Towed Vehicle Techniques for Pavement Slipperiness Testing in the United States," Proceedings of the First International Skid Prevention Conference, Virginia Council of Highway Investigation and Research, Charlottesville, Virginia, p. 257, August 1958.*
109. *V. E. Gough and D. W. Badger, "Tyres and Road Safety," World Highways, Vol. 17, No. 10, p. 47, October 1966.*
117. *V. E. Gough, "Practical Tyre Research," SAE Trans., Vol. 64, p. 310, 1956.*
118. *J. E. Gray and F. A. Renninger, "The Skid Resistant Properties of Carbonate Aggregates," National Research Council, Highway Research Record 120, pp. 18-34, 1965.*
123. *J. A. Greenwood and D. Tabor, "The Friction of Hard Sliders on Lubricated Rubber: The Importance of Deformation Losses," Proceedings of the Physical Society (London), Vol. LXXI, p. 989, 1958.*
128. *K. A. Grosch, "The Relation Between the Friction and Viscoelastic Properties of Rubber," The Proceedings of The Royal Society (London), Vol. 274, Ser. A, pp. 21-39, 1963; Reprinted in Rubber Chemistry and Technology, Vol. 37, pp. 370-385, 1963.*
143. *R. R. Hegmon, "The Contribution of Deformation Losses to Rubber Friction," Rubber Chemistry and Technology, Vol. 42, pp. 1122-1135, 1969.*

147. *C. Hofelt, Jr., "Factors in Tires that Influence Skid Resistance, Part V, Effect of Speed, Load Distribution and Inflation," Proceedings of the First International Skid Prevention Conference, Virginia Council of Highway Investigation and Research, Charlottesville, Virginia, pp. 173-187, August 1958.*

150. *K. E. Holmes and R. D. Stone, "Tyre Forces as Functions of Cornering and Braking Slip on Wet Road Surfaces," Symposium, Handling of Vehicles Under Emergency Conditions, Inst. Mech. Engrs., Auto. Div., Paper No. 6, pp. 81-101, January 8, 1969; Also RRL Report LR254, Road Research Laboratory, Ministry of Transport (U.K.), Crowthorne, Berkshire, England, 1969.*

152. *W. B. Horne and T. J. W. Leland, "Influence of Tire Tread Pattern and Runway Surface Condition on Braking Friction and Rolling Resistance of Modern Aircraft Tire," National Aeronautics and Space Administration, NASA TN D-1376, 1962.*

153. *W. B. Horne and R. C. Dreher, "Phenomena of Pneumatic Tire Hydroplaning," National Aeronautics and Space Administration, NASA TN D-2056, November 1963.*

157. *W. B. Horne, "Results from Studies of Highway Grooving and Texturing at NASA Wallops Station," Pavement Grooving and Traction Studies, National Aeronautics and Space Administration, NASA SP-5073, p. 425, 1969.*

164. *J. D. Huffington, "The Frictional Method of Measuring Contact Area Between Surfaces and Its Relation to the Surface Microtopography of Some Abraded Metals and Polymers," Wear, Vol. 12, p. 343-356, November 1968.*

165. *J. D. Huffington, "Friction and the Microtopography of Some Surfaces," Proceeding of the Royal Society (London), Ser. A, Vol. 279, No. 1377, pp. 161-169, May 1964.*

167. *G. Jones, "The Skidding Behavior of Motor Vehicles," Proceedings, Automotive Division, Institute of Mechanical Engineers, p. 65, August 1962.*

172. *J. D. Kelley, "Factors Affecting Passenger Tire Traction on the Wet Roads," SAE Paper No. 680138, Automotive Engineering Congress, Detroit, Michigan, January 8-12, 1968.*

173. *W. F. Kern, "Coefficient of Wet Friction of Tire Treads," Rubber Chemistry and Technology, Vol. 40, p. 984, 1967.*

188. *H. W. Kummer and W. M. Meyer, "New Theory Permits Better Frictional Coupling Between Tire and Road," Proceedings, 11th International Automobile Technical Congress (FISITA), Munich, Germany, 1966, Publication Verein Deutscher Ingenieure, Duesseldorf, Germany, B11, 37p.*

189. *H. W. Kummer and W. E. Meyer, "Tentative Skid-Resistance Requirements for Main Rural Highways," Highway Research Board, National Cooperative Highway Research Program Report 37, 80p. 1967.*

191. *F. T. W. Lander and T. Williams, "The Skidding Resistance of Wet Runway Surfaces With Reference to Surface Texture and Tire Condition," RRL Report LR 184, Road Research Laboratory, Ministry of Transport (U.K.), Crowthorne, Berkshire, England, 1968.*

195. *R. C. Leathers and R.I. Kingham, "Skid Studies at the AASHO Road Test," Highway Research Board, Special Report 66, pp. 47-58, 1961.*

196. *T. J. W. Leland, "Some Parameters Affecting the Performance of Pneumatic Tires on Wet Pavement," Proceedings of 9th Stapp Car Crash Conference, University of Minnesota, pp. 383-402, October 20-21, 1965.*

198. *S. A. Lippmann, "Structural Mechanisms of Tires Leading to the Development of Steering Forces," SAE Automotive Engineering Congress (Preprint 794C), January 1964.*

200. *R. D. Lister, "Retention of Directional Control When Braking," SAE International Automotive Engineering Conference, Detroit, Michigan (Preprint 963A), January 1965.*

203. *D. I. Livingston, "Physics of the Slipping Wheel. III. Slip Under Camber and Other Forces in Pneumatic Tires," Rubber Chemistry and Technology, Vol. 43, No. 5, pp. 981-994, 1970.*

204. *D. I. Livingston and J. E. Brown, Jr., "Physics of Slipping Wheel. II. Slip Under Both Tractive and Lateral Forces," Rubber Chemistry and Technology, Vol. 43, No. 2, pp. 244-261, 1970.*

211. *K. C. Ludema and Chul Soo Lee, "Influence of Road Surface Texture on Tire-Road Interface Traction Limits," ASME Paper 69-Lub-20, 1969.*

222. *G. Maycock, "Studies on the Skidding Resistance of Passenger Car Tires on Wet Surfaces," Proceedings, Automotive Division, Institute Mechanical Engineers, Vol. 180, Part 2A, No. 1, pp. 122-157, 1965-66.*

225. *J. K. Meades, "The Effect of Tire Construction on Braking Force Coefficient," RRL Report LR 224, Road Research Laboratory, Ministry of Transport (U.K.), Crowthorne, Berkshire, England, 1969.*

230. *Harold L. Michael, "Effect of Pavement Type and Composition on Slipperiness – A summary of Research in Indiana," Proceedings of the First International Skid Prevention Conference, Virginia Council of Highway Investigation and Research, Charlottesville, Virginia, p. 485, August 1958.*

235. *D. F. Moore, "Drainage Criteria for Runway Surface Roughness," Journal, Royal Aeronautical Society, Vol. 69, pp. 337-342, May 1965.*

238. *D. F. Moore, "A Theory of Viscous Hydroplaning," International Journal of Mechanical Sciences, Vol. 9, pp. 797-810, 1967.*

239. *D. F. Moore, "An Elastohydrodynamic Theory of Tire Skidding," 12th International FISITA Congress, Barcelona, Spain, May 1968.*

245. *R. A. Moyer, "Skidding Characteristics of Road Surfaces," Proceedings, Highway Research Board, 13th Annual Meeting, pp. 3-48, December 1933.*

247. *R. A. Moyer, "Effect of Pavement Type and Composition on Slipperiness-California Experience," Proceedings of the First International Skid Prevention Conference, Virginia Council of Highway Investigation and Research, Charlottesville, Virginia, p. 469, August 1958.*

250. *F. P. Nichols, Jr., J. H. Dillard and R. L. Alwood, "Skid Resistance Pavements in Virginia," Highway Research Board, Bulletin 139, pp. 35-39, 1956.*

262. *M. A. Parshin, B. Yurkouski, I. M. Karpinskaya, A. T. Epifasteu and L. D. Slyudikou, "Study of Adhesion of Car Tires to the Road Surfaces (Translated by P. G. Williams)," Soviet Rubber Technology, No. 3, March 1966.*

275. *A. D. Roberts and P. D. Swales, "The Elastohydrodynamic Lubrication of Highly Elastic Surface," British Journal of Applied Physics, Ser. 2, Vol. 2, pp. 1317-1327, 1969.*

278. *B. E. Sabey, T. E. Williams and G. N. Lupton, "Factors Affecting the Friction of Tires on Wet Roads," SAE Paper No. 700376, 1970 International Automobile Safety Conference Compendium, pp. 324-340, 1970.*

283. *B. E. Sabey and G. N. Lupton, "Photography of the Real Contact Area of Tires During Motion," RRL Report LR 64, Road Research Laboratory, Ministry of Transport (U.K.), Crowthorne, Berkshire, England, 1967.*

284. *B. E. Sabey and G. N. Lupton, "Measurement of Road Surface Texture Using Photogrammetry," RRL Report LR 57, Road Research Laboratory, Ministry of Transport (U.K.), Crowthorne, Berkshire, England, 1967.*

286. *B. E. Sabey, "Road Surface Texture and the Change in Skidding Resistance with Speed," RRL Report LR 20, Road Research Laboratory, Ministry of Transport (U.K.), Crowthorne, Berkshire, England, 1966.*

295. *A. R. Savkoor, "The Relation of the Adhesional Friction of Rubber to the Friction Between Tire and Ground," FISITA 11th International Automobile Technical Congress, Munich 1966.*

296. *A. R. Savkoor, "Some Aspects of Friction and Wear Arising from Deformation, Slip and Stresses at the Ground Contact," Wear, Vol. 9, p. 66, 1966.*

306. *A. Schallamach, "A Theory of Dynamic Rubber Friction," Wear, Vol. 6, pp. 375-382, October 1963.*

315. K. H. Schulze and L. Beckmann, "Friction Properties of Pavement at Different Speeds," Symposium on Skid Resistance, ASTM Spec. Tech. Publ., 32G, pp. 42-49, 1962.

320. Paul J. Serafin, "Michigan Experience with Different Materials and Designs on the Skid Resistance of Bituminous Pavements," Highway Research Board, 49th Annual Meeting, Washington, D.C., January 12-16, 1970.

325. J. W. Shupe and R. W. Lounsbury, "Polishing Characteristics of Mineral Aggregates," Proceedings of the First International Skid Prevention Conference, Virginia Council of Highway Investigation and Research, Charlottesville, Virginia, pp. 509-529, August 1958.

338. Karl W. E. Stinson and Charles P. Roberts, "Coefficient of Friction Between Tires and Road Surfaces," Proceedings, Highway Research Board, 13th Annual Meeting, pp. 49-69, December 1933.

339. W. J. Stutzenberger and J. H. Havens, "A Study of the Polishing Characteristics of Limestones and Sandstone Aggregates in Regard to Pavement Slipperiness," Highway Research Board, Bulletin 186, pp. 58-81, 1958.

342. D. Tabor, "The Importance of Hysterisis Losses in the Friction of Lubricated Rubber," Proceedings of the First International Skid Prevention Conference, Virginia Council of Highway Investigation and Research, Charlottesville, Virginia, p. 211, August 1958.

364. D. E. Weller and D. P. Maynard, "The Influence of Materials and Mix Design on the Skid Resistance Value and Texture Depth of Concrete," RRL Report LR 334, Road Research Laboratory, Ministry of Transport (U.K.), Crowthorne, Berkshire, England, 1970.

368. N. W. West and T. F. Ross, "A Study of Road Friction in New South Wales," Proceedings, Australian Road Research Board, Vol. 2, Pt. I, pp. 478-495, 1964.

386. W. O. Yandell, "The Effect of Surface Geometry on the Lubricated Sliding Friction and Polishing," Australian Road Research, Vol. 3, No. 10, June 1969.

THE PHYSICS OF TIRE TRACTION: A REVIEW OF CURRENT THEORY

A. L. BROWNE

General Motors Research Laboratories, Warren, Michigan

SUMMARY REMARKS

When we were in the planning stage for this Symposium, some individuals thought that the physics of tire traction was too narrow a subject to be the topic for a two day meeting. If anything, we certainly have demonstrated that two days is too brief a time in which to attempt to thoroughly treat this multifaceted problem.

In the first session, we focused on a general description of tire traction phenomena discussing the different requirements necessary for good traction performance in the principal operating environments. Operating environment meant not only the prevailing weather conditions – whether the pavement is dry, wet, or snow covered – but also the dynamic state of the traction demand situation – for instance whether the tire is experiencing slip or has nonzero camber and cornering angles at the time that it is required to provide traction.

In the opening paper by Dr. M. G. Pottinger, we were reintroduced to the persistent dilemma of ranking tires as to their wet traction performance. The research of recent years has repeatedly pointed out that these rankings can switch depending on vehicle speed, pavement surface texture, such tire properties as tread design, and the depth of the water film.

The present goal as was expressed in this first paper is to find some representative quantity that gives an average performance value or average ranking of the tire under the gamut of wet traction conditions. Ultimate Performance Rating is proposed as meeting this demand. Operational Severity is an additional useful concept that was introduced. It gives a qualitative display of the roles that many of the important factors play in affecting traction performance.

Aside from Pottinger's paper, certain general observations can be made about wet traction. Past research has given us sufficient insight into the wet traction problem to enable us to alter the shape of the traction-speed curves. Thus, an important question that should be asked is what features of these curves are most important? Should we make changes that offer increases in traction performance at high operational severity while causing decreases in traction at low severity? Or should we adopt those that do the reverse? We need to establish guidelines as to how to evaluate these alternatives.

The paper by Mr. R. W. Yeager dealt with wet traction conditions of high operational severity. However, the results of this study are highly significant for the wet traction problem in general – they are the first to clearly reveal the details of the interactions between the different tread elements and the fluid. Because of this study, we can gain an understanding of the roles payed by the different elements of the tread pattern in influencing wet traction.

Another important aspect of the wet traction problem that was emphasized in both the paper by Yeager and that by Pottinger is the critical importance of water depth to tire traction performance. This is even more important in light of the fact that there are no reliable ways to measure and control the fluid depths within the accuracy needed to insure reliability of traction test data. This is especially true in the case of thin fluid films of thickness of the order of the size of the pavement surface texture.

Working in the tire traction area it is sometimes easy to succumb to thinking that the only significant problems lie in how to get rid of the interface contaminant such as water or snow in order to restore intimate dry contact between tire and road. However, changes in the dynamic operational environment of a tire can have equally significant effects on traction performance as those due to the presence of an interface contaminant.

In the papers by Professor L. Segel and by Dr. H. B. Pacejka, the problem of traction dynamics was highlighted. In the paper by Segel, the study of dry traction was broken into the mechanics of shear-force generation to which he addressed himself and the study of the detailed mechanics controlling the friction coupling at the tire-road interface. He reiterated that operating conditions such as amount of slip affect the maximum available traction components, and indicated that tire geometry, type of construction, tread wear and pavement surface texture can greatly modify the traction values for these different operating conditions. Pacejka took this a step further by analyzing the tire response to varying force inputs. The implication that can be drawn from these papers is that it is not only important for us to know how factors such as tread pattern design and carcass construction can affect traction directly, but also to learn about the equally important effects that they can have on traction by modifying the mechanics of traction force generation.

The closing paper in the first session by Dr. A. L. Browne was on the snow traction performance of tires. A recent letter from Ralph Moyer, who has been actively involved in the area of tire traction research for the past 40 years, contained the following quote: "In my opinion, the area of study in skid prevention in which we now lack reliable test data and an understanding of the basic facts, is concerning traction and skid prevention under snow and ice conditions."

The paper that was presented at this Symposium on snow traction addressed itself to the lack of reliable test data and to the lack of an understanding of the basic facts concerning snow traction. A comprehensive theory of snow traction was presented based on fundamental conjectures about the mechanisms that are involved in and the nature of the traction producing interactions between the tire and the snow. It also defined what essential efforts must be made in snow traction research if investigators are to start obtaining useful and meaningful results.

The second session was devoted to a study of the fundamental mechanisms of rubber friction and their contribution to tire traction.

The opening paper by Dr. K. A. Grosch restated the hope that correctly designed laboratory experiments which measured the adhesional part of the viscoelastic behavior of the tread compound could be used to predict tire traction performance. He stated that this was generally true for operation on ice, and that results were confirmed by extensive outdoor testing.

The paper also emphasized the important well-known point that there is no one tread compound that is best for all operating conditions, even from just the standpoint of tire traction. Grosch's work is oriented toward creating order out of this very difficult situation.

In the paper by Dr. A. Schallamach, a closer look was taken at the possible nature of adhesive friction. The previously stated "waves of detachment" model for sliding on smooth dry surfaces was solidified with an interesting film and mathematical development.

In the third paper, Dr. A. D. Roberts addressed himself to the difficulties in establishing zones of adhesive contact when a lubricant is present. As he stated, when a fluid is added to the interface zone the picture is drastically changed. During rolling without slip on smooth surfaces, fluid bells or globules are trapped in interface zones. Such bells have an extremely high resistance to collapse due to effects such as the electrical double layer, and thus, can greatly reduce the area in which adhesional bonds can act. He further reported that the presence of slip on such smooth surfaces will not in itself destroy the thin film, the regions of entrapped fluid in the minute contact zones then take on the shape of the elastohydrodynamic horseshoe.

As a comment on this study, it clearly demonstrates the critical role of pavement surface texture that is small in scale compared to the size of a liquid bell.

In the final paper of the second session, Professor K. C. Ludema addressed himself to the deformation component of rubber friction contributing to tire traction during lubricated contacts.

He has presented a basic model for asperity-rubber interaction in the presence of a lubricant. This is designed to serve as a building block for a theory of traction under mildly wet conditions where the tread pattern and pavement surface texture are effective in rapidly removing the bulk of the fluid from the interface zone. The model approximates reality best under operation on polished pavement asperities. The authors have also offered evidence supporting two conclusions: one that skid numbers at low speeds may be influenced by the non-wettability of the rubber by water; and two, that the interacting surfaces are always separated by a water film even in dry contacts. Questions exist as to when this water film will be thin enough as to have negligible effect on the interface interaction.

The third session focused on the role of the tire in the production of traction.

In the first paper, Dr. R. F. Peterson discussed the effect of the tire tread compound. He expressed the view that rubber compounding has the greatest impact on traction during wet operation.

The dilemma of the many conflicting demands placed on the rubber compound was carefully discussed and a description was given of the many compound ingredients that can affect wet traction performance.

Another aspect of the tire important to traction development was discussed by Dr. R. N. Kienle in the third session: the tire tread pattern. This talk gave us some insight as to the mathematical techniques that are used for the design of tread patterns for improved wet traction performance. The following observations seem to be appropriate in light of that discussion.

We seem to have a pretty good idea of what we would like tread patterns to do – provide good traction on wet and snow-covered pavements. However, if today's multiplicity of tread designs is any indication, there does not seem to be very good agreement as to the best way to reach our goal.

From comparison of wet traction tests especially under commonly found severe wet performance conditions (such as 60 mph during a moderate rain) to which the SAE tests do not correspond, it is obvious that many commonly used tread patterns really are quite ineffective in spite of their complex appearances. Almost more seems to be known about the effective design of tread patterns from a noise and wear standpoint than for wet traction.

Maybe we have focused our attention a little too long on preventing bad effects of tread pattern incorporation on the other tire performance aspects and not enough on trying to make the tread pattern effective at doing well what we put it on the tire for in the first place.

Several approaches for effective tread design have been offered at this meeting. Most notable among these are first the use of glass plate tests to study just how the individual features of the tread pattern affect the fluid flow and second, the beginnings of different hydrodynamic approaches to the problem. These are encouraging. However, let us hope that they are only the beginning of a concerted effort at the design of effective tread patterns.

The comments that were made in the review presentation by Dr. V. E. Gough present us with some important challenges.

His advice as to the importance of insuring a lasting, dual road surface texture should be seriously considered by all highway departments.

He has challenged tread designers to attempt a mathematical approach to the design of effective tread features for increased wet traction.

Finally his indictment of many of those who are theorizing about the physics of tire traction on dry pavements is most notable: "Any theory ascribing frictional grip or tangential force transmission of polymer to road surface as being entirely due to energy loss within (the plane of) or on the surface of the polymer . . . will not stand a moment's critical inspection . . . Clearly friction is not caused by losses at the tyre to ground interface – (it must be) due primarily to a form of surface-to-surface adhesion capable of transmitting traction forces without necessitating an energy loss mechanism."

The subject of tire carcass and size effects on tire traction should be mentioned in this review even though it was not the subject of a formal paper.

The last several years have seen a proliferation of tire sizes and constructions. As we know, as the tire size and construction change, so do many aspects of its performance. Tire traction is one performance characteristic that can be significantly altered by changes in tire carcass construction and size. Now we are entering a period in which this proliferation in tire sizes and types of construction is being questioned. It is obvious that drastic measures at standardization will occur in the future. Choices are being made. It is important that we now make a major effort to critically explore the effects of tire size and construction on tire performance – especially tire traction – so that these choices will be the right ones.

In the final session our concern shifted to an attempt to define more clearly the complex interrelationship between pavement surface texture and tire traction.

Mr. R. Schonfeld gave further elaboration to his proven experimental procedure for predicting a pavement's skid resistance by photo interpretation of its surface texture. A critical aspect of this technique was the identification of six (and only six) geometrical structure parameters of the pavement surface texture that are important to tire traction. The success of this experimental approach in predicting skid resistance should not be treated lightly. Those attempting to mathematically model the pavement surface texture in terms of its influence on tire traction would be wise to look closely at Mr. Schonfeld's six parameters of geometrical structure.

The effect that pavement surface texture will have on tire traction is very strongly linked to factors such as the tread material and pattern, the presence of an interface contaminant like water, and differences in operating conditions such as speed and load besides being dependent on the nature of the texture itself.

Dr. W. O. Yandell discussed the way in which traction is affected by the interaction of the pavement surface texture and the vehicle load.

Mr. R. Bond, Professor G. Lees, and Dr. A. R. Williams addressed themselves to the problem of optimizing the amounts of pavement micro- and macro-texture in terms of total tire performance. Minimum and maximum levels of micro- and macro-texture were recommended for good traction performance without undue penalties in terms of increased wear and noise or inferior ride characteristics. These recommended levels, of course, are closely linked to the tire tread design and must, therefore, vary within certain ranges which the authors also specified. The most important part of this work may be, however, that the authors have implemented their findings by constructing actual pavement surfaces to their specifications. The results of this program may, thus, have great positive influences on pavement construction in the future.

In conclusion, enough issues have been raised at this Symposium to show us that tire traction is far from a closed subject. In fact the number of points that have been actively debated, clearly indicates that sufficient important work remains to be done in this area to challenge all of our minds over the next many years. Through all of your efforts, we have come a long way. However, we have certainly demonstrated that our knowledge of tire traction is still growing and must continue to grow for many years before we will be able to say that we know all that should be known.

SYMPOSIUM SPEAKERS AND CHAIRMEN

SYMPOSIUM CHAIRMEN

D. F. Hays

A. L. Browne

SESSION I: TIRE TRACTION – A MANY FACETED PROBLEM

H. C. A. van Eldik Thieme

M. G. Pottinger

R. W. Yeager

L. Segel

SESSION II: FUNDAMENTAL ASPECTS OF RUBBER FRICTION

D. F. Moore

K. A. Grosch

A. Schallamach

A. O. Roberts

SESSION III: TIRE TRACTION – THE ROLE OF THE TIRE

S. K Clark

R. F. Peterson

SESSION III: (Continued)

R. N. Kienle

H. B. Pacejka

SESSION IV: TIRE TRACTION – THE ROLE OF THE PAVEMENT

K. C. Ludema

W. O. Yandell

R. Schonfeld

A. R. Williams

SESSION II PANEL

SESSION III PANEL

SESSION IV PANEL

IN MEMORIAM

Vernon Eric Gough

1908 – 1973

Vernon Eric Gough passed away after a long illness on December 27, 1973. During his many prolific years of research, he made substantial contributions to our knowledge of the physics of tire traction. To him we owe a great debt; and certainly he is an individual who shall be missed.

PARTICIPANTS

* Denotes Session Chairman
** Denotes Speaker

I. A. Abu-Isa
Polymers Department
Research Laboratories, GMC

W. G. Agnew
Technical Director
Research Laboratories, GMC

W. A. Albers
Societal Analysis Department
Research Laboratories, GMC

C. V. Allen
Proving Ground, GMC
Tire & Wheel Development Engineering
Milford, MI 48042

C. A. Amann
Engine Research Department
Research Laboratories, GMC

D. D. Anderson
Proving Ground, GMC
Tire & Wheel Development Engineering
Milford, MI 48042

G. G. Balmer
Federal Highway Administration
HRS-12 (FHRS)
Washington, D.C. 20590

M. R. Barone
Engineering Mechanics Department
Research Laboratories, GMC

C. Beauregard
Ford Motor Company
P.O. Box 2053
Bldg. 5, Room 1222
Dearborn, MI 48121

P. S. Beebe
Fluid Dynamics Department
Research Laboratories, GMC

W. Bergman
Ford Motor Company
P.O. Box 2053, Bldg. 5
Dearborn, MI 48121

J. B. Bidwell
Executive Director
Research Laboratories, GMC

K. D. Bird
Calspan Corporation
4455 Genesee Street
Buffalo, New York 14221

C. E. Bleil
Physics Department
Research Laboratories, GMC

L. S. Bonderson
Transportation & Urban Analysis Department
Research Laboratories, GMC

P. Boyd
National Highway Traffic Safety Administration
Safety Systems Lab/Research Institute
Attn: N43-20, Room 5210
Washington, D.C. 20590

J. J. Brockman
Electrical Engineering Department
Research Laboratories, GMC

T. J. Brown
Pontiac Motor Division, GMC
Pontiac Engineering Department
Pontiac, MI 48053

**A. L. Browne
Engineering Mechanics Department
Research Laboratories, GMC

A. V. Butterworth
Mathematics Department
Research Laboratories, GMC

K. L. Campbell
Firestone Tire & Rubber Company
Development Department
Akron, Ohio

J. D. Caplan
Executive Director
Research Laboratories, GMC

W. L. Carpenter
B. F. Goodrich Company
500 S. Main Street
Akron, Ohio 44318

C. I. Carr
Uniroyal Research Center
Oxford Management & Research Center
Middlebury, Connecticut 06749

A. Y. Casanova
National Highway Traffic Safety Administration
400 7th Street, S.W.
Washington, D.C. 20590

P. F. Chenea
Vice President
Research Laboratories, GMC

A. Chiesa
Pirelli Milano Italy
Industrie Pirelli Spa
Milano, Italy

B. E. Clapson
Avon Rubber Company
Bath Road
Melksham
Wiltshire, United Kingdom

*S. K. Clark
University of Michigan
Department of Applied Mechanics and Engineering Science
201 W. Engineering Bldg.
Ann Arbor, MI 48104

A. G. Clary
Highway Research Board
2101 Constitution Avenue
Washington, D.C. 20418

H. R. Clemett
Ford Motor Company
Product Development
Room 1222, Bldg. 5
P.O. Box 2053
Dearborn, MI 48121

J. S. Collman
Gas Turbine Research
Research Laboratories, GMC

T. P. Cote
Chevrolet Motor Division, GMC
30003 Van Dyke
Warren, MI 48090

W. B. Crum
Ford Motor Company
P.O. Box 2053
Bldg. 5, Room 1222
Dearborn, MI 48121

J. Dames
Institut für Strassen-und Verkehrswesen
Technische Universität Berlin
Strasse des Af. Juni A35
1 Berlin A2, Germany FR

R. E. Davies
B. F. Goodrich Tire Company
Dept. 6142, Bldg. 10E
500 S. Main Street
Akron, Ohio 44318

F. R. Davis
General Tire
Gilchrist Road
Akron, Ohio 44309

J. Eagleburger
Goodyear Tire & Rubber Company
1144 E. Market Street
Akron, Ohio 44316

I. R. Ehrlich
Stevens Institute of Technology
Davidson Laboratory
Castle Point Station
Hoboken, NJ 07030

W. T. Emmett
Firestone Tire & Rubber Company
1200 Firestone Parkway
Akron, Ohio 44317

R. F. Forbes
Overseas Operations, GMC
9-135 GM Building
Detroit, MI 48202

M. Fox
General Motors South African
P.O. Box 1137
Port Elizabeth, South Africa

D. L. Fry
Physics Department
Research Laboratories, GMC

R. M. Gallaway
Texas Transportation Institute
Texas A & M
College Station, Texas 77840

M. W. Gallogly
Ohio State University
2070 Neil Avenue
Columbus, Ohio 43210

D. M. Gearig
Chevrolet Motor Division, GMC
Room 3-140
30003 Van Dyke
Warren, MI 48090

R. W. Gibson
Library
Research Laboratories, GMC

**K. A. Grosch
Uniroyal ETDC
51 Aachen
Postfach 410
German Federal Republic

B. D. Gujrati
University of Michigan
2046 E. Engineering Building
Ann Arbor, MI 48104

L. R. Hamkins
American Motors Corporation
14250 Plymouth Road
Detroit, MI 48232

E. A. Hanysz
Electronics & Instrumentation Dept.
Research Laboratories, GMC

D. F. Hays
Engineering Mechanics Department
Research Laboratories, GMC

R. Hegmon
Federal Highway Administration
HRS-12
Washington, D.C. 20590

F. Herzegh
2884 Woodbury Road
Shaker Heights, Ohio 44120

R. Hickling
Fluid Dynamics Department
Research Laboratories, GMC

H. C. Hodges
Hodges Transportation Inc.
Nevada Automotive Test Center
P.O. Box 234
Carson City, Nevada 89701

G. T. Hodgkinson
Dunlop Tire and Rubber Company
P.O. Box 1109
Buffalo, New York 14240

R. N. Hollyer
Physics Department
Research Laboratories, GMC

J. C. Holzwarth
Technical Director
Research Laboratories, GMC

D. S. Howarth
Electronics & Instrumentation Dept.
Research Laboratories, GMC

J. W. Hutchinson
University of Kentucky
Department of Civil Engineering
207 Anderson Hall
Lexington, Kentucky 40506

G. A. Ilkka
Polymers Department
Research Laboratories, GMC

F. E. Jamerson
Physics Department
Research Laboratories, GMC

T. A. Johnson
Proving Ground, GMC
Noise & Vibration Laboratory
Milford, MI 48031

T. M. Johnson
Proving Ground, GMC
Noise & Vibration Laboratory
Milford, MI 48031

W. D. Kegg
The General Tire & Rubber Company
One General Street
Akron, Ohio 44329

**R. N. Kienle
Uniroyal, Inc.
6600 E. Jefferson Avenue
Detroit, MI 48232

T. King
The General Tire & Rubber Company
Experimental Testing Laboratory
One General Street
Akron, Ohio 44329

W. F. King
Engineering Mechanics Department
Research Laboratories, GMC

R. L. Klimisch
Physical Chemistry Department
Research Laboratories, GMC

E. Kontos
Uniroyal, Inc.
6600 E. Jefferson Avenue
Detroit, MI 48232

M. K. Krage
Electronics & Instrumentation Dept.
Research Laboratories, GMC

M. R. Ladd
Buick Motor Division, GMC
E. Hamilton Avenue
Flint, MI 48505

J. G. Larson
Physical Chemistry Department
Research Laboratories, GMC

T. C. Laurell
Cadillac Motor Car Division, GMC
2860 Clark Street
Detroit, MI 48232

D. Livingston
Goodyear Tire & Rubber Company
142 Goodyear Blvd.
Akron, Ohio 44316

J. M. Long
B. F. Goodrich Company
500 S. Main Street
Akron, Ohio 44318

*,**K. C. Ludema
2046 E. Engineering Building
University of Michigan
Ann Arbor, MI 48104

Y. Maeda
Tire Technical Center
Yokohama Rubber Company
150 Shinshuku, Hiratsuka-shi
Kanagawa-Ken 254
Japan

D. C. Mahone
Virginia Highway Research Council
P.O. Box 3817 University Station
Charlottesville, Virginia 22903

T. J. Mao
Polymers Department
Research Laboratories, GMC

J. F. Martin
Calspan Corporation
Box 235
Buffalo, New York 14221

W. T. Mason
Fluid Dynamics Department
Research Laboratories, GMC

F. E. Matyja
General Tire
685 Longacre Drive
Akron, Ohio 44319

W. E. Meyer
Pennsylvania State University
Research Building B
University Park, PA 16802

R. Miller
General Tire & Rubber Company
One General Street
Tire Development, Plant II
Akron, Ohio 44329

J. B. Montagne
Service F Michelin
63000 Clermont-Ferrand
France

*D. F. Moore
National University of Ireland
(University College)
Department of Mechanical Engineering
Upper Merrion Street
Dublin, Ireland

R. Moore
Proving Ground, GMC
Tire & Wheel Development Engineering
Milford, MI 48031

R. A. Moyer
University of California
109 McLaughlin Hall
Berkeley, California 94720

N. L. Muench
Technical Director
Research Laboratories, GMC

A. H. Neill, Jr.
National Highway Traffic Safety Administration
Safety Systems Laboratory/RI
Attn: 43-20
Washington, D.C. 20590

D. W. Nicholson
Goodyear Research
142 Goodyear Boulevard
Akron, Ohio 44316

O. Nordstrom
National Swedish Road & Traffic Research Institute
Statens Vag-Och Trafikinstitut
Drottning Kristinas Vag 25
114 28 Stockholm
Sweden

H. Oberhaus
A. Opel A.G.
D-6090 Russelsheim
W. Germany

K. P. Oh
Engineering Mechanics Department
Research Laboratories, GMC

W. C. Olander
Electronics & Instrumentation Dept.
Research Laboratories, GMC

L. J. Oswald
Fluid Dynamics Department
Research Laboratories, GMC

**H. B. Pacejka
Delft University of Technology
Vehicle Research Laboratory
Mekelweg 2
Delft, The Netherlands

K. G. Peterson
Proving Ground, GMC
Milford, MI 48031

**R. F. Peterson, Jr.
Uniroyal Management and Research Center
Middlebury, Connecticut 06749

A. Pfeifer
Adam Opel, A.G.
Russelsheim, Germany

B. D. A. Phillips
Lanchester Polytechnic Coventry
Mechanical Engineering Department
Coventry, Warwickshire
England

R. Pochart
American Motors Corporation
14250 Plymouth Road
Detroit, MI 48232

**M. G. Pottinger
B. F. Goodrich
9921 Brecksville Road
Brecksville, Ohio 44141

J. Potts
Firestone Tire & Rubber Company
Firestone Research Laboratories
1200 Firestone Parkway
Akron, Ohio 44317

R. Quick
GM of Canada, Ltd.
Chassis Engineering Department
Park Road South
Oshawa, Ontario, Canada

G. Randall
GMC Truck & Coach Division, GMC
660 S. Boulevard
Pontiac, MI 48053

O. J. A. Repelaer Van Driel
International B. F. Goodrich
Europe Research and Development Center
P.O. Box 208
Enschede, The Netherlands

C. I. Reynolds
Uniroyal, Inc.
6600 E. Jefferson
Detroit, MI 48232

**A. D. Roberts
Cambridge
24 Montague Road
Cambridge CB4 1BX
England

S. R. Sacia
Goodyear Tire & Rubber Company
Dept. 462D
1200 E. Market Street
Akron, Ohio 44316

**A. Schallamach
73 Longmore Avenue
Barnet, Herts
En5 1La
England

F. K. Schenkel
Fluid Dynamics Department
Research Laboratories, GMC

N. A. Schilke
Fluid Dynamics Department
Research Laboratories, GMC

**R. Schonfeld
Government, Ontario, Canada
1281 Wilson Avenue
Downsview, Ontario M3M1J8
Canada

N. W. Schubring
Electronics & Instrumentation Department
Research Laboratories, GMC

D. J. Schuring
Calspan Corporation
Buffalo, New York 14221

J. H. Schutz
Chrysler Corporation
CIMS 418-36-04
P.O. Box 118
Detroit, MI 48231

W. P. Schwager
Firestone
1200 Firestone Parkway
Akron, Ohio 44317

**L. Segel
University of Michigan
Highway Safety Research Institute
Huron Parkway & Baxter Road
Ann Arbor, MI 48105

W. L. Shepard
Proving Ground, GMC
Tire-Wheel Development Engineering
Milford, MI 48042

J. Scowcroft
Motor Vehicle Manufacturers Association
320 New Center Building
Detroit, MI 48202

P. Shoemaker
Uniroyal, Inc.
4500 Enterprise Drive
Allen Park, MI 48101

H. A. Smith
Highway Research Board – NCHRP
2101 Constitution Avenue, NW
Washington, D.C. 20418

F. Smithson
Proving Ground, GMC
Tire-Wheel Development Engineering
Milford, MI 48042

L. F. Spaine
Highway Research Board
2101 Constitution Avenue, NW
Washington, D.C. 20418

W. B. Straub
The Firestone Tire & Rubber Company
1200 Firestone Parkway
Akron, Ohio 44317

R. Straughan
Olympic Tyre & Rubber Company
P.O. Box 1, West Footscray
Victoria, 3012, Australia

S. Takahasi
Bridgestone Tire Company of America, Inc.
16921 South Western Avenue
Gardena, California 90247

Y. Takami
Toyo Rubber Industry
2685 Shoreline Dr., A-1
Akron, Ohio 44314

J. Tanner
NASA Langley Research Center
Hampton, Virginia

H. D. Tarpinian
Uniroyal, Inc.
6600 E. Jefferson
Detroit, MI 48232

*H. C. A. van Eldik Thieme
Delft, University of Technology
Laboratory of Vehicle Engineering
Mechanical Engineering Department
Technische Hogeschool
2 Mekelweg
Delft, The Netherlands

R. F. Thomson
Executive Director
Research Laboratories, GMC

J. T. Tielking
University of Michigan
Highway Safety Research Institute
Huron Parkway & Baxter Road
Ann Arbor, MI 48105

R. W. Topping
B. F. Goodrich Company
9921 Brecksville Road
Brecksville, Ohio 44141

W. A. Turunen
Fluid Dynamics Research Department
Research Laboratories, GMC

T. Umeda
The Yokohama Rubber Company, Ltd.
1530 Church Road
Montebello, California 90640

N. K. Van Allen
Library
Research Laboratories, GMC

H. Verdier
Michelin
Service F Michelin
63000 Clermont-Ferrand
France

J. Viner
Federal Highway Administration
Structures and Applied Mechanics
HRS-12
Washington, D.C. 20590

J. D. Walter
Firestone Tire & Rubber Company
Central Research Laboratory
1200 Firestone Parkway
Akron, Ohio 44317

J. R. Watson, Jr.
Federal Highway Administration
Implementation Division (HDV-22)
Washington, D.C. 20590

N. M. Wang
Mathematics Department
Research Laboratories, GMC

P. Weiss
Polymers Department
Research Laboratories, GMC

E. F. Weller
Electronics & Instrumentation Department
Research Laboratories, GMC

E. A. Whitehurst
Ohio State University
Transportation Research Center
475 Hitchcock Hall
2070 Neil Avenue
Columbus, Ohio 43210

D. R. Whitney
Environmental Activities Staff, GMC
Vehicular Noise Control – Room 248-A
GM Technical Center
Warren, MI 48090

I. D. Wilken
Fluid Dynamics Department
Research Laboratories, GMC

**A. R. Williams
Dunlop Ltd.
Fort Dunlop Tyre Research Dept.
Erdington, Birmingham B23
England

F. L. Williams
Physical Chemistry Department
Research Laboratories, GMC

R. L. Williams
Environmental Science Department
Research Laboratories, GMC

R. L. Wilson
Motor Vehicle Manufacturers Association
320 New Center Building
Detroit, MI 48202

**W. O. Yandell
University of N.S.W.
School of Highway Engineering
Box 1 P.O.
Kensington
N.S.W., Australia

**R. W. Yeager
Goodyear Tire & Rubber Company
Dept. 460G, Plant 1
1144 E. Market Street
Akron, Ohio 44316

L. K. Yu
General Tire & Rubber Company
Corporate Research & Development
One General Street
Akron, Ohio 44329

AUTHORS AND DISCUSSORS INDEX

Bold Face Type refers to first page of papers and Symposium and Session Summaries.

AUTHORS OF CITED REFERENCES

Numbers in parentheses indicate the numbers of the references when these are cited in the text without the names of the authors.

Numbers set in **boldface** refer to the first page of a contributor's paper.

SUBJECT INDEX